KT-551-041

AS

Information
Communication
Technology

for Advanced Level

SECOND EDITION

Julian Mott and Anne Leeming

ASSESSMENT and
QUALIFICATIONS
ALLIANCE

Hodder & Stoughton

A MEMBER OF THE GROUP

Order: please contact Bookpoint Ltd, 130 Milton Park, Abingdon,
Oxon OX14 4SB. Telephone: (44) 01235 827720,
Fax: (44) 01235 400454. Lines are open from 9.00–6.00,
Monday to Saturday, with a 24 hour message answering service.
Email address: orders@bookpoint.co.uk

A catalogue record for this title is available from
The British Library

ISBN 0 340 772441

First published 1998

Impression number   10   9   8   7   6   5   4   3
Year   2003   2002   2001

Cover photo from Telegraph Colour Library

Typeset by Wearset, Boldon , Tyne and Wear.
Printed in Great Britain for Hodder & Stoughton Educational,
a division of Hodder Headline Ltd, 338 Euston Road, London
NW1 3BH by The Bath Press, Bath.

Contents

Introduction

This book aims to cover the AQA A and AS level information technology syllabuses by including IT theory, case studies, bullet-point summaries for quick reference, practice and revision questions, and coursework guidance. We gratefully acknowledge the permission of AQA to reproduce exam and revision questions in this book.

The Core section of the book covers the AS sections of the syllabus which constitute the examinable part of the AS Level syllabus. The Extension section covers the A2 sections, which together with AS make up the examinable part of the A Level syllabus.

A project chapter gives advice on how to tackle projects for this syllabus suggesting possible projects and what should be included.

The book is also suited to students of GNVQ Advanced IT.

Copyright holders of photographs reproduced in this book:
Telegraph Colour Library: Cover
Life File Photographic Library: Pages 6, 139, 150, 171, 186, 191, 192, 197, 214, 223, 224, 240, 256
British Telecom: Page 142
PA News Photographic Library: Page 4

What is information?

Knowledge, information and data

Computers are *automatic data processors*.

Information Technology means collection, storage, processing and dissemination of information using computers. What is the difference between information and data?

Raw facts and figures are **data**.

Organised facts and figures are **information**. Often information and data are regarded as the same. They are not the same; information must have a *context*, which makes it understandable.

Knowledge is knowing how to use information to make decisions.

For example:

12, 16, 15, 13

has no context. It is *data*.

The price of 100 widgets as follows:

Company A	£12.00
Company B	£16.00
Company C	£15.00
Company D	£13.00

has a context. It is *information*.

Deciding to order widgets from Company A because they are the cheapest is using information to make a decision. It is *knowledge*.

Computers store data, for example a list of numbers representing sales in a shop. Computer output should be information, such as sales information that can enable decisions to be made about what stock to order.

Information is important. Without information we have no knowledge, so we cannot make decisions.

Computers have enabled us to store and process more data.

Data is derived from different sources. It can be input directly and indirectly – a computer reading a bar code in a supermarket, account details being read directly from the magnetic strip on a credit card, a computer automatically reading the numbers on the bottom of a cheque and information from an automatic weather station being down-loaded into a computer – are all examples of deriving data directly.

A VDU operator taking data from a piece of paper and typing it into a database is an example of data being entered indirectly. Indirect entry is more likely to lead to mistakes, due to human error.

We expect information to be **reliable** and **accurate**. It must be **up-to-date**, **complete** and **precise**. It must be **comprehensible** (understood by the recipient).

Conversely, we must try to avoid **inconsistent**, **misleading** or **incomprehensible** information.

If data is entered incorrectly, for whatever reason, (for example the wrong data has been supplied, it has been mistyped or been given deliberately wrong) the information output will be incorrect. This is referred to as **GIGO – Garbage In, Garbage Out.**

Input, process and output

Input means entering **data** into the computer. It is data as it does not have a context at this stage, for example it might be the number from a bar code or reading from a sensor.

Processing means manipulating this **data** into **information** in a form understandable to the user. This might be by looking up the details of the product whose code has been entered or by performing a calculation on numbers that have been entered.

Output means presenting this information **to the user.** It must be in a form the user can understand, so it must have a context. It is information. It could be printed, displayed on a screen or in another form. It might be audible such as an alarm telling the user to evacuate the building.

Feedback means using the output of a computer to influence the input. A customer using a self-scanning machine in a supermarket can see the total cost of the goods scanned so far. This **output** may be used to help the customer decide what else they can afford to buy. This is feedback.

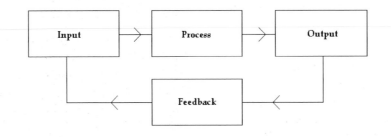

A pelican crossing, controlled by computer, has a push button device which pedestrians push when they wish to cross the road. When a pedestrian pushes this button it is **input**. A signal is sent to the computer. This is **data.**

The computer calculates how long the red and green light phases should be in each direction, according to instructions in its program. This is **processing** data into a form understood by the user – the changing of the lights

The lighting of the 'wait' light by the push button, changing of the lights from green to red, turning on the green man and the audible bleep are the **output**. They provide **information** to the users.

A motorist seeing the red light will decide to stop, using the output information to make a decision – this is **knowledge**.

Another pedestrian arriving a few moments later will see that the red man (output) is now lit, preventing her from crossing. She presses the button. This is input affected by the output – **feedback.**

The value and importance of information

Information, particularly exclusive information to which no other organisation has access, is **a valuable commodity.**

Marketing departments of businesses often build up mailing lists of potential customers. The names and addresses will be stored on a computer **database**, possibly with other information about the subject, such as how much they earn. The company then uses this mailing list to send out direct-mail using a *mail-merge*; a program that merges names and addresses into a letter. The mail-merge can be selective; it selects only certain names according to certain criteria. The criteria might be a certain post-code, within a certain age range or in a certain earnings bracket.

A new car showroom is opening shortly. The owners may approach a company specialising in supplying data and buy from them a database of potential customers living nearby who will be invited to a test-drive or a promotional event. *Selling data like this is perfectly legal.* Information from the database is valuable as it can help the business.

Sales information in a supermarket is very valuable as it is used to help decide what to order, whether to promote an item or whether to sell an item at all. Using this information means supermarkets can order stock more accurately and be more efficient.

Computer data is so valuable that the vast majority of businesses go bankrupt if they lose the data stored on their computers.

■ 'WHERE DID THEY GET MY NAME FROM?'

A mailing list can be built up from the electoral register. You can visit the local council offices and get a copy of the electoral register: a list of all the adults living in the area together with their addresses. It is available free or for a nominal charge for photocopying. However, you may also buy this information on computer disk. This is more expensive but has the advantage that you don't have to type all the names and addresses into the computer.

Direct-mail

We all receive direct-mail, usually advertising sent by companies to selected names taken from a computer database. Three billion direct-mail letters are sent in Britain every year: the number has more than doubled over ten years. Direct-mail now represents one sixth of all mail in this country. It is often called junk-mail and thrown away without being opened.

You might not like getting direct-mail (junk-mail); you might not even open it before throwing it away. However direct-mail is obviously a very successful marketing strategy for businesses; otherwise businesses would not use it. Direct-mail is all about **targeting** – reaching particular groups of people who are more likely to purchase their goods or services.

Direct-mail companies would suggest that this is an efficient technique, which enables them to reach only those likely to buy, for example

- those who have bought similar products before
- people who have a particular interest
- those who come from a certain socio-economic grouping
- those living near a particular shop

Storing data on customers in a database enables businesses to target specific customers and to analyse purchases to predict likely consumer demands.

This means that they can maintain appropriate stock levels and operate efficiently so their prices will be competitive.

The Mailing Preference Service (MPS) – the end of junk-mail?

The Mailing Preference Service (MPS) is a non-profit making body, set up in 1983 to foster good relations between direct-mail users and the general public, by giving you the option not to receive junk-mail. The MPS slogan is **'Helping you to ensure the mail you receive is the mail you want'**.

Direct-mailers don't want to waste valuable time and money by mailing to people who simply aren't interested. So the direct-mailing industry funds the work of the Mailing Preference Service by paying a levy whenever they use the Royal Mail volume mailing service.

■ WHAT CAN YOU DO?

You can register with MPS by writing to them at FREEPOST 22, London W1E 7EZ. Everyone at your address who shares your surname will no longer receive junk-mail.

After a few months, you will notice a reduction in the postal advertising you receive. Your name remains on file for five years. You will still receive mailings from companies with whom you have done business in the past, as well as from some small local companies.

In practice many people will not register, as they are worried that they will not get some information on some product they actually want. However over 400 000 people have registered. There is also a Telephone Preference Service, a Fax Preference Service and an E-mail Preference Service being set up.

Getting information illegally

Information can be so valuable that reputable companies may take illegal steps to spy on their rivals. In 1991 Richard Branson's Virgin Atlantic Airways accused British Airways (BA) of using such illegal 'dirty tricks' to gain access to information on Virgin's computer files, particularly the names and addresses of Virgin's business customers. Branson alleged BA then contacted these customers offering big discounts and other incentives to switch from flying Virgin to flying BA.

In March 1992 Branson sued BA for libel over suggestions that he had invented the claims. In January 1993 BA apologised unreservedly to Branson and paid Virgin Atlantic over £1.5 million in costs and damages in an out-of-court settlement.

The Sunday Times revealed on 10 January 1993 what British Airways had been doing.

The 'dirty tricks' campaign that BA tried to keep hidden

Fresh evidence of how British Airways waged a secret war against rival Richard Branson's Virgin Atlantic Airways has been uncovered by the *Sunday Times* Insight team.

Had the trial gone ahead it would have revealed details of BA's 'dirty tricks' campaign. In autumn 1990 a team of BA computer operators was told to access information on Virgin's passengers and flights.

The commercially confidential information, obtained via the BA booking system (BABS), allowed BA to monitor the number of passengers on each Virgin flight, their class of seat and details of lucrative corporate accounts.

Soon after the secret operation was set up, Virgin passengers were cold-called at home. They were offered free supplementary tickets, flights on Concorde and even triple air miles to switch to BA flights.

In January 1991 Branson complained to the European Commission about alleged unfair tactics by BA. The allegations were strenuously denied by BA. But behind the scenes BA staff were quietly told to destroy evidence.

Harvey Elliott in The Times 12 January 1993:

Branson wins BA apology for libel

British Airways yesterday apologised 'unreservedly' and agreed to pay £610 000 libel damages to Richard Branson, chairman of Virgin Atlantic, bringing to an end one of the most extraordinary disputes in aviation history.

Christopher Clarke, QC, counsel for BA, had told the court that allegations made against Mr Branson were 'wholly untrue'. His clients undertook not to repeat them and apologised unreservedly for the 'injury' caused to the reputation and feelings of Richard Branson and Virgin Atlantic and in particular for attacking the 'good faith and integrity' of Mr Branson.

'The investigation which BA carried out during the course of this litigation revealed a number of incidents involving their employees which BA accept were regrettable and gave Richard Branson and Virgin reasonable grounds for concern,' Mr Clarke said.

'Industrial espionage' like this to obtain information from a competitor is not unusual. Accurate information would be very useful to a rival company. The value of commercial information depends on accuracy, potential user and intended use.

How good is information?

I phone up a record store to see if they have a CD in stock. They look at the stock file on their computer and see that they have one left. I go to the store but when I get there it is not in stock. Why?

➤ They could have sold it.

➤ It could have been stolen.

➤ The computer may have been wrong.

A more likely possibility is that the computer information is out-of-date. Details of each sale goes into a transaction file – a file of today's sales. At the end of the day this file is used to up-date the stock file; each sale will result in the reduction of the number in stock of the given item. The stock file is only up-to-date at the end of each day.

Figure 1.1 Richard Branson

Information must be **up-to-date**. It is no good having a mailing list when the addresses are no longer correct. If people receive junk-mail for the last resident, they just put it in the bin. They don't tell the company who sent it.

The electoral register is up-dated every year in October. It is expensive to do this, as surveyors may have to go out and visit each house. Keeping such information as up-to-date as this is time consuming. Each entry must be checked and alterations typed in, which is slow and can lead to mistakes. However if the information is not up-dated, it is not as useful.

Date stamping information

Perishable food in a shop has to have a date stamped on it after which it is no longer safe to eat. Similarly it is a good idea to date stamp an information file, so the user knows how up-to-date it is. For example, if the record store uses an old version of its stock file to order new stock, it will not order the right items.

Reports and other print-outs from a database should be date stamped (and possibly time stamped) as well. A supermarket computer can print out a report of the number in stock of each item. This information varies as products are sold, so it is important that the report is stamped with the time and date when it was produced.

A user may have two or more versions of the same file on disk. Date and time stamping will show which is the most recent.

Making sure information is good

We can take steps to ensure information is good enough for the purpose required:

- checking data when it is entered (validation and verification – see Chapter 8)
- date stamping
- using a system that up-dates files more quickly, if not immediately, if up-to-date information is vital

The law requires that personal data is as up-to-date and accurate as possible (see Chapter 2).

Coding information

If large amounts of data are being processed, data items are often replaced by codes, for example goods in a shop have a bar code number, books have an International Standard Book Number. (ISBN). The code is used by the person entering the data into the computer. However if the data item is printed out, it will be printed in full. Thus at a supermarket till, the bar code of the item is entered into the computer but the name of the item is displayed on the till.

To encode data means to convert information or data into a code for processing. To decode data means to convert back into a form which can be understood.

Examples of common codes used to replace items of data include

- gender as M or F
- post codes: SO9 identifies an area of Southampton – SO9 5NH is the university. If it is written on an envelope, it can be converted into a series of dots at the sorting office for automatic sorting.
- date of birth, for example 06 02 81 stands for 6 February 1981
- bank branch sort-codes like 60-18-46

Reasons for using codes include

- the code is short and easy to enter
- codes takes up less storage space on disk
- it is easy to check that the code is valid (validation – see Chapter 8)

Using codes can lead to problems. Some codes are mnemonic and easy to remember, for example on airline baggage LHR means London Heathrow, FRA means Frankfurt. However you might not recognise an ISBN like 0 340 71180 9. Can you be sure it is the correct code?

Value judgements

A value judgement is when you given a value to something. It is entirely your opinion and may differ completely from someone else's opinion.

If someone says, 'I get up at 7 o'clock each morning,' this is not a value judgement but a statement of fact.

If they say, 'I get up *early* each morning,' this is a value judgement.

A milkman used to getting up at 3.30 wouldn't describe 7 o'clock as *early*. A student used to getting up at 11 o'clock might describe 7 o'clock as *very early*.

Value judgements and IT

Businesses are interested in our opinions. Do we like the taste of a product? Is the packaging attractive? Is it too expensive? These are some of the questions a market researcher might ask members of the public before a new product is launched. Information from the research will be used to determine the price, the image and even whether the product is produced at all.

All these questions involve value judgements. A lot of data is gathered and IT used to process it into a form understood by the decision makers of the business. It is much easier to gather and process data if the customer is given fixed answers.

Loss of precision due to coding value judgements

Market research like this can lead to a loss of precision. It is likely that the customer will be given quite a limited choice of responses. For example the question might be: This new product will cost £1.65. Is it:

- much too cheap
- too cheap
- about right
- too expensive
- much too expensive

What happens if the customer thinks it is a little bit too expensive? Do they choose 'about right' or 'too expensive?'

In a survey of taste customers might be asked to give a score from the following:

*5 – very pleasant   4 – pleasant   3 – satisfactory
2 – unpleasant   1 – very unpleasant*

The survey's average score was 3.6. What does this mean? Does it mean most people thought the taste was satisfactory or pleasant? Does it mean more people liked the taste than disliked it? Is the score accurate if customers only had 5 choices to choose from?

Direct Mail Case Study

▶ Porsche Cars, Reading

Porsche Cars Great Britain (PCGB) imports all Porsches into Britain and owns five Porsche dealerships. Sales are small compared with volume manufacturers such as Ford or Vauxhall. PCGB know that the people most likely to buy one of their cars are people who have bought one in the past.

This means that PCGB spend 80 per cent of their marketing budget on direct-mail. They have built up a large database of 22 000 current Porsche owners, over half of all Porsche owners in the country. These owners receive a copy of *Marque*, PCGB's magazine. This is an important part of their marketing strategy. Industry sources suggest that sixty per cent of customers will buy from you again simply because you keep in touch.

Keeping a database like this is quite legal but PCGB must comply with the Data Protection Act (see Chapter 2). □

Figure 1.2 A Porsche car

▶ Summary

☐ Data means raw figures.

☐ Information means data in a context which gives it a meaning.

☐ Information can have a very high commercial value.

☐ Data is only valuable if it is correct and up-to-date.

☐ Ensuring data is up-to-date can be time consuming and costly.

☐ It is possible to buy information already in computer readable form, for example on disk.

☐ Some organisations have used illegal practices to try to gain commercial information.

☐ Information may be out-of-date by the time it has been processed.

☐ Direct-mail is an effective, if somewhat unpopular, way of targeting when promoting a product.

☐ You have the right not to have it if you wish.

☐ Information is often coded when it is stored on computer.

☐ Coding information reduces the accuracy and may make it meaningless.

Data Questions

1 Explain why it might be better for a radio station to get a copy of a mailing list on computer, rather than just put an advert in the magazine. (2)

2 A hotel asked customers to score their service in the hotel according to this system.

> **1 – Excellent   2 – Good   3 – Average**
> **4 – Bad   5 – Poor**

The average mark for 100 customers was 1.8.
Are these statements true?
All our customers think our service is good or better.
Our average score is good to excellent.
Our customers think we are consistently good.
Give two reasons why this is not always a reliable way of storing information. (5)

3 A supermarket stock control computer system up-dates its stock levels every evening based on that day's sales. List two possible consequences of the supermarket using out-of-date data. (2)

4 What is meant by the term GIGO? Give an example. (2)

5 With the aid of an example, describe **one** problem which may occur when coding a value judgement. (2) *NEAB 1996 Paper 1*

6 Encoding value judgements can have the effect of reducing its accuracy or meaning. This becomes evident when the data is retrieved and used. Explain, with the use of two appropriate examples, why this may happen. (4) *NEAB Specimen Paper 1*

7 Describe, with the aid of examples, the difference between knowledge, information and data. (6) *NEAB Specimen Paper 1*

8 Computers use information and data

a) What is meant by the term 'data'? (1)

b) What is meant by the term 'information'? (1)

c) Give an example which clearly shows the difference between 'data' and 'information' (2) *NEAB 1996 Paper 1*

9 A local bookshop uses an interactive computer database system to store stock details and answer customer enquires regarding the availability of books. An enquiry for a particular book shows that there is a single copy remaining in stock. However on searching the shelves (and stockroom) the book cannot be found. Give possible reasons for this apparent discrepancy. (4) *NEAB Specimen Paper 2*

10 Study the information about the MPS – an organisation funded by the direct-mail industry to try to help people avoid junk-mail.

a) Give three reasons why the industry is keen to help people who don't want junk-mail.

b) Give one reason why you might register with MPS and one reason why you wouldn't.

c) Give three reasons why businesses use direct-mail.

11 Travelling sales representatives working in the UK can make extensive use of company credit cards to pay for goods and services. A company credit card is one that is issued by a company to its representative. All charges and information relating to each transaction are sent directly to the company.

a) List **four** items of data which are captured each time the card is used. (4)

b) Other than payment information, suggest **one** other potential use for information which can be derived from this data. (2) *NEAB 1997 Paper 1*

12 Information processing is concerned with:

- Input

- Processing

- Output

- Feedback

a) Briefly describe these **four** elements of information processing, using a diagram to illustrate your answer. (6)

b) Explain the difference between 'knowledge' and 'information'. (2) *NEAB 1998 Paper 1*

13 Low quality information can be misleading, distorted or incomprehensible. This type of information is of little value to the decision maker. The output of good quality information is costly and dependent upon many factors.

    a) Identify **three** factors which affect the *quality* of information. (3)

    b) State **two** factors which affect the *cost* of providing good quality information. (2) *NEAB 1998 Paper 1*

14 State **two** factors that affect the value of information and give an example of each one. (4) *NEAB 1999 Paper 1*

15 Many market research firms use questionnaires as a means of gathering raw data for companies about the popularity of their products.

    a) Explain why Information Technology is widely used in market research. (4)

    b) Once the data has been collected, it can be used to give the clients information about their products. Explain the difference between information and data in this context. (4) *NEAB 1999 Paper 1*

Keeping information secure, private and correct

Security and integrity of data

Data stored on computer is vital to the success of any business. The loss of computer files is an extremely serious problem for any organisation, so it is vital that businesses take steps to protect the security and integrity of their data.

Security of data means keeping data safe from physical loss. This could be due to accidental damage, for example natural hazards such as flooding and fire, or it could be damage caused by hardware failure, for example when a tape gets caught up in a drive and is destroyed. The loss of data could be intentional, for example theft by a competitor, unauthorised access (hacking), destruction by viruses or terrorism.

Integrity of data means the correctness of the stored data. Data may be incorrect because of errors in data transmission (caused by background noise on the line), input errors (data typed in wrongly), operator errors (for example an out-of-date version of the file has been loaded), program bugs, hardware breakdown, viruses or other computer crime.

Privacy of data means keeping data secret so that it cannot be accessed by unauthorised users. In the 1980s a hacker accessed the Duke of Edinburgh's e-mail box after finding a system password left on the screen. Since then, the Computer Misuse Act 1990 has made unauthorised access to computer material a criminal offence.

If a computer user orders goods by e-mail, they need to give their credit card number and expiry date. This is private information, which someone else might use to order goods fraudulently if the information was not kept secret.

Methods of maintaining security

There are many security procedures that organisations should use to maintain the security of their data.

Physical security

The most obvious way is to lock the door to any computer installation. The lock can be operated by a conventional key, a 'swipe' card or a code number typed into a key-pad. Any code must be kept secret. Staff should not lend keys or swipe cards to anybody else. Locks activated by voice recognition or fingerprint comparison offer alternative, but expensive, stronger methods of security.

Additional security measures could include computer keyboard locks, closed circuit television cameras, security staff and alarm systems. Passive infra-red alarm systems to detect body heat and movement are commonly used, as they are reliable and inexpensive.

Computer systems with terminals at remote sites are a weak link in any system and must be fully protected. Disk and tape libraries also need to be protected, otherwise it would be possible for a thief to take file media to another computer with compatible hardware and software.

Staff and authorised visitors should wear identity cards, which should not be copyable and should contain a photograph. These are effective and cheap. These security methods are only effective if the supporting administrative procedures are properly adhered to, for example doors must not be left unlocked and security staff should check identity cards.

The security measures used by an organisation will reflect the value of the data stored and the consequences of data loss, alteration or theft. Financial institutions like banks need to have the very highest levels of security to prevent fraud.

Software security

Access to files must be controlled by passwords, which have to be keyed into the terminal in response to a series of questions displayed on the screen. There will be different levels of permitted access for different users depending on their needs.

Some areas of the disk will be write-protected for everyone except the network manager. For example, only the network manager should be allowed to install new and delete old software; add new and delete old users; delete old files, copy files and set files to read-only or read-write. To access very sensitive information, users will need to know several passwords.

'Hackers' are people who, acting illegally for fun or fraud, specialise in breaking through the software protection to gain access to an information system's files. The passwords used should therefore be carefully chosen, kept secure (memorised and not divulged) and changed frequently.

Using people's names, for example, is not a good idea as it may allow entry by guessing or by trial and error. Some computer systems only allow passwords which are not in a dictionary, as 'hackers' may use programs that try every word in a dictionary until they get access.

The system should be aware of repeated unsuccessful attempts to gain access, as this is likely to be an unauthorised user. The network can be set up so that a computer is disabled after three wrong passwords have been entered and only able to restart after a certain time has elapsed. The network manager should be alerted by a message at the server if a number of wrong passwords have been entered in a short time.

A network access log can be kept. This keeps a records of the usernames of all users of the network, which station they have used, the time they logged on, the time they logged off, which programs they have used and which files they have created or accessed.

Figure 2.1 shows part of a school's access log and could be used to detect unauthorised access. The first

excerpt shows the automatic back-up taking place at 3.00 in the morning.

The second excerpt shows computers being turned on and logging on to the network at 8.29 a.m.

The third excerpt shows machines being turned on and users logging on at the start of the school day.

Security procedures

It is important that users follow standard administrative procedures for maintaining security. Passwords should be changed regularly and should never be left written down on scraps of paper. Whenever computer users leave their terminal, for example to see a colleague or to go to the toilet, they must log off so that unauthorised users cannot gain access. Other simple procedures may include:

- using a floppy drive lock to prevent someone using a floppy disk to steal data

- locking the keyboard to prevent unauthorised use

- using a virus check before using floppy disks from an unknown source

- not allowing unauthorised personnel to use your PC, even at home

- backing-up regularly

Figure 2.1 A network access log tells the network manager who has used the network

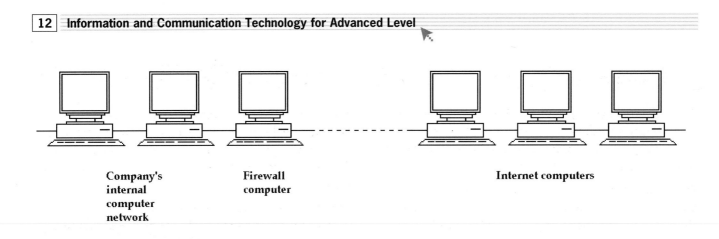

Company's
internal
computer
network

Firewall
computer

Internet computers

Firewalls

Maintaining security is much harder if a company's computer system is connected to a public network like the Internet. As many businesses today use the Internet to make information available to other parts of the businesses or to suppliers, this could present a major security risk. This information may be sensitive to the business and they would not want it getting into the hands of competitors.

'Firewalls' can be used to protect the business's computers from intruders. A firewall is a single security point through which all traffic must pass. The firewall can use passwords to control traffic and can also log details of all access.

A firewall allows organisations to manage and control access easily, greatly reducing the risk of network break-in and the destruction or theft of data.

Virus protection: prevention, detection and repair

A virus is a program which can reproduce itself. A virus on a hard disk of an infected computer can reproduce itself on to a floppy disk. When the floppy disk is used on a second computer, the virus copies itself on to this computer's hard disk. This copying is hidden and automatic and the user usually is unaware of the existence of the virus – until something goes wrong.

Thousands of viruses exist with damage varying from the trivial to the disastrous. The most common virus in the world, *Form,* makes the speaker beep when you press a key on the 18th day of each month. It does not damage the hard disk. The *Jerusalem* virus is more serious. It deletes a program you try to run

on Friday 13th. The *Dark Avenger* virus is very dangerous as it corrupts the hard disk and back-up copies.

Viruses can be prevented by not allowing users to bring their home floppy disks to use on the system, or to take the company's disks home to use on their own PC. Systems can be set up only to allow specially formatted disks, so that users cannot use their home computer disks.

Viruses can be detected and damage repaired using Anti-Virus Toolkit software. This sort of software is widely available and can detect and repair thousands of viruses. Whenever an infected disk is placed in the computer's drive, a warning message appears on the screen. Up-dates of this software are produced every month as new viruses are detected.

Staff may have to put floppy disks into a 'sheep-dip' station before use on the organisation's other computers. This station is fitted with the latest virus detectors. Organisations may also use PCs without floppy disk drives to avoid virus problems.

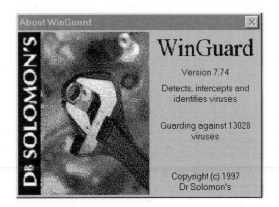

Figure 2.2 Anti-virus software can load automatically and scan for viruses

Virus damage

The computer firm Apricot was infected by one of the Black Baron's viruses *Pathogen*. They had to shut down many of their computers and found that 22 out of 60 machines had been infected.

The company Microprose estimated that one of the Black Baron's viruses cost them half a million pounds after they had to spend more than 480 staff hours checking more than a million files. The Black Baron was eventually sentenced to 18 months in prison. Details of how he operated are given later in this chapter in 'Other computer security laws'.

Encrypting messages

Encryption is a method of scrambling or coding messages so that someone who intercepts it, for example by breaking into a cable, cannot understand or change the message. It should have little effect on performance of the computer system, and may contain other benefits, such as data compression. Data encryption methods are used to protect important and confidential information during transmission from one centre to another.

An example of encryption is the banks' Electronic Funds Transfer (EFT) system. Banks and other financial institutions transfer vast sums of money electronically. These transfers are protected by the latest data encryption techniques.

The simplest of all of the methods of encrypting data uses a translation table. Each character is replaced by a code character from the table. However this method is relatively straightforward for code breakers to decipher.

More sophisticated methods use two or more tables. An example of this method might use translation table 'A' on all of the 'even' bytes, and translation table 'B' on all of the 'odd' bytes. The use of more than one translation table makes code-breaking relatively difficult.

Even more sophisticated methods exist based on patterns, random numbers and using a key to send data in a different order. Combinations of more than one encryption method make it even more difficult for code-breakers to determine how to decipher your encrypted data. Encryption schemes can still be broken, but making them as hard as possible to break is the job of a good cipher designer.

Good back-up procedures

Even the best security may still be broken. Could a computer installation survive a fire that destroyed all its equipment? New equipment could be bought but the data could only be recovered if back-up files were kept.

Back-up files are kept for just such an eventuality. They are kept in a fire proof safe or on another site, which is unlikely to be damaged.

Magnetic tape files automatically create a new master file when they are up-dated using batch processing. The old master file is then retained as a back-up. Usually three generations of old files are kept, just in case. This procedure provides a built-in security system. (See Chapter 9.)

Magnetic disk files are usually up-dated by overwriting the existing data. It is necessary to take regular back-up copies as processing proceeds. The frequency of backing-up will depend on the nature of the data. Sales data for a supermarket, which affects orders and deliveries, will be backed-up hourly, if not more frequently. User data, such as passwords and user names, need only be backed-up every week.

If the data files are lost or corrupted, the data can be recovered by using programs which restore the data from the back-up files.

Vetting of employees

Security procedures are only as good as the employees using them. One of the most common breaches of a company's security is through its own employees. We have already seen how one business's information may be valuable to a competitor. Industrial espionage does exist in the cut-throat competitive world of big business. It is not unknown for employees to be bribed to provide information to a rival. Data may be altered or erased to sabotage the efforts of a company. Employees working in sensitive areas must be totally reliable. They will often need to be vetted before appointment. Strict codes of conduct exist for employees. Anyone found to have breached the organisation's regulations is likely to be instantly dismissed.

Methods of maintaining integrity

There are many ways in which the integrity of data can be affected. Many of the measures to ensure data security will also maintain data integrity. However data can be corrupted accidentally, for example through transmission errors, and businesses need to be aware of this. Again the methods of maintaining integrity will reflect the importance of accurate information.

Errors in transmission can usually be detected by transmitting a *check sum* with the code for each character. The check sum counts up the number of bytes to be sent. This can be compared with the number of bytes received. They should, of course, be the same.

It is also possible to send the same message back to the sender after it has been received. The sender can compare the message received with the original. If there are any differences the message can be sent again until there are no errors.

No human being can guarantee being 100 per cent accurate, so data may be incorrect due to keyboard errors. Keyboard errors can be reduced by validation and verification. (See Chapter 8.) Automatic methods of input like OMR are more reliable than typing in data.

Rights of individuals
The Data Protection Act

It is not just companies who want to ensure information is secret and correct. What about the individuals whose details are stored in the data files? The rights of individuals and their confidential records are protected by law in the UK. This legislation, called the **Data Protection Act 1984 (DPA)**, attempts to ensure data held on computer is secret and correct.

The Data Protection Act is concerned with the automatic processing of personal information. This usually means information kept on computer, but also includes word-processors, smart cards, document image processors, punch-card processors and telephone logging systems.

What does the law say?

Anyone holding personal information on computer files must register with a government official called the **Data Protection Registrar**. The following must be registered:

- what the personal data is used for
- the sources used to obtain it
- people to whom it may be disclosed
- countries where the data may be sent

The role of the Registrar

The Data Protection Registrar is responsible for administering the Act. The Registrar must keep a list (the register) of those who keep personal information on computer and for considering complaints from individuals who feel information about them is being misused. The Registrar may take legal action against anyone considered to be breaking the law.

The Registrar also encourages good practice by publicising how organisations should operate, drawing attention to possible problems and acting as Ombudsman or referee in disputed cases.

Principles of the Data Protection Act

Personal data stored on computer must

- be obtained and processed fairly and lawfully
- be held only for lawful purposes which are described in the register entry
- be used or disclosed only for those purposes
- be adequate, relevant and not excessive in relation to their purpose
- be accurate and up-to-date
- be held no longer than necessary
- be surrounded by proper security
- be provided, on request, to the individual concerned who, where appropriate, has the right to have personal information corrected or erased

(*from* Data Protection Registrar. *DPA 1984: The Guidelines*. November 1994)

Exemptions from the Data Protection Act

The following do not need to be registered:

- Personal data for family affairs or recreational purposes such as lists of club members
- Personal data used only for calculating wages and pensions
- Personal data used for distributing information to the data subjects, for example a mailing list
- Personal data which the law already requires the user to make public
- Personal data which must be secret to safeguard national security or fight crime
- Personal data not stored on computer

Individuals' rights under the DPA

- Individuals have the right to see information held about them (subject access) on payment of a fee of up to £10. Response must be within 40 days.
- Individuals can complain to the registrar if they feel the law has been broken. The registrar may prosecute. Any mistakes must be corrected or deleted.
- Individuals can sue for compensation through the courts if they feel they have been damaged in some way by the misuse of information held on computer.

Is it legal to sell personal data?

It is perfectly legal for companies to sell data to other companies and they frequently do. Before selling, they may wish to draw up an agreement which might say:

- that both companies must be registered under the DPA
- what measures the buying company will take to keep the information secure
- whether the buying company is allowed to sell the data on to someone else
- how long the buying company can keep the data
- the buying company will not store data irrelevant to them

■ ARE COMPUTER BUREAUX COVERED BY THE ACT?

A computer bureau processes data for other people. It does not control the content and use of the data, but it must still comply with the Data Protection Act, in particular the seventh principle concerning the security of personal data.

■ WHAT ABOUT OTHER COUNTRIES?

Other countries like the USA have 'Freedom of Information Acts' which allow the individual to see any personal information stored in their own files, except where national security is thought to be threatened.

■ CRITICISMS OF THE ACT

As the use of information technology increased in the 1980s and 1990s, many people felt that the act did not go far enough. Among their particular concerns were:

- The Internet, which hardly existed when the Act was passed, was not covered.
- There were no controls over paper based records on computer.
- The Data Protection Registrar had no real power to investigate.
- When filling-in forms members of the public had to 'opt-out' if they did not want their details passed on to other users. This often meant ticking a tiny box.
- There were too many exemptions, such as police or MI5 information.

In 1997 the Data Protection Registrar, Elizabeth France, urged the Prime Minister, Tony Blair, to end MI5's exemption from the Data Protection Act after it was discovered that MI5 was storing out-of-date, irrelevant information about people including the Home Secretary, Jack Straw and the former Prime Minister, Edward Heath.

The 1998 Data Protection Act

The European Union Data Protection Directive of 1995 aimed to strengthen Data Protection laws and harmonise them throughout Europe. The 1998 Data Protection Act made the directive law in the UK, to be phased in by 2001.

One of the problems with the 1984 Act was that it gave the Data Protection Registrar very limited powers. If you were not registered and broke the requirements of the Act, the only thing the Registrar could do was to make you register.

The new Act greatly strengthens the powers of the Data Protection Registrar, who will be called the Commissioner when the Act comes into force. Among the changes brought in by the Act were:

- Registration is replaced by simple notification.

- The Data Protection Commissioner can enforce the law with unregistered firms.

- Personal data stored on the Internet and sent by e-mail is covered by the Act.

- Data users must gain the consent of data subjects for the gathering and use of personal information.

- Data will be open for scrutiny both by the Data Protection Commissioner and by any member of the public, whether registered or not.

- The penalty for breaking the law can be a fine and a criminal conviction.

- It is the responsibility of the data user to stop personal data being hacked, lost, damaged or stolen.

- The law is harmonised throughout Europe.

- Data cannot be exported outside the European Union unless the country it is going to has 'adequate' data protection legislation. When the new Act came into force only one country in Asia (Hong Kong) had any data protection laws.

- Paper-based records are included for the first time: the deadline for them is 24 October 2007.

The Home Office has estimated the cost to British businesses as £837 million.

Other computer security laws

Hacking is a crime under **The Computer Misuse Act, 1990**. This Act aims 'to make provision for securing computer material against unauthorised access or modification; and for connected purposes.'

Three new offences were created:

- unauthorised access to computer material

- unauthorised access with intent to commit or facilitate commission of further offences

- unauthorised modification of computer material

In the first category, you are committing an offence if you try to access to any program or data held in any computer and the access is unauthorised and you know at the time that is the case. The maximum penalty is six months in prison and a £5 000 fine.

Persistent hacking is included in the second category, which can lead to up to five years in prison.

The third category of altering data can also lead to six months in prison.

In the summer of 1994 Stephen Fleming, a temporary employee at British Telecom, gained access to a computer database containing the telephone numbers and addresses of top secret Government installations. Mr Fleming, who worked at BT for two months, found passwords written down and left lying around offices and used them to call up information on a screen and copy it. Police interviewed Mr Fleming and threatened prosecution under the first category of the Computer Misuse Act for gaining unauthorised access to computer data. BT said that security would be tightened.

In November 1995 Christopher Pile, who called himself the Black Baron, became the first person convicted under the Computer Misuse Act. Pile created two viruses named *Pathogen* and *Queeg* after characters in the BBC sci-fi comedy Red Dwarf. The viruses wiped data from a computer's hard drive and left a Red Dwarf joke on screen which read: 'Smoke me a kipper, I'll be back for breakfast ... unfortunately some of your data won't'.

Pile, of Plymouth, Devon, pleaded guilty to five charges of gaining unauthorised access to computers, one of inciting others to spread the virus and five of making unauthorised modifications and was jailed for 18 months.

Daily Telegraph, Saturday 2 November 1996

Organ group president is fined for list on computer

By Maurice Weaver

The honorary president of the Association of Organ Enthusiasts was fined £50 and ordered to pay £683 costs yesterday for keeping his membership list on his home computer.

Trevor Daniels, 52, told magistrates in Towcester, Northants, he did not realise that he had to register and pay a £75 fee to the Data Protection Registry. Admitting two breaches of the Data Protection Act, Daniels told the court that he had spent hours keying in details of organ enthusiasts who subscribe to his bi-monthly magazine.

He said: 'I didn't think names and addresses were sensitive information. I am sorry for what I have done.'

Jan Jellema, prosecuting for the Data Protection Registrar, said: 'As a result of information received, an investigator visited Mr Daniels' home. He was fully co-operative and showed him his computer system. During the interview he was frank and accepted he was not registered. He said he was unaware that he needed to be registered and that he had no idea he was contravening the law.'

Daniels, a father of five, works as a professional musician giving organ recitals around the country. He pleaded guilty on behalf of two small businesses he runs with his wife, Rosemary, from their home in Highfields, Towcester. Both are linked to the non-profit making Association of Organ Enthusiasts and were using information in the database.

One organisation, the Association of Organ Enthusiasts Festivals Limited, runs holidays for organ enthusiasts. The other is called the Association of Organ Enthusiasts Publications which publishes the bi-monthly magazine. After yesterday's court hearing he said he suspected that he might have been reported by a competitor in the organ world.

He said: 'I think the whole Data Protection Act is a con. One minute you can have something written down on paper but as soon as it goes on computer it becomes sensitive. It is crazy. I don't think this law protects you one bit. It is just another excuse for a tax. You have to pay £75 to register and the fact of the matter is that once you are registered you can do what the hell you like with the information.'

Note: Although personal data stored for recreational purposes such as lists of club members is exempt from the Act, Mr Daniels was storing information on behalf of a limited company. The 1984 Data Protection Act does not distinguish between multi-national companies and small companies.

Lack of security

Case Study

Danny Hughes is an A level student who had a holiday job on the production line at Betta Biscuit plc. One lunch-time Danny decided to explore the factory and found his way into the computer room. There was no-one about. Danny sat down at a terminal and typed in a few usernames with no luck. Then he noticed a birthday card on the desk; to Bob with love from Jane. Danny typed in the username BOB and was asked for a password. He typed in JANE and to his surprise it was accepted!

A menu appeared on the screen. Danny chose payroll. He could load up the payroll information of all the employees. Danny loaded the file of his friend Chris and cut his hourly pay by half. Two workers came in and saw Danny, but no-one said anything. Danny logged off quietly and slipped out of the room undetected.

▶ Summary

The following may be causes of data loss:

☐ natural hazards such as flooding and fire

☐ damage caused by hardware failure

☐ theft by a competitor

☐ unauthorised access (hacking)

☐ viruses

☐ terrorism

☐ errors in data transmission

☐ input errors

☐ operator errors

☐ program bugs

Unauthorised access to a system may:

☐ provide vital information to competitors

☐ result in the deliberate corruption of the data

☐ allow fraudulent changes to be made to data by employees or others

☐ result in loss of privacy for individuals if the information is confidential

Loss of data may be disastrous to the organisation.

Data loss can be prevented by

☐ sticking to agreed administrative procedures

☐ physical methods, for example locks

☐ software methods, for example passwords

☐ anti-virus software

☐ encryption

☐ keeping back-up files

☐ protecting the data from accidental or deliberate corruption – maintaining data integrity

☐ vetting new employees

To detect any unauthorised access or changes to the information system:

☐ user actions should be logged and monitored

☐ users should be identifiable

☐ the files should be capable of being audited

Computer users who use or process information about people (personal data) must register with the Data Protection Registrar. The information must

☐ be obtained and processed fairly and lawfully

☐ be held only for lawful purposes which are described in the register entry

☐ be adequate, relevant and not excessive, accurate and up-to-date

☐ be held no longer than necessary

☐ be surrounded by proper security

Security questions

1 Read the Case Study 'Lack of security' on page 17.

   a) Suggest as many steps as you can that Betta Biscuit plc should take to improve their security.

   b) Who has broken the law, Danny or Betta Biscuit?

2 Describe three ways a computer user should guard against loss of data due to fire and flood.

3 Explain the difference between security and privacy of data.

4 Some organisations suggest that you should change your password every week and not use a word in the dictionary as a password. Explain why this advice is given.

5 A national distribution company advertises its products by sending personalised letters to thousands of people across the country each year. This type of letter is often known as 'junk mail'. The distribution company purchases the list of names and addresses from an agency.

   a) Describe **two** ways in which the use of information technology has increased the use of 'junk mail'. (4)

   b) The company wishes to target letters to people who are likely to buy its products. How might this be done? (2)

   c) A person receiving this type of mail writes to the company to complain that it is acting illegally under the terms of the Data Protection Act. Give **three** statements the company may use in its reply to show that it is operating within the terms of the Act. (3) *NEAB 1996 Paper 1*

6 Charlie Reid wishes to buy a washing machine on hire purchase. He is refused credit as he has a bad credit rating on the computer. He asks why and he is told that his son, who lives at home, has a bad credit rating. When he asks why his son has a bad credit rating, he is told that his son used to live at an address 100 miles away in Lancashire. Another resident at the Lancashire address owes his bank £1000. He complains to the bank that they've broken the Data Protection Act. They deny that they have. Explain whether you think the Data Protection Act has been broken or not.

7 An on-line information retrieval system holds confidential personal data.

   a) What precautions should be taken to

      i) minimise unauthorised access

      ii) detect unauthorised access? (4)

   b) Why might different users be given different access privileges? (2)

   c) Explain how the data should be protected from corruption. (4) *NEAB Specimen Paper 1*

8 A multi-national organisation maintains an information technology system which holds a large amount of vital and sensitive data.

   a) Describe **three** steps which should be taken to protect the data against deliberate theft or corruption. (6)

   b) Describe **three** steps which should be taken to protect the data against accidental loss. (6) *NEAB 1996 Paper 1*

9 A common way of permitting different levels of access to on-line files is the use of passwords. Once a password has been input the user may be allowed to perform a number of different actions upon the data within the files, dependant on the level of access given by that password. Describe **four** of these possible actions. (4) *NEAB 1997 Paper 1*

10 A college maintains an extensive database of its full-time students. The database contains personal data, the courses students attend, and higher education or employment applications.

   a) Describe how the college might keep the personal data of the students up to date. (3)

   b) The college wishes to sell the personal data to a local sports retailer. An agreement is to be written between the college and the retailer. Describe **three** issues, relating to the data, that should be included in this agreement. (3) *NEAB 1997 Paper 1*

11 In 1997 the Data Protection Registrar, Elizabeth France, threatened to take two utility companies to court for

breaking restrictions in the Data Protection Act. The companies were sending promotional leaflets and advertisements to their customers. The official notice from the registrar to the two companies said they should stop using the customer list for advertising. Write briefly to one of the companies explaining which principle of the Data Protection Act may have been broken.

12 A company equips its sales staff with portable notebook computers. The IT department feels that a set of 'procedures' is required to ensure the integrity of the data and software held on the notebooks. Suggest **four** different items that the company might include in its set of procedures. (4) *NEAB 1997 Paper 1*

13 Mr Daniels was very critical of the Data Protection Act. (See press cutting: 'Organ group president is fined for list on computer' on page 17.)

   a) Is it inconsistent that the DPA does not include data stored on paper at present?

   b) Can you do 'what the hell you like with the information' once you are registered?

   c) Does this law offer any protection or any other advantages or is it really 'just another excuse for a tax?'

   d) Should the Data Protection Act distinguish between multi-national corporations and small organisations?

14 The Computer Misuse Act defines three types of offence. With the aid of examples, describe each of these **three** types of offence. (9) *NEAB 1997 Paper 1*

15 A publishing company administers its business by using a database system running on a network of PCs. The main uses are to process customer orders and to log payments. You have been asked about backup strategies and their importance.

   a) Give **two** reasons why it is essential that this company has a workable backup strategy. (2)

   b) State **five** factors that should be considered in a backup strategy, illustrating each factor with an example. (10)

   c) Despite all the precautions, some data might still be lost if there was a system failure. Give **two** reasons why this might be the case. (2) *NEAB 1998 Paper 2*

16 The term 'data protection' covers the maintenance of the integrity, quality and ownership of data handled by information technology systems. There are many ways to protect data, and there is also legislation to ensure that data is kept private and secure.

Discuss the Data Protection Act 1984, including reference to:

- objections to the Act;

- the information that should be recorded when registering with the Office of the Data Protection Registrar;

- the likely future of the Act and the consequences for data owners/users as a result of the EU Directive on Data Protection.

Quality of language will be assessed in this question. (15) *NEAB 1998 Paper 1*

17 Different levels of access can be provided for on-line files which permit users to perform a number of different actions upon the records within the files. Give **four** of these possible actions. (4) *NEAB 1999 Paper 1*

   a) State the **three** levels of offence under the Computer Misuse Act of 1990. Illustrate each answer with a relevant example. (6)

   b) Describe **four** separate measures that can be taken to prevent accidental or deliberate misuse of data on a stand-alone computer system. (8) *NEAB 1999 Paper 1*

18 Explain, using examples, the distinction between security and privacy as applied to data held in a computerised information system. (4) *NEAB 1999 Paper 2*

Social impact of IT

Computers have already had a great effect on society, for example in the fields of employment, leisure and communications. This trend is likely to increase as processors get faster, computer memory gets bigger and their price is comparatively less. In this chapter we look at some of the ways IT has affected the way we live and work.

IT and business

Much of business today is unrecognisable compared with business in the days before computers. The electronic office is an obvious example of the effect of IT on business. A modern office is likely to have a computer on every desk. The work carried out in offices is generally the receipt, processing, storage and despatch of information. The computer can do all these things more efficiently than traditional methods.

The arrival of fax, e-mail and satellite links has completely changed how businesses and some individuals communicate. Businesses now advertise their fax numbers and e-mail addresses prominently. It would seem logical that this would affect the level of traditional communications, particularly letters sent by post. However the number of these letters has still increased reflecting the increased need for communication.

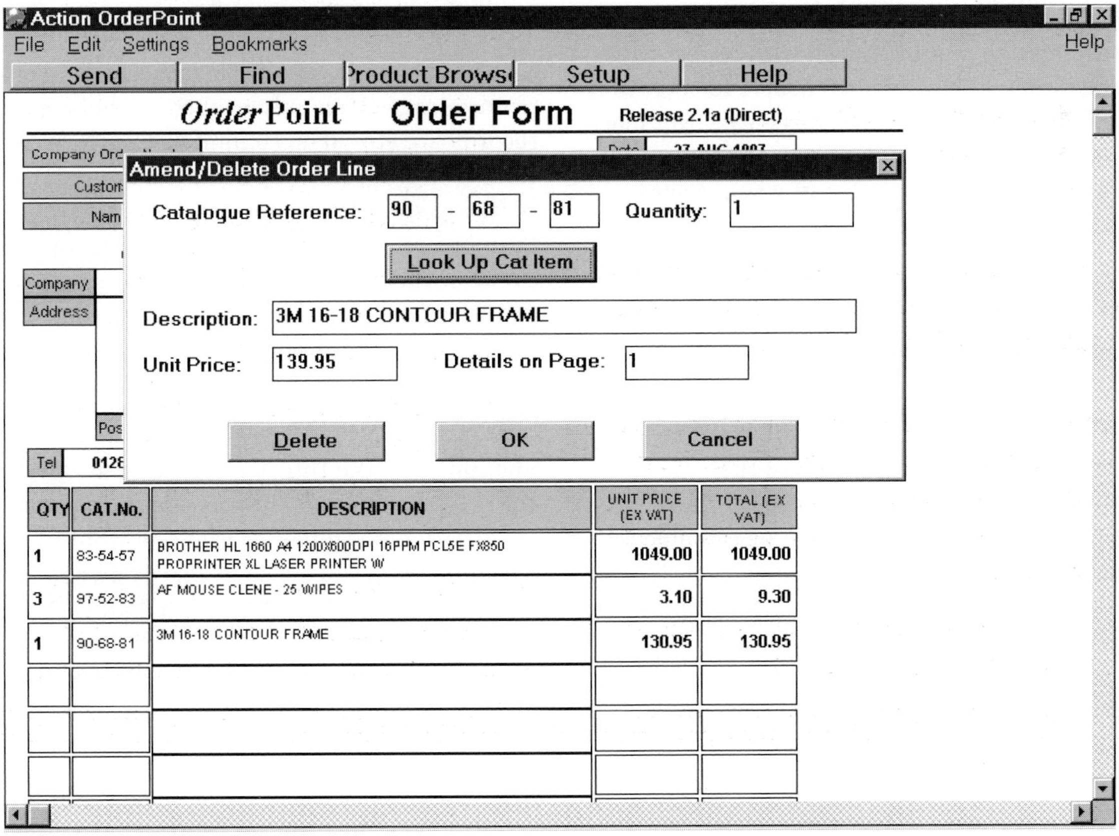

Figure 3.1 An on-line order form

Products can be ordered on-line using the Internet. Electronic Data Interchange (EDI) is a system of sending orders, paying invoices and sending information electronically. (See pages 35 and 36.) Money can be transferred electronically to pay for goods and services. These are all examples of *e-commerce* (electronic commerce), which is revolutionising how businesses order and sell products.

IT and manufacturing

Many industries now use Computer-Aided Manufacturing (CAM). For example, motor cars are assembled by computer-controlled robot welders; Benetton has been able to produce new clothes more quickly as a result of using IT (see Case Study 1 on page 26).

The quality of these computer-manufactured products is more consistent and higher, leading to greater reliability and increased productivity. Working conditions are often cleaner. More automation offers the workforce the prospect of shorter working hours and more leisure time but it always doesn't work out as predicted. There may be further reductions in the workforce.

Products can also be manufactured with fewer staff. A brewery canning line needed 15 people to operate it before computerisation. Now it needs only 2, yet more beer is canned and there are fewer breakdowns. Benetton's warehouse needs 16 staff when it once needs 300.

Obviously, some jobs have been lost. Workers have been made redundant or redeployed to other jobs. Skilled workers may have seen their skills replaced by automatic machines.

The development of IT has severely affected the number and role of people employed in the manufacturing industry. But if these industries had refused to change, they would not have been able to offer value for money in a competitive market. In the early 1960s the British motorcycle industry failed to adapt to cheap but reliable competition, mainly from Japan. As a result the industry died out completely. Industries that do not modernise will lose out to competition from those who have modernised. Even more jobs will be lost.

However computers have created jobs in the manufacturing of hardware, and also in sales, repair and maintenance, technical support and consultancy sectors. New products are now being manufactured thanks to computers, such as CDs, videos, microwaves and satellite dishes.

IT and commerce

The number of transactions carried out by banks has grown so rapidly that they could now not operate without computers. Banks transfer money electronically. For example, most workers are paid directly into their bank account by computer.

Cash is not as important as it used to be. We don't need to carry as much cash as before thanks to credit and debit cards. If we run out of cash, we can visit an ATM at any time. Some people have predicted a cashless society (see Case Study 2 on page 27). Home banking means that you can check your account and pay bills from home.

IT has also affected our shopping habits in other ways. Credit cards mean it is possible to shop and pay for your goods without leaving home. Cable and digital TV shopping channels and the Internet have provided new ways of finding out what to buy instead of the traditional catalogues.

Many supermarkets offer loyalty cards to customers to encourage them only to shop at their store. All records are stored on computer. Computerised stock control has enabled supermarkets to reduce costs by reducing stock levels and stock even more products than before. These trends have contributed to supermarkets getting bigger and bigger, improving facilities for shoppers but leading to the decline of small corner shops and traditional town centres. Phone cards for public telephones have resulted in less theft from phone boxes.

Shopping via the Internet is taking an increasing share of the market. It is not just hi-tech and multinational companies that use what is often called e-commerce to sell goods. A family owned butchers from Yorkshire, Jack Scaife, used an Internet site (*www.jackscaife.co.uk*) to sell its bacon, sending deliveries all over the world. Soon e-commerce was bringing in £200 000 worth of sales from a site that cost £250 a year to maintain.

However companies advertising on the Internet are more likely to be large, multinational companies, particularly those involved in high-tech industries.

Software suppliers can even send programs and files by e-mail. Customers can order by e-mail giving their credit card numbers. (See Amazon Books, Chapter 4)

There are problems however. E-mail is not completely secure. It is difficult for a business to check that it is really your card number or for a user to find out if a business is genuine (particularly if it is not even in this country). Some web browsers often have the facility to encrypt your details, for example to increase security if you are ordering goods on-line.

By the year 2000, 28 million Americans had purchased goods on-line.

> ### Loyalty cards
>
> **Case Study**
>
> Retail chains like Tesco, Sainsbury, Boots and WH Smith offer loyalty cards to shoppers. Card-holders receive vouchers to spend in the store; the more they spend the more vouchers they get.
>
> It sounds a good deal for the customer as they get money back and a good deal for the shops as it encourages the customer to go back to that store. But is there something more sinister behind these cards?
>
> Stores can use these cards and data from sales to build up a huge database on customers. This might include not just how much we spend but what we spend it on and when we go shopping.
>
> This could be used in direct mail marketing. Customers at a store like Boots, who have bought cosmetics may be told of the latest products and offers. Customers who have never bought cosmetics need not receive these details.
>
> This targeting could be seen as more efficient. Stores need not waste money sending mail to people who aren't interested. Other people may see it as threatening. 'They know exactly what I have bought from them in the last year. They know more about my shopping habits than I do,' said one customer.

IT and medicine

Computers are used in the administration of hospitals and doctors' surgeries, storing patients' records. Pharmacists keep records of customers and their prescriptions. This makes the administration much easier but leads to obvious concerns about inaccurate data and security of data.

Some hospitals are now experimenting with storing medical records on smart cards kept by the patient and taken with them every time they visit a doctor, dentist, pharmacist or hospital. The smart card can store a complete medical history and can be up-dated at the end of each visit.

IT also helps in the diagnosis and monitoring of patients' illnesses. Expert systems can be set up to help in diagnosis by asking questions about symptoms and using the answers to draw conclusions. Computer controlled ultra-sound scanners enable doctors to screen patients very accurately. X-ray film is being replaced by on-screen digital pictures. Computers can be used for continuous monitoring of patients' bodily functions such as blood pressure, pulse and respiration rates.

IT offers many opportunities for people with disabilities, particularly those who have difficulty communicating. There are various computer adaptations available for people who can't use a mouse or keyboard or who can't see a normal monitor too well. On-screen keyboards allow users who can use a mouse access to the computer, providing point-and-click access to standard keyboard letters, whole words and communication phrases. Speech recognition means it is not necessary to be able to operate a keyboard or a mouse to use a computer. Output can be to large screens, 'spoken' or in the form of Braille to help users with poor eyesight.

Many organisations specialise in providing free computer training for people with disabilities. The British Computer Society has a Disabled Access Group providing information and support for people suffering from deafness, blindness, learning difficulties, partial sight, mental health or physical disabilities.

IT in the home

Information technology has already changed our lives at home. New gadgets such as automatic washing machines, dishwashers and microwaves have made our lives easier. Our standard of living has improved as goods are made more cheaply but to a higher standard.

Computers can be used to control and protect our environment by controlling heating, ventilation, hot water and security devices like alarms and closed circuit TV cameras. We can use computer communications to check our bank account, order goods via the Internet and even do our work at home.

Computers have changed people's leisure activities. Video recorders, satellite and cable TV and have changed viewing habits; over 25 per cent of British homes have a PC and this is expected to double over the next few years; home computers and games consoles offer a new form of entertainment.

IT and education

Computers are of course common in schools, colleges and universities, but IT has improved education in other ways. Intelligent tutoring systems enable the computer to give the student information, ask questions, record scores and work at the individual's pace.

Computer communications provide new opportunities for distance learning. Students can send in their work to their tutors by e-mail. Videoconferencing may be used for lectures, enabling two-way communication and discussion. This is particularly useful in remote and under-populated areas.

Computer Based Training (CBT) is a sophisticated way of learning with help from IT. A simulated aircraft cockpit used for pilot training is an example of CBT, which is cost-effective and less dangerous than the real thing.

Martin Byron is studying to be a teacher, undertaking a Post Graduate Certificate in Education course at the Open University. His tutor, Kate Staples, lives 80 miles away. Martin has to send assignments to Kate regularly. Kate then marks the assignments and sends her comments back to Martin.

To help students communicate with their tutor, all the students receive an Apple Mac personal computer at the start of their course. Martin sends in his assignments by e-mail.

Martin has found the system very helpful. It has helped him maintain deadlines, as he cannot say that he has lost work in the post and e-mail is date stamped. He can easily send part of assignments to Kate to make sure that he is on the right track. Communication between the student working throughout the day in a school and the tutor working in a college is much easier. Martin has also improved his IT skills.

Changes in employment patterns

Technological developments since the Industrial Revolution over two hundred years ago have led to changing patterns of employment. The Industrial Revolution led to the building of factories that resulted in a shift in the population from the countryside to towns.

Inventions since then such as the steam engine, mechanical looms, the internal combustion engine, the typewriter and the telephone have each led to changes in employment patterns and working practices. The horse and cart has been replaced by motor transport, coal has been replaced as a domestic fuel leading to considerable changes at work.

The development of the computer has also changed patterns of work, but it is different because it has affected nearly every part of industry and commerce.

Some skills have disappeared completely. For example, in the printing industry, typesetting used to be a skilled operation using hot metal. It was performed by print workers who had undergone a seven-year apprenticeship. Now it can be done in the office using a desk-top publishing program and a standard PC. This has resulted in greater job flexibility and the breakdown of the traditional demarcation lines between printers and journalists.

The introduction of computers into the newspaper industry was a major cause of industrial unrest in the 1980s. When News International, who own *The Sun*, *News Of The World*, *The Sunday Times* and *The Times*, moved printing from Fleet Street to Wapping in 1986, 5200 workers went on strike in an attempt to defend their jobs, working conditions and union recognition.

The workers were dismissed for breach of contract. Workers picketed the new plant at Wapping every night for 13 months in an unsuccessful attempt to get News International to re-instate the workforce.

Automatic telephone exchanges have cut the number of personnel needed. Football pools checkers are no longer needed as the job can be done automatically.

Some jobs may have changed little, such as gardeners or delivery drivers, but they still may be affected by such inventions as computer-controlled greenhouses or automated stock control. Other jobs such as supermarket check-out operator, bank clerk or secretary have changed considerably. This has usually meant that existing staff have had to retrain to use IT.

Changing employment locations

In many industries improved communications mean that job location is no longer so important. Telecommuting means that in some jobs it is possible to work from home. Small information technology businesses may prefer to be based in a rural 'telecottage' rather than in a city office, with its problems of congestion and pollution.

The biggest private employer on the Channel Island of Alderney is a company called SportingBet who take bets over the Internet: (http://www.sportingbet.co.uk/) No longer does the betting shop need to be near the customer. SportingBet have brought much needed employment to a small island and as Alderney is outside the United Kingdom tax area it can offer tax-free betting.

As future developments in the computer industry take place, work patterns will continue to change. Telecommuting may lead to many more people working from home, changing work patterns again. There is a danger of society splitting into the haves and the have-nots. Those with IT skills, who will be in demand and have well-paid jobs. Those with no IT skills will find it difficult to gain employment.

Drawbacks of IT

The development of IT is not entirely good news. As well as possible loss of employment opportunities, there are other disadvantages too.

Many people are concerned about the amount of personal data held on computer databases and the potential for lack of privacy resulting from misuse of this. The Data Protection Act (see Chapter 2) was passed to try to ensure that the data is accurate and protected by appropriate security. However it does not apply to the police in the fight against crime or security organisations such as MI5. Other people feel threatened by suggestions of a national identity card (see Case Study 3 on page 28).

Computers have led to increased opportunities for criminal activity such as fraud or theft of data (see Chapter 6). Computers have also led to health and safety problems (see Chapter 7). It is true to say that we

are prepared to put up with these disadvantages as our society today is dependent on information technology.

Dependence on computers

Our lifestyles today depend on IT. Computers offer the facility to process data quickly which could not be done by any other method. IT is vital to so many aspects of modern life that it is impossible to imagine life without the computer. Failure of a computer system would be potentially disastrous for a business.

We have seen that a study in the USA showed that businesses never recover from a loss of computers for ten days or more. For this reason businesses must have appropriate back-up systems to ensure that no data is lost and that alternative hardware is available in case of breakdown. Many large businesses have more than one computer centre in different locations so that if one fails or experiences power failure, the other can take over.

In 1996 the world's largest Internet provider, America Online (AOL), experienced a 19-hour breakdown, while installing new software to allow more callers to use the system. Businesses using AOL could not send or receive e-mail, severely disrupting large users. Newspapers such as the *New York Times* and the *Chicago Tribune* who receive copy by e-mail were unable to do so.

Fortunately the breakdown did not last too long.

Pressure groups and IT

Pressure groups are organisations trying to press for a change in the law or our attitudes. Some pressure groups try to prevent some of the worst pitfalls of the development of IT, for example trade unions try to protect the pay and conditions of their members whose jobs are under threat. They also will be concerned with health and safety issues. Liberty (formerly the National Council for Civil Liberties) are concerned about threats to individual liberty, for example by storing personal data on computer or by forcing individuals to carry an ID card. Liberty are campaigning against censorship of the Internet.

Pressure groups use IT to store records of their members and communicate with them, for example by producing newsletters. Lots of pressure groups have their own web sites, which is a cheap way of informing the public of their existence and their aims. Anyone interested can send them an e-mail.

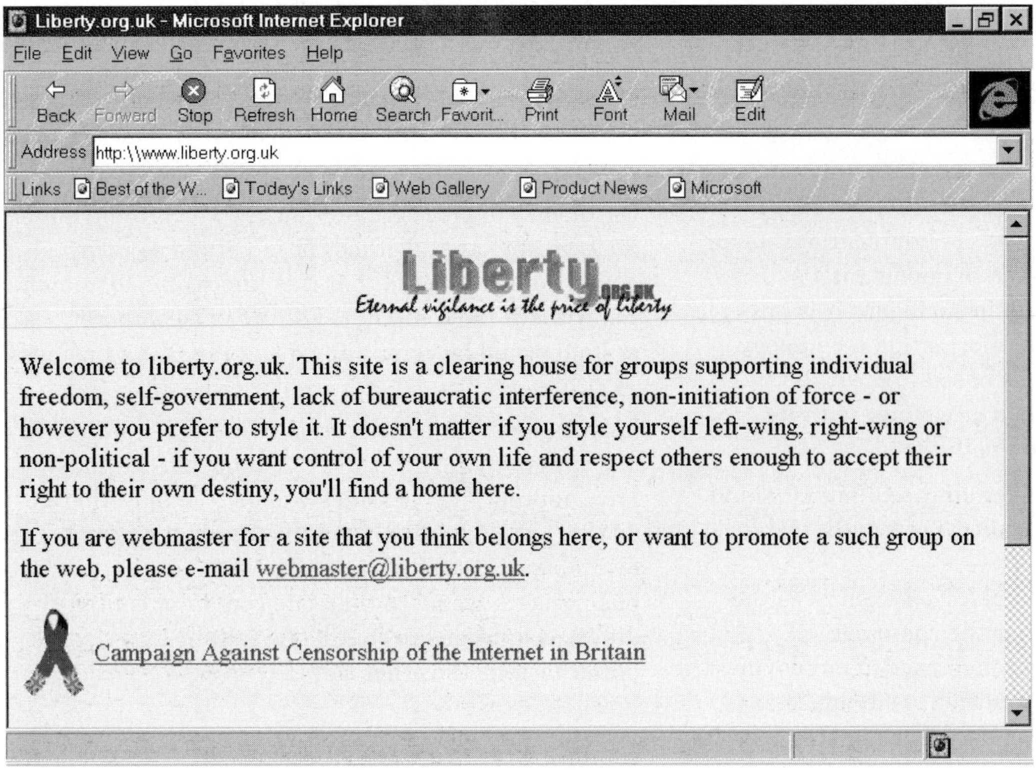

Figure 3.2 The web site of Liberty

Benetton

Case Study 1

Founded in 1965, Benetton is a highly successful Italian fashion company that manufactures and retails its own range of clothing through a world-wide chain of shops.

COMPUTER-AIDED DESIGN

Benetton makes use of CAD systems to produce a template for items such as a pair of jeans in a range of sizes. The CAD system automatically calculates the best way to lay the templates on the fabric so as to minimise the wastage of materials used and cuts the time taken to produce these templates from 24 hours to two hours.

Benetton is linking its CAD system to its computer-aided manufacturing (CAM) systems. This enables designs created on the CAD systems to be transferred directly to computer-controlled knitting machines, increasing the flexibility of the system still further.

PRODUCTION AND WAREHOUSING

The warehouse is totally automated and requires **16 people** to run it, consisting of **8** maintenance personnel, **3** warehousemen, **2** computer operators, **1** general director, **1** director of computer operations and **1** distribution director. It is estimated that a similar warehouse organised in the traditional way would need about **300 workers**.

THE INFORMATION SYSTEM

Benetton has two mainframe computers linked to a network of personal computers. This system is used for general administrative tasks such as payroll, stock control, and invoicing, and has links to the other locations involved in the design, manufacturing, warehousing, and retailing of garments. They use a wide area network to link the Benetton main computer system with the company's agents and factories.

Benetton has 75 agents throughout the world, responsible for the retail operations in his or her area. The agents use a personal computer to link into the communications network and will send orders using e-mail. The system collects information from the agents, up-dates the agent's product and price files and confirms orders.

This network provides management with up-to-date information on sales in each market, making planning much easier. The network also reduces the time taken for orders to reach headquarters, making scheduling of manufacturing easier and supply faster. □

Case Study 2

The cashless society

There have been many developments in Information Technology that are leading to a society without cash. These include:

- credit cards where computers store financial details
- cheques which are processed by computers using MICR
- direct debit used to pay regular bills is generated by computer
- wages and salaries are paid by electronic transfer and not in cash
- phone-cards can be used for telephone calls
- smart cards containing a microchip can be used for automatic debit and credit (see below)
- electronic fund transfer connects the shop with the banks' computers

In 1995 a company called Mondex, jointly owned by the NatWest Bank, the Midland Bank and British Telecom, launched an experiment in Swindon hailed as the start of the cashless society. Local people were issued with a Mondex card, an 'electronic purse' that can be charged with cash and then used to pay for goods and services for a monthly fee of £1.50. Mondex took five years to develop at a cost of 'tens of millions of pounds'. Swindon was chosen because it is seen as a perfect example of an average town. After three months, just 4 per cent of householders had the card and a total of only £250 000 had been spent using Mondex.

In 1996 a rival company, Visa International, one of the world's largest credit card providers, started a similar experiment in Leeds. Visa Cash, a chip-based plastic card, allowed users to make everyday purchases of small items such as a newspaper or a pint of milk, without having to scrape around for the right change. Holders of the Visa Cash card can 'load' the card's electronic chip from their bank account, up to a limit of £100, at any of 3000 specially designed automated telling machines.

Mondex reported 13 000 users in Swindon after 15 months, well below target, although trials had been more successful in Hong Kong, where more than 20 000 customers and 400 stores signed up for the cash card at two shopping malls. Mondex was also being tested in Canada and in Australia. Visa International said its trial would be among 70 000 Leeds residents in 1997. A spokesman for Visa said the card was unlikely to lead to the complete demise of cash, but they hoped it would ultimately eliminate cash from many smaller everyday transactions, typically under £5.

Advantages of cashless society
- no need to carry cash
- can shop from home by phone or mail order
- less chance of being mugged

Disadvantages of cashless society
- some transactions are very small amounts of money
- small traders unable or unwilling to install equipment
- some people prefer cash
- cash sales are usually quicker ☐

■ CASHLESS ON THE UNDERGROUND?

In 1997, London Transport announced plans for a smart card ticket project costing over £1 billion. The project call PRESTIGE, the Procurement of Revenue Services Ticketing Information Gates and Electronics, aims to replace tickets with contact-less smart cards on the London Underground tube and bus network by 2001.

Passengers using the system will charge up smart cards at terminals in London Underground stations. Money will automatically be deducted from the cards as passengers walk through electronic barriers. ☐

Case Study 3

National Identity Cards

In recent years many people have proposed the introduction of a National Identity Card system to tackle rising crime. The cards could store information in a magnetic strip or be smart cards, storing information in a chip inside the card.

In 1994 the then Home Secretary Michael Howard proposed introducing a voluntary ID card. He said the new card might serve as a bank card, driving licence, social security card and kidney donor card. 'No one would be forced to get one but I believe in time the vast majority would'. He said that an ID card would help stop under-age drinking, teenage smokers and young children hiring adult videos.

Another benefit of a national scheme could be 'enabling people to dispense with bulky passports when travelling around Europe, and handy proof of identity for commercial transactions.'

In 1995 a Gallup poll said 74 per cent were in favour of ID cards while 62 per cent thought they should be compulsory.

Polls can be misleading. When the Australian government tried to introduce a National Identity Card in 1986, the initial polls showed that the majority of the population were in favour. But when it came to implementation a year later, a public outcry led to the shelving of the plans.

The Data Protection Registrar warned about the dangers of a voluntary identity card becoming compulsory by stealth. For example if to get a driving licence you had to have an ID card, then the identity card would have become compulsory for most people. A report for ministers said that fraudulent use of smart cards could be contained at a very low level. Measures proposed to counter fraud included using hand-writing and fingerprint identities.

Many people think that ID cards represent a threat to individual liberty. Jack Straw, then the Shadow Home Secretary, said Labour would oppose compulsory cards as 'alien to the British tradition' of individual liberty. He was not opposed to voluntary ID cards in principle, but the case had yet to be proved. ☐

Figure 3.3 Could all these cards be replaced by a single card?

Case Study 4

Electronic information retrieval

Traditionally we have received information on printed media, such as newspapers, magazines, books and encyclopaedias. Now we can get this information from electronic sources; viewdata, teletext, the Internet, CD-ROMs.

Electronic newspapers such as teletext already exist, are free and up-to-date. They reduce paper waste and the need for deliveries, cutting pollution and congestion. The user only needs to access the areas he wants. However they have poor graphics and are limited by the size of the screen. They have only one size of text. They may have a long access time and as yet, have not affected newspaper and magazine sales

An encyclopaedia on CD-ROM offers the opportunity to search through the whole encyclopaedia for a particular word and to copy text and graphics into a word-processor or a DTP package. The CD-ROM can easily be up-dated as it is stored on computer and is cheap to produce. A CD-ROM takes up very little physical storage space.

The Internet offers millions of pages of information and the opportunity to send and retrieve e-mail.

As with any technology there will be those who want to be ahead of the rest, if only to show off. Others will be frightened of the new technology and try to avoid it. There is again the danger that society will split into two: those who have access to the new technology and those that don't.

You may have heard on the radio or television, 'Fax us now on ...' Only those who have a fax can take part. Those without a fax feel left out. This is now beginning to happen with the Internet as more organisations advertise their web sites.

New jobs are springing up in creating CD ROMs and Web pages. Other jobs in printing encyclopaedias are likely to reduce or even disappear.

Books and newspapers are unlikely to disappear. They are relatively cheap to produce, convenient to use, easy to carry around anywhere and not dependent on technology. ☐

Direct Mail Case Study 5

The CD-ROM encyclopaedia brings death of a salesman

By A.J. McIlroy

For years his hard-sell reputation made him the butt of music hall jokes. But now the door stepping days of the Encyclopaedia Britannica salesman have gone forever. The end came yesterday when Encyclopaedia Britannica International, the Chicago-based owners of the world's oldest continually published English-language encyclopaedia, closed down its Britain and Ireland direct sales operation 'because it cannot compete with the electronic age'.

The 70-strong team could not talk their way around the advent of the CD-ROM and the Internet. They found it difficult to persuade a family to spend up to £3,000 on 32 heavy volumes, when the CD-ROM version costs just £125. Fifteen months ago there were 225 salesmen in the team.

Daily Telegraph, 13 January 1998

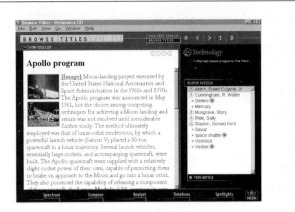

Figure 3.4 The CD-ROM version of the Encyclopaedia Britannica costs just a fraction of the price of the now-defunct paper version. It includes sound, videos and links to the Internet.

Summary

Computers have had an enormous impact on society in business, manufacturing, commerce, medicine, education and in the home.

☐ Computerisation has led to the loss of some jobs. Other jobs have been created. Employment patterns have changed throughout history and will continue to change. Employment locations are also changing.

☐ Developments such as fax, e-mail and on-line databases have changed the way businesses operate.

☐ International banks transfer vast sums (about $2 trillion a day) electronically. The importance of cash has been reduced.

These developments have raised concerns about dependence on computers, security of sensitive information, suggestions of a national identity card, threats to individual liberty, the potential for fraud and the polarisation of society into the IT literate and the IT illiterate – the haves and the have-nots.

Computers and society questions

1 'The development of Information Technology has resulted in increased unemployment.' Discuss this statement including specific examples. Include in your discussion:

- examples of jobs affected or lost due to Information Technology

- examples of new jobs created by Information Technology

- examples of jobs which have not been affected at all.

- deskilling caused by IT (20)

2 'Identity Cards are unnecessary and a threat to the British way of life.' Discuss this statement including reasons for introducing ID cards and for not introducing them. (15)

3 'In the future, magazines and newspapers will be replaced by information retrieval systems.' State whether or not you agree with this claim and give **three** reasons to support your answer. (3) *AEB Computing Specimen Paper 2*

4 Describe briefly **three** different systems in which computers are involved in the payment for purchased goods or services. In each case describe clearly the role of the computer. (3)

5 Do you think that computers will eventually eliminate the need for cash transactions? Justify your answer. (2) *AEB Computing Specimen Paper 2*

6 Suggest **one** benefit and **one** danger in each of the following applications of computers.

   a) The screening of patients before they see a doctor. (2)

   b) The use of computers to teach school students (2)

   c) The use of computer systems to suggest the sentence to be served by a convicted criminal. (2) *AEB Computing Specimen Paper 2*

7 There is considerable public anxiety over computer data banks and their implications for an individual's right to personal privacy.

Suggest **three** different reasons why such anxieties persist. (3)

8 Individuals and organisations have become so dependant upon IT systems that the consequences of their failure could be catastrophic to the individual or the organisation.

Give **two** different examples of types of IT system for which failure would be catastrophic. In each case explain why the failure could prove to be catastrophic. (4) *NEAB 1997 Paper 1*

9 'The development of Information Technology has had significant effects on society, individuals and organisations.' With the aid of specific examples discuss this statement. Include in your discussion:

- the effects of Information Technology on society at large;

- the effects of Information Technology on employment and work methods;

- the effects of Information Technology on individuals. (20) *NEAB Specimen Paper 1*

10 Briefly describe **two** social impacts and **two** organisational impacts commonly identified as a result of introducing computerised information systems into business organisations. (8) *NEAB 1998 Paper 1*

11 Many organisations and individuals are concerned about the 'Year 2000 problem' or the 'Millennium Bug'.

Discuss this issue. Particular attention should be given to:

- what is meant by the Millennium Bug;

- how and why it occurred;

- what is being done to tackle the problem;

- the effects on individuals and organisations of the failure of information systems.

Illustrate your answer with specific examples. (20) *NEAB 1999 Paper 1*

Computer communications and the Internet

Computer communications involve the transfer of messages from one place to another using a computer. Two or more computers can easily be connected to each other. If they are in the same building they can be connected by a local area network. If they are further apart they can transfer data using a phone line and a modem, a device that today is often standard when you buy a computer.

Data can be transmitted by radio waves (via communications satellites), electrical currents in wires or light pulses on fibre optic cables. Computer communications have grown greatly in recent years as computer processing speeds have increased, modems have become more efficient and transmission methods have improved. Modern fibre optic cables have a greater bandwidth than old copper cable, which means they can carry a large number of messages simultaneously and transmit data more quickly. They are very tough, very high quality and transmission is fast enough to make possiblethe sending of graphics and video images.

These developments in communication, sometimes referred to as **informatics** or **telematics**, include fax, electronic mail, teletext, viewdata, private networks, public networks like the Internet, videoconferencing and telecommuting.

It is no wonder that the phrase 'information super-highway,' first coined by US Vice-President Al Gore, referring to computer communications networks spanning the globe and sharing information, entered the English language in the early 1990s.

Facsimile transmission (Fax)

A fax machine uses telephone lines to transmit copies of an original document. The fax machine scans the document, encodes the contents and transmits them to another fax machine that decodes the data and automatically prints a copy of the original.

The document is sent as a graphic and therefore takes longer to send than a text file. Obviously you can only send faxes to someone with a fax machine but fax machines are now extremely common in business – in fact it is rare for a business not to advertise its fax number.

Transport telematics

Transport telematics is the use of information communications technology in transport. This may be relatively simple, such as providing traffic information on the Internet or having large information signs on motorways to warn drivers of possible delays.

It may be more sophisticated such as an in-car computer that receives details of traffic hold-ups and plans the best route or road pricing systems that automatically collect data on cars travelling on a toll road.

Electronic mail (e-mail)

E-mail has expanded enormously, partly due to the popularity of the Internet. Computer users paying a few pounds a month to an Internet service provider are provided with an e-mail address as well as access to the Internet. You access both services using a modem which connects to a local telephone number, reducing phone charges.

With e-mail software it is very easy to send e-mail to anyone else who has an e-mail address anywhere in the world for the cost of a local phone call. Many businesses now advertise their e-mail addresses. An e-mail address is usually of the form:

sally.miggins@computerland.co.uk

All addresses are in lower case letters. Words are separated by full stops. The uk at the end is the only indication of the geographical location of this address.

Faxes arrive automatically but not e-mail. E-mail users have to check their mail-box to see if they have any mail. If they forget to check, e-mail isn't very quick!

E-mail can also be sent within an organisation on a local area network.

Documents sent by e-mail can be loaded by the receiver without the need to re-type them. Journalists send copy to their newspapers by e-mail. The copy can then be imported in the newspapers desk-top publishing system. Information sent by fax would have to be re-typed, wasting time and introducing the possibility of errors.

■ CAN MESSAGES GET DISTORTED?

There is a danger that a poor quality line or background noise can distort a message sent by computer.

Today telephone lines, especially the fast ISDN (Integrated Service Digital Network) lines, are very high quality. Modems are also very reliable. However it is possible to verify a message if it is automatically returned to the sender. The returned message can then be compared with the original to make sure that they are the same.

Fax modems

Recently there has been **a convergence of technology**. New fax modems can be used to send faxes as well as e-mail. However the computer can only receive faxes if it is left turned on in receive mode.

E-mail or fax?

Sending a letter may be too slow. A phone call does not give hard copy. An office in another time zone, for example in Australia or the USA, may be closed. How can a business send a message? What are the advantages of using e-mail and fax?

	Fast?	Hard copy?	Who can receive it?	Can load directly into the computer?	Easy?	Cost?
E-mail	Yes faster than a letter	Yes	Only those with an e-mail address	Yes For example a newspaper can place an article sent by e-mail directly into a desk-top publishing program	Must dial-up to get it, but can dial-up from anywhere with a lap-top	A few pounds per month. All calls at local rate.
Fax	Yes but a fax sends a message as a picture. Therefore it takes longer than sending e-mail which sends the message as text	Yes	Only those with a fax machine or computer with fax/modem	No a newspaper would have to re-type a faxed article.	Yes	Aprroximately £200 for machine. Cost of call depends on distance.

Internal e-mail is also suitable for memos within a business using a LAN network. It is easy to use, involves no paper and costs nothing but in some offices it is replacing personal contact. Employees are sending e-mails to the next door office instead of going to see the recipient personally.

The telephone is still a very useful means of communication, particularly where personal contact is involved and an immediate answer is required.

Voice mail

Voice mail offers subscribers an alternative to the answering machine, saving telephone messages in the subscriber's own personal voice mailbox. When callers call the subscriber's number, if there is no answer or if someone is already on the phone, the call is automatically forwarded to the voice mailbox, where up to 100 messages can be stored.

Voice mail can offer callers a menu of options, leading to another menu, a mailbox, or an announcement. For example, a computer company hotline menu may have three options:

1 software products

2 hardware products

3 requests for more information

Callers can choose the menu item using the buttons on a touch-tone phone.

Subscribers can retrieve your messages from their voice mailbox at any time using any touch-tone phone, simply by calling the provider and keying in their ID number and passcode. They can change their passcode at any time.

Subscribers normally pay a monthly charge for this service.

■ ADVANTAGES OVER AN ANSWERING MACHINE

- Voice mail is always turned on.
- There is no equipment to buy or maintain.
- Voice mail can store messages even when your line is busy.
- Voice mail can respond to more than one caller at a time.
- Voice mail uses digital sound for good quality recording.
- No one can hear your messages unless they know your passcode.

Teletext

Teletext is an electronic information service which can be viewed on specially adapted televisions. The user will have to pay about £50 extra for a Teletext TV, which is operated using the TV's remote control. Teletext can be used to view information such as news, weather forecasts, traffic information and sports results provided by the television companies. Teletext is cheap and easy to use and can be read by over 30 million people in the UK, with an average of 18 million people using the service each week.

Viewdata

Viewdata means an interactive information system using a computer, modem and telephone lines. There are a few networks that provide viewdata to the general public. PRESTEL provided by British Telecom was the world's first public viewdata service, starting in 1979. It enabled users to send e-mail and to access and send information, for example home banking or holiday booking.

Similar services exist in the USA (Prodigy) and Japan (Captain).

Teletel is an on-line information service provided since 1980 by France Telecom, who gave 6.5 million users a free Minitel terminal. Originally Teletel was designed to let users look up telephone numbers and replace directory enquiries. Now it has been extended to include e-mail, information and shopping, including buying Eurostar tickets.

The terminal connects to phone lines, increasing phone use. Teletel is accessed around two billion times a year. It was obviously a good investment for the company!

Campus is a British educational information service for schools and colleges operated by BT. It provides information, for example on university places, and also offers e-mail.

Videoconferencing

Videoconferencing, sometimes called teleconferencing, means being able to see and interact with people who are geographically apart. Two or more people can be connected to each other. The equipment needed includes a high specification PC, a video camera usually positioned above the monitor, microphone and loud speakers, a high speed modem and high speed communication line, for example an ISDN line.

The use of videoconferencing enables business meetings and interviews to take place avoiding the expense and time of travel. The equipment can be expensive and at present the image quality is still not as good as on television or video but continues to develop as hardware improves.

Case Study

Videoconferencing

South Australia covers a large geographical area including many remote, rural areas. Videoconferencing has been used in Further Education there since 1990, having been developed to bring courses to small rural centres by connecting learners with their teachers and other learners. Videoconferencing aims to meet the demands for courses which has been impossible to answer previously.

The equipment includes a range of cameras and microphones to make the rooms flexible but simple to use. It has been fitted into classrooms with the minimum of intrusion, to try to make the facilities look like the conventional situation and to make the users as comfortable as possible.

All the rooms are fitted with cameras for the teacher and class, as well as a video recorder, which can be used to either play back a tape, or record a session. Control of the rooms is left entirely to the teacher and there are no other staff present during the session to help them.

It is now possible to undertake courses in accountancy, automotive studies, business studies, horticulture, hospitality, languages, law, sign language for the deaf and tourism studies via videoconferencing. Class sizes reflect the sizes in normal classes, with about 15 to 25 students on average. Three or four centres may be connected together to combine enough students for a class.

The busiest rooms are in use for up to 45 hours for their busiest weeks, providing about 1000 different sessions during the year, totalling around 16 000 hours a year throughout the state. ☐

What is EDI?

EDI (Electronic Data Interchange) is a means of transferring information such as invitations to tender, letters, orders and invoices electronically via the telephone network. It allows the computers in one organisation, to 'talk' to the computers in their supplier's organisation, regardless of computer manufacturer or software type.

EDI cuts down the paper mountain. Although all large organisations and most smaller ones use computers, it is true that the vast majority are still essentially paper based.

For example, a simple order is raised on one computer, printed, mailed, then received by the supplier; who re-keys into another computer.

The process is expensive, time-consuming and is prone to postal delays and errors. EDI changes all that. Acting as a giant, efficient electronic mailbox, it collects the orders directly from one company's computer and sends them to the suppliers' computer. It cuts out printed mailings and removes re-keying, minimises the margin for error, saving days in the processing cycle.

EDI Case Study 1

EDI at Nissan

Nissan's Sunderland plant started production in 1986. Rapid increases in production levels meant that the paperwork generated soon reached large proportions with as many as 15 000 delivery notes each week.

The labour costs of dealing with all this paperwork and the associated mailing costs were excessive. There was also the potential for human error. Following an investigation Nissan started to use EDI in 1989.

Almost immediately there were savings in labour and mailing costs, a shortening of the time for delivery information to reach suppliers and a reduction in the level of human errors. EDI is used to transmit delivery requirements to Nissan's logistics partner, Ryder Distribution Services, who in turn use EDI to send delivery data to Nissan. The volume of mail to suppliers has been reduced by 90–95 per cent.

EDI at Tallent Engineering

EDI Case Study 2

Tallent Engineering manufacture car chassis products such as axles. They supply a number of European car makers including Nissan in Sunderland, Ford in Belgium and Rover at Longbridge.

The Tallent plant at Newton Aycliffe operates a Just-in-Time (JIT) stock system, holding only eight hours worth of finished goods. Tallent deal with vast volumes of information every day such as orders and delivery information. They use EDI to manage all of this information.

Tallent have gained the following benefits from using EDI:

- Competitive advantage
- Demand updated in hours not days
- Reduction in lead times
- Schedule and daily JIT requirements managed by EDI
- Cost savings on data input.

Network questions 1

1 A multi-national company is considering the use of 'teleconferencing'.

a) What is meant by the term 'teleconferencing'? (3)

b) List the minimum facilities required to enable 'teleconferencing' to take place. (4)

c) Discuss **two** advantages and **two** disadvantages to the **firm** of using 'teleconferencing' as compared to traditional methods. (4) *NEAB 1997 Paper 1*

2 Give **two** advantages and **one** disadvantage of a firm using electronic mail as a method of keeping in touch with its large number of travelling salespersons. (3) *NEAB Specimen Paper 1*

3 'The development of communication systems has enabled individuals, organisations and society to operate on a global basis.'

Discuss this statement. Include in your discussion:

- specific examples of facilities and/or tasks that make use of communication systems;
- specific examples of applications that make use of these facilities and/or tasks;
- the communication technology and/or techniques that have enabled this development. (20) *NEAB Specimen Paper 1*

4 Give four reasons why is it easier for schools and colleges to use e-mail for exam entries (4)

5 A large company has introduced a communication system which includes electronic mail. This system will be used both for internal use within the company and for external links to other organisations.

a) Describe **two** features of an electronic mail system which may encourage its use for internal communication between colleagues. (2)

b) Contrast the use of an electronic mail system with each of fax and the telephone. (6)

c) Describe **two** functions the communication system might have, other than the creation and reception of messages. (4) *NEAB 1996 Paper 1*

6 Facsimile and computer based electronic mailing systems are different forms of message systems.

a) For each of these systems, describe **two** of the facilities offered. (4)

b) Discuss the relative strengths and weaknesses of each of these systems. (8) *NEAB 1998 Paper 1*

7 Many businesses now use EDI to communicate with their suppliers

a) What is meant by EDI? (3)

b) Give four advantages of introducing EDI (4)

The Internet

The Internet (International Network) is a large public computer network. It was established in the 1960s for US research establishments. It is now available to businesses and the general public.

To access the Internet, a home user needs a computer, a telephone line, a modem and communication software. This software will include a browser for looking at web pages and e-mail software. She will need to subscribe to an Internet Service Provider (ISP). She may have to pay extra for time spent logged on to the Internet. The software may be provided by the ISP who usually has a local phone number which allows access to the Internet for the cost of a local call.

Once set up, the software will dial the number and make the connection to the Internet for the user at the click of a button.

The World Wide Web (WWW) allows Internet users to access multimedia pages of information. This information can include text, images, sound and video. An Internet subscription allows users to:

- view web pages to find out information
- save these pages for reference off-line
- set up web pages for others to look at
- have an e-mail address
- send and receive e-mail

The Internet is truly world wide as it can be accessed from anywhere in the world. Sending e-mail to someone in another continent is the same as sending it to the next street. In fact you may not know which country you are loading a web page from or sending e-mail to. The latest estimates say there are 40 million users. The number of web pages available increased from 130 000 in 1993 to 30 million in 1997. Over 12 million users had visited the Buckingham Palace (**http://www.royal.gov.uk/**) web site within two months of it being set up! No wonder the 'Net' is big business.

With so many sites, a lot of the pages are bound to be very specialised and not very interesting. Conversely with so much information available, some of its is bound to be useful. The main problem is finding it.

Search engines such as Yahoo and Lycos enable users to search the Internet for selected key-words. This is commonly known as surfing the net. The search engines provide the information required. A search on one key-word may provide thousands of links. Finding the right ones can be like finding a needle in a haystack!

How fast is the Internet?

Some web pages on the Internet are graphics intensive and may take a long time to load. Many factors affect the rate at which data is transmitted. These include

- Modem speed – even if you have the very latest modem, the person you are communicating with may have an old, slow modem. You can only work at the speed of the slower modem.

- Type of line – ISDN lines transmit data at least four times as quickly as standard telephone lines.

- Number of other users – if the server you are trying to access is very busy, it will slow down the rate at which you can down-load data.

You often have to spend a long time waiting for pages to load, particularly if they are graphics intensive and your computer, modem and line are not of the highest specification.

Browser wars

The two most common Web browser programs are Netscape *Navigator* and Microsoft *Internet Explorer*. Netscape recognised the potential of the World Wide Web before Microsoft and for many years their software had the greater market share.

To improve its market share, Microsoft started to insist that PC manufacturers included *Internet Explorer* along with *Windows 95* on new computers. As virtually every new PC used *Windows 95*, it also had a 'free' copy of *Internet Explorer*. PC makers could include Netscape's *Navigator* as well, but that would have involved paying an extra licensing fee.

Netscape complained to the US Department of Justice that Microsoft was using the dominance of *Windows 95* to gain an unfair advantage in Web browsers. Microsoft argued that *Windows 95* is an 'integrated' product containing *Internet*

Explorer. They said the browser was an extension not a separate product, such as windscreen wipers on a car. Wipers can be sold separately, but they're an integral part of any new car.

However most experts said *Internet Explorer* was a separate product. In 1997 Judge Thomas Penfield Jackson of the US District Court in Washington agreed and banned Microsoft from forcing PC manufacturers to include *Internet Explorer*. This was obviously good news for Netscape, but Microsoft were already planning a new version of *Windows, Windows 98*, with fully-integrated browsing capabilities.

What is on the net?

On the 'Net' you can find advertisements from companies, government statistics, football club information, details of people's research, pictures of pop stars and the lyrics of their songs. You name it, it's probably there. You can order clothes from catalogues, food and wine from supermarkets using your credit card.

Many newspapers publish editions on the Internet. Users can search to find references to any subject they choose. University researchers put details of their research on to web pages.

Internet Case Study 1

Amazon.com: the world's biggest bookstore

Why are some British bookshops concerned about competition from a shop 6000 miles away? Because Amazon.com from Seattle, USA are selling books over the Internet. Amazon.com offer two and a half million titles whereas the largest physical bookstores offer only about 175 000 books. *Time* Magazine has described Amazon.com as 'the best place to do holiday shopping without leaving your house'.

Amazon.com are open 24 hours a day, 365 days a year and send books to anywhere in the world. Customers pay by credit card. Overheads are so low that they can give discounts of up to 40 per cent on some books. Microsoft chairman Bill Gates said in 1996 that he buys books at Amazon.com. Perhaps he's attracted by the low prices!

Potential customers can use search tools to find exactly the book they want in the company's extensive database, quickly and easily. Amazon.com can use e-mail to market their goods to customers who sign up for receiving e-mails.

No wonder small bookshops in Britain, who have already lost trade to supermarkets, should be worried.

Will these shops lose trade as the Internet expands? Some booksellers remain confident. They feel the Internet is over-hyped, believing Net users will need to spend a lot of time finding the book they want. They feel that their customers like browsing in their shops. Bookshops have a nice atmosphere. Customers visit a bookshop while passing and can view the purchase before they buy. You cannot get that same knowledge by browsing electronically. ☐

Amazon.com opened a British site in 1998.

Internet Case Study 2

Banks introduce Internet services

In 1997 the Royal Bank of Scotland, Nationwide Building Society and Lloyds TSB all announced that they would introduce Internet banking services, such as checking your balance, mini-statements, gas, electricity and credit bill payment and even overdraft applications via the Net.

They chose the Internet as it is becoming the most common platform for users.

Other banks, including Barclays and First Direct, plan to use a private dial-up connection as they are more cautious about the security of the Internet. Lloyds however are confident about the Internet's security, using a 'proven' 128-bit encryption method. ☐

Commercial use of the Internet has developed so rapidly that it has spawned a whole new jargon, including such words as e-commerce, e-tailers (electronic retailers). e-business, e-shopping, e-banking and even e-greetings.

Policing the Internet

The global nature of the Internet and the anonymous nature of the pages make policing very difficult. The Internet is alleged to contain pornographic pictures and offensive and racist literature. Some Internet Service Providers try to 'firewall' these pages, that is not allow access to them, but millions of new pages are being created making the

What is a URL?

URL stands for Uniform Resource Locator or address where to find a particular web page.
http://www.royal.gov.uk/index.html is the URL for the Buckingham Palace web site index.

What is HTTP?

Internet addresses all begin with http. This stands for HyperText Transfer Protocol. This is the standard protocol (or rules for exchanging data) that computers must use to identify the Internet address. This means that different makes of computer can all access the Internet as long as they use this protocol.

How do you know where to look?

Finding information on the Internet is not always easy. Many companies publish their web site address (URL) and you can access the information by typing it in the white box

Searching for information

If you do not know the address, you have to use a search engine to find key-words. This is sometimes called 'surfing the net.' As there are millions of pages, this can take a very long time and you still might not find what you want.

Favourites

You can save the address of frequently visited sites as a 'favourite' to avoid having to remember the URL. Web pages often include highlighted words which are links to other web pages. By clicking on the highlighted word you load the next page.

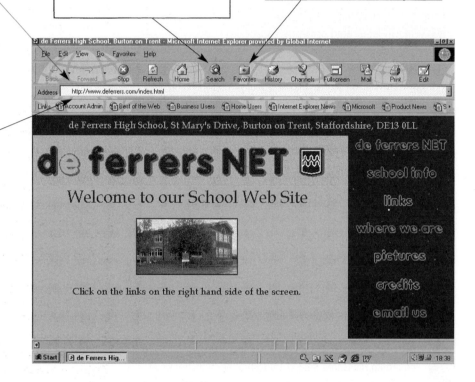

How do you set up web pages?

Pages are set up in a special language called HTML (Hyper Text Mark-Up Language). HTML is relatively simple but there are now several computer programs on sale commercially that help you set up your own web pages.

Web pages often include counters to show how many users have visited the site.

```
<HTML>
<HEAD>
<TITLE>de Ferrers High School, Burton on Trent</TITLE>
<META NAME="keywords" CONTENT="Burton on Trent Staffordshire "></HEAD>
<FRAMESET cols="579,*" BORDER=0 FRAMEBORDER=0 FRAMESPACING=0>
<FRAME SRC="main.html" NAME="main" SCROLLING="auto"
MARGINWIDTH="0" MARGINHEIGHT="0">
<FRAME SRC="menu.html" NAME="menu" SCROLLING="auto"
MARGINWIDTH="0" MARGINHEIGHT="0">
<BODY BGCOLOR="FFFFFF" TEXT="" LINK="CCCCFF" ALINK="FFFFFF"
VLINK="CCCCFF">
</FRAMESET>
</html>
```

From the Nationwide banking web-site

How secure is Online Banking?

When you register for Online Banking we'll give you a customer number and a PIN (Personal Identification Number). Both are different from your account number and cash machine PIN, and you'll need to use them to identify yourself whenever you go online. When you first use Online Banking you will be asked to give an additional password of your choice, which you will need whenever you log on to Online Banking. As long as you keep these numbers and your password secret, only you can sign on to Online Banking.

Figure 4.1 In 1999 30 Tesco stores offered e-shopping, which made up 7 per cent of sales in these shops

Figure 4.2 The amazon.com Web Site is the world's biggest bookstore

preventing of access difficult and prosecution practically impossible.

Different countries have different laws on such matters. If someone in country A publishes information that is legal in country A but illegal in country B, have they broken the law if someone in country B accesses their pages?

There is some concern that buying goods over the Internet may lead to problems. How do you know if the organisation selling the goods is genuine or a fraud? What if your credit card details are intercepted and used by someone else to pay for goods?

Case Study

▶ Mr Evans

The dangers of using the Internet for buying and selling goods were shown up in the case of 80-year-old Alec Evans. Mr Evans was puzzled when the purchase of more than £200 of goods he did not recognise on his Abbey National Visa card bill. He found out from his local Abbey National branch that the goods had been purchased over the Internet. Yet Mr Evans did not own a computer and had never made a purchase over the Internet.

Abbey National immediately refunded Mr Evans but did not know how the misuse of his card number could have happened, suggesting that he might have been overheard giving out his number by someone who later used the number fraudulently. Visa says that the Internet is not a safe place for financial transactions.

Despite the possible dangers, shopping over the Internet, sometimes called **e-commerce**, is growing rapidly. Dell Computers claim their web site generates $3 million worth of sales every day. ☐

Intranets

An intranet is an internal network, for example, within a company, including information pages and electronic mail facilities. It uses the same software as the Internet and can be connected to the Internet. A company could use this to provide employees with information – for example schedules for the day, stock information, orders due for delivery – as well as internal e-mail.

Lina Saigol, The Guardian, 9 August 1997

Virtual bargains now in store

Cybershopping is no longer just a passing fad. Amid the hype surrounding the Internet, companies are rapidly establishing a web presence from which to sell their wares.

By the end of the year, an estimated 500,000 consumers in Britain will participate in virtual shopping, now worth an estimated £1 billion – big business for banks and retailers.

Insurance, mortgage and savings rates can be found floating in cyberspace, as can investment advice and share dealing services.

Cybershopping is convenient, open 24 hours and goods are often cheaper than in stores because there are no overhead costs. Couch potatoes can have fresh groceries delivered to their door for a £5 fee by Tesco, with a choice of 20 000 products.

Fashion victims can browse for clothes at their leisure. Great Universal Stores, the mail order giant which owns Kays and Marshall Ward catalogues, has set up more than 15 virtual shops. Music lovers can buy CDs direct from the US making significant savings and bookworms can down-load an extensive selection of books. The Net also accounts for a growing proportion of travel sales as on-line reservations can be cheaper.

Safety on the Net

- To protect your credit card number when paying for goods, check your browser is secure and equipped to encrypt and scramble purchase information. If you don't have encryption software, fax your order.

- Shop with companies you have heard of. If in doubt ask for a paper catalogue to be sent in the post.

- Check the company's refund and return policies prior to placing an order.

- Never give out your Internet password

- Be cautious if you are asked too many personal details on a transaction

- The same laws that protect you when you shop apply to cyberspace shopping, and if you pay by credit card you are protected by the Consumer Credit Act.

Case Study

▶ Electronic Arts

Electronic Arts is a successful software company based in Silicon Valley, California, known for some of the best-selling computer games in the world. EA uses an internal web network, or intranet, to make important information available to its 2300 employees and to enhance teamwork.

EA wanted a system that would allow employees to exchange information in an easy and cost-effective way and could be used by different platforms. Using a Web browser makes it very easy to access information, which can include text, graphics, sounds and videoclips.

EA believe it is a lot easier to access information electronically. It's all in one place and up-to-date. Workers are better-informed and so make better decisions.

Intranet newsgroups can discuss projects over the intranet, enabling EA employees to collaborate to tackle a project regardless of where they are based.

The benefits of Electronic Arts' intranet include:

- information sharing – employees can access all information in one place

- team collaboration – newsgroups enable 'virtual' teams to work together ☐

Extranets

Companies can extend their corporate intranet systems into an extranet. Extranets are direct network links between two intranets, often between two businesses such as a retail company and their supplier.

The retailer can access their suppliers' stock records, check stock levels and delivery times before ordering directly via the extranet.

These links can use either the Internet using encryption or private leased secure lines. Connections to consumers and to business prospects are directly over the Internet – often encrypted for privacy.

Remote Databases

It is now possible to book airline tickets, holidays, train tickets, hotel rooms and even theatre seats from travel agents all over the country. These travel

▶ Darnton Tiling, Leeds

Extranet Case Study

Darnton carry out tiling work in offices, factories and homes. They need to know if their suppliers can deliver the tiles for a particular job quickly. Darnton has an Internet link with tile manufacturers H&R Johnson Tiles, of Stoke-on-Trent.

Darnton access the manufacturer's home web page and by entering a user name, password and PIN number can find out what is in stock, when new stock will arrive and place orders. The Internet also allows Darnton to check on tile prices and take advantage of any offers. ☐

agents can connect to the holiday company's computer using a WAN and find out which holidays are available. When the customer has decided on the holiday, it can be booked electronically and the holiday company's database up-dated.

This is an example of a remote database. The computer communication enables the customer to know exactly what is on offer and have their booking confirmed immediately.

Computer networks

A **network** is two or more computers linked together. A network enables users to share resources. They may share **hardware**; peripherals such as printers, disks or CD-ROM drives; **software** or **data files**.

The network may be in one room or building. This is called a **Local Area Network** or **LAN**. It will use its own dedicated cables.

The network may be spread over a wide area, even in different countries, linked by telecommunications systems such as phone, satellite or cable. This is a **Wide Area Network** or **WAN.**

Network gateways are used to connect LANs together to form wide area networks.

Networks provide opportunities for communication, particularly the sharing of data files and electronic mail. However connecting your computer to someone else's computer creates problems in terms of security. These problems might be caused by

- remote users – a WAN system that can be accessed from off the site will not be as secure as one than can only be accessed on-site such as a LAN system;

- unsupervised users – users accessing a WAN system from home will not be supervised in the same way as if they were in an office with other employees;

- unauthorised access – if a WAN system can be accessed legitimately from a remote computer, there is a danger that an unauthorised user may be able to gain access;

- difficulty in tracking down abuse – it is not easy to trace where unauthorised access has been made.

It is important that a network offers **different levels of access**:

The network manager will have unlimited access to all areas and drives. The manager will need greater privileges than ordinary users, for example in order to install new software, add and delete users, set up menus, and so on. The manager's programs obviously must be protected by passwords.

Users, however, need read/write access to a dedicated area of the disk where their files are saved, read-only access to some areas (for example where software is loaded from) and no access at all to other areas (for example other user areas).

Why network?

Networks are complicated to install and manage but they offer many advantages:

- they allow users to share information without the need to swap floppy disks

- they allow users to share software – this may be cheaper and quicker to install

- they allow users to share peripherals such as printers – this should be cheaper too

- they can have better security than stand-alone machines – for example there is no need for disk drives that may lead to virus problems if an infested disk is used

- they provide a good way of sending data accurately so that it does not have to be typed into the computer again – for example examination board entries, newspaper stories, National Lottery tickets, record sales, credit card sales

When buying a network of say 20 machines, you can buy a network licence to run a program such as

▶ Levels of access

Allan Bromley is the network manager for a small plastics manufacturing company. The company's financial records and payroll details are stored on this network. Unauthorised personnel are not allowed to access this data.

- Allan and other authorised users can inspect the data.

- Authorised users add to the data, for example when someone starts work at the company.

- Authorised users can edit the data, for example when all employees get a pay rise.

- Authorised users can delete some of the data, for example when an employee leaves.

- Allan backs up all the data every night on to magnetic tape.

- Allan safeguards the data by locking the previous day's back-up tape in a fireproof safe. □

Microsoft Office. This will be cheaper than buying 20 copies Microsoft Office which you would need to do if you had 20 stand-alone machines.

You will only need to install the software once. With 20 stand-alone machines the software will need to be installed twenty times. This would take a lot of time.

Start-up costs may look more expensive as you will need to buy the network server, the cabling and the network software. However when you consider the software costs, networks should work out cheaper.

You can also share data more easily. This can only be done by swapping disks when using stand-alone machines. This increases the chances of getting viruses. Networks do not need floppy disk drives.

■ THE INCREASE IN WIDE AREA NETWORKS

There has been a great increase in the use of WANs over the past ten years due to

◢ the need to share information

◢ the increase in information available

◢ the fall in cost, so more computers are used, storing more information

◢ LANs can easily be connected to each other via WANs

◢ the increased speed of transmission

Examples of the use of WANs include airline, theatre and hotel booking systems; home banking; bank ATMs; the Internet; the National Lottery; videoconferencing; stock control; e-mail.

■ PROBLEMS WITH NETWORKS

There may be problems with networks particularly with unauthorised or illegal access. This is particularly true is the case of WANs. Anyone with a home computer and a modem can, in theory, get into other computer systems connected to telephone lines if they know all the passwords. It is vital that all security measures are taken to prevent unauthorised access (see Chapter 2).

■ RADIO NETWORKS

Network stations can now be connected by infra-red transmitters (if they are in the same room) which work in the same way as a TV remote controller, or by microwaves (over a longer distance). This saves the cost of installing cabling.

Network topologies

The **topology** means the shape of the network – how the computers are connected together. Different networks require different topologies.

If a central computer controls the network it is called the **host**. Each terminal on a network is called a **node**. A **repeater** is a signal amplifier. It is necessary on a long cable or sometimes at every node.

1 Star network (for example bank cash dispensing machines or the Camelot computer)

In a star network, the host is at the hub or the centre of the network. All nodes are connected directly to it. The host switches between all the nodes in turn. This type of network is good for communications to and from the host, so it is ideal for WANs like bank cash dispensers. Each cash dispenser is connected directly to the bank's main computer.

Specialist network hardware

Repeater

Transmitted signals deteriorate as they travel distances until they reach a stage when they are unrecognisable. To prevent this happening, a repeater is installed between segments in the network. A repeater inputs a signal, regenerates it, and outputs it in boosted form.

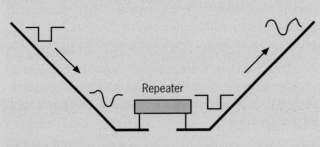

Figure 4.3 Repeater

Bridge

A bridge is a device which allows two network topologies to be linked, thus extending the network. This may become necessary if a network has reached a maximum size, either in the number of terminals covered or in the distance covered. The topologies of the linked networks do not have to be of the same type: a bridge can link a star to a bus network.

A bridge examines all information packages that it receives and only passes across from one network those that are destined for nodes on the other network. Two smaller networks, each with its own file server with access to software and data files, would generate less traffic than one large network as most data transfer would be confined to one network. Only occasionally would data cross the bridge.

An intelligent bridge can enhance security by only allowing messages from designated work stations to cross the bridge.

Backbone

If a number of networks are to be linked, a backbone can be used. This is a length of cable to which the networks are linked, each via a bridge. Each segment would have its own server. Users would have access to other networks and their servers via the backbone.

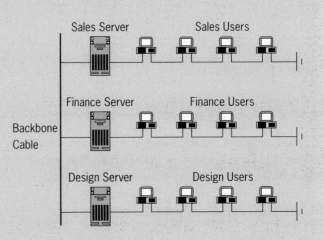

Figure 4.4 Backbone

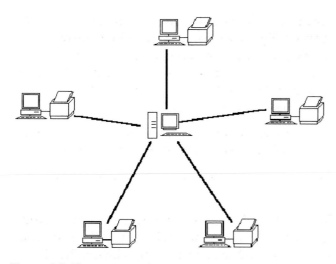

Figure 4.5 The star network

2 Ring network (LANs)

In a ring network, all the nodes are in a loop, which can extend up to five kilometres in diameter. It is therefore only suitable for local area networks. Data travels in one direction round the ring. This means high transmission speeds are possible.

There is no host and none of the nodes controls the network.

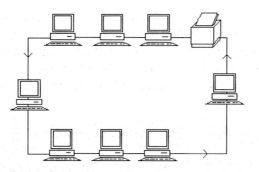

Figure 4.6 The ring network

3 Bus network

In the bus network, sometimes called an Ethernet network, data travels in both directions along a single cable called the bus. All the nodes are in a line. As data travels both ways, it is much slower than the ring network.

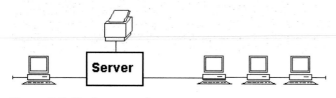

Figure 4.7 The bus network

Networks: Server or peer-to-peer?

Some simple networks have no server. These are called peer-to-peer networks where all stations have equal rights and access. As there is no central server, individual copies of software must be stored on each station. Similarly there is no central hard disk so users save to a local disk drive.

On a server based network, user menus are set up centrally. A local menu system offers more flexibility. Server based networks are complicated to install and maintain but backing-up a central hard disk on a server is much easier than backing-up many local ones.

As all stations on a peer-to-peer network have equal access rights, such a network cannot have complete security. It would not be suitable for use in a school for pupils to use!

As a server is very expensive to buy, a small peer-to-peer network would probably be much cheaper than a small server network. A peer-to-peer network would be ideal in a small office where four PCs need to be networked to share data.

Network Case Study 1

The National Lottery

The National Lottery, run by Camelot, sells tickets in around 35 000 retail outlets. Tills in all the retailers are connected to Camelot's wide area network either by cable or by satellite. As lottery tickets are sold, details of the numbers chosen are entered by optical mark reading (OMR).

The data is transmitted to Camelot's computer centre in Rickmansworth, Hertfordshire. The network needs to be very sophisticated to cope with the large volumes of sales (particularly early in the evening before the draw is made) which have reached over 50 000 transactions a minute. Camelot say that the network has been designed to cope with considerably more traffic than this. □

Telecommuting (also called teleworking)

Telecommuting means using IT to work from home (or another place away from the office). It has been made possible by advances in technology and networking, such as fibre optics, faster modems, fax, satellite systems, internal e-mail and teleconferencing.

▶ Debit cards

Over 16 million people in Britain have a debit card, often called a Switch card. These cards can be used in nearly a quarter of a million retail outlets throughout the country.

When a card holder presents their card at the till, it is 'swiped' through a terminal. The terminal can check the card against a list of stolen cards. If the purchase is over a certain amount, the terminal connects to a wide area network called SwitchNet to make sure that the card holder has the money in their account and to authorise the transaction. This takes less than five seconds in nearly all cases.

Details of the sale are stored in the terminal. Every night the sales data is sent to the bank's computer and the list of stolen cards is up-dated, using the wide area network.

Each month debit cards are used for over 50 million purchases worth £1.5 billion. Use of debit cards is rising rapidly and mainly replacing cheques. ☐

▶ Marks and Spencer

Marks and Spencer use IT for stock control. When you buy an item at M&S, an electronic message is sent to the central computer via a wide area network which calculates what stock is needed and plans the following day's deliveries.

The company's seven distribution depots know each night exactly what has been sold in each branch and can therefore supply exact replacements.

The company benefits from lower stock levels. It no longer needs to keep large amounts of stock in the warehouse of each store. This reduces the amount of capital tied up in stock and enables the company to react more quickly to the fickle demands of fashion. Customers benefit from a larger range of goods on sale. ☐

Working from home has been common for a long time in some industries, such as sales representatives, telephone sales and the self employed. Now telecommuting has extended home-working to other industries.

It is now possible for authors, journalists, computer programmers, accountants and word-processor operators to do their work by telecommuting. British Airways flight booking takes place in Bombay, proving you don't even need to be in the same continent. Much computer programming and testing is done in Asia for British companies, taking advantage of cheaper labour costs.

It is common for telecommuters still to spend part of their working time at the office, and part at home.

Telecommuting demands new skills and training is essential.

There are obvious advantages for the employee such as flexible working hours and avoiding the rush hour, travel costs and some child care problems. For the employer, telecommuting saves the cost of office provision – desks, chairs, space, heating, car parking space. The computer can record the amount of time spent at work, which leads to good productivity levels

The disadvantages for the employee include the lack of the social side of work. Many people make friends through their job. The work is all individual with teamwork only at a distance. Telecommuters may have problems in finding quiet at home. The employer cannot know if they are they really working or working productively.

In the long term, will large scale telecommuting have effects on society? It should be more environmentally friendly and perhaps the end of cities and offices as we know then today. It may lead to a feeling of never being off the job and result in more stress at home.

Building society introduces teleworking

In 1997 Britannia Building Society launched one of the UK's most ambitious telecommuting projects. The building society has implemented a policy for its text creation department – what used to be called the typing pool.

Dictation of letters is done over the phone, stored and then transmitted to the teleworker's home. The completed documents are typed at home and submitted by e-mail. Britannia say that the system works really well and it is easy to monitor the work rate, and error rate, of the teleworkers.

Britannia, which claims to be one of the first UK employers to do this, is also considering offering teleworking to some programmers and developers.

CMC

(Details from *Explain IT,* October 1996)
Telecommuting means using information technology with communications systems in the form of phone, fax and e-mail to enable people to work away from an office: some teleworkers work from home while others use small offices sometimes called telecottages.

BT estimates that there are around 1.3 million teleworkers in Britain, made up of 650 000 self employed persons, 150 000 company employees and 500 000 mobile workers such as sales staff.

Telecommuting certainly sounds attractive: no more commuting to and from the office and no more petty work politics. But is life really better for the teleworker?

The Complete Multimedia Company (CMC) is based at Capons Farm in Sussex. The company is run by Rosemary and Martin Mares. Set up in 1994, CMC provides sound recording and editing facilities for companies producing CD-ROMs and also supplies creates pages for the Internet's World Wide Web.

The company started with just a single PC and a telephone line: 'We soon realised that we needed a fax, because almost everyone we spoke to would ask us for our fax number. Now we've got three PCs, a fax and e-mail,' said Rosemary.

One of the advantages of telecommuting is flexibility. 'I normally start work at 7.45 a.m., but today I didn't get in until after nine because I had to take my daughter to the hospital,' Rosemary said. 'If I had a regular employer, they probably wouldn't like it if I took time off because my child was sick.'

She added, 'One problem is making people aware that although you work in the countryside, you are a professional organisation and can undertake any work and do a damn good job. Some people expect to see plush offices with a pretty receptionist, and when they don't, they think: this company's a bit small.'

Teleworkers who work in their own homes often find it difficult to switch off. 'We try not to talk about work in front of our daughter, and we'll wait until she's in bed. Martin will work from 9 a.m. to 5 or 6 p.m. and then go home and see Nat. But after she's gone to bed, he could be at his desk until 2 or 3 in the morning. If you work for yourself, you've got to be prepared to work hard.'

But despite the problems over never being away from work, Rosemary says teleworking is a good way of earning a living. 'There is something special about working for yourself, and the environment here is so nice. The sheep wander into our office, and during the lambing season, we bottle feed the lambs. You can't do this in an ordinary office.' □

Summary

E-mail or fax?

☐ more people have fax machines than e-mail addresses but e-mail is expanding

☐ you can send files, for example text, graphics by e-mail but not by fax

☐ there is no need to re-type documents sent by e-mail

☐ fax arrives automatically; you must log-in to get e-mail

☐ e-mail is a local phone call

A network is a number of computers linked together allowing users to share data:

☐ a local area network (LAN) will be in one building or neighbouring buildings

☐ a wide area network (WAN) is spread over a wider area linked by satellite or cable

Effects of the Internet

☐ users can e-mail other users

☐ access to millions of pages of information with excellent research facilities

☐ can be accessed from anywhere usually for the price of a local phone call

☐ dangers of pornography and racist literature

☐ it is impossible to police

☐ Internet shopping may affect traditional shops

Network questions 2

1 Describe two differences between a Local Area Network (LAN) and other networks such as a Wide Area Network (WAN). (2)

2 Give five reasons why Camelot use a WAN for collecting data from shops. (5)

3 The following are applications which use either a wide area or a local area network or combination of both. For each, justify which network type is most suitable.

   a) Cash-dispensing and account-inquiry facility for a national building society. (2)

   b) Accounting and stock-control system for a department store, using point of sale terminals. (2)
   AEB Computing Specimen Paper 2

4 A school is investing in 16 computers. They need to choose a between network or 16 stand alone machines. Describe the advantages and disadvantages of each. (8)

5 With the aid of diagrams, describe a ring network and a bus network. (4)

6 A local surgery uses a number of stand-alone computer systems to manage patient records, appointments, staff pay and all financial accounts. The surgery manager is considering changing to a local area network.

   a) Compare the relative advantages of stand-alone and local area network systems. (6)

   b) Describe, with the aid of diagrams, **two** alternative network topologies. (6) *NEAB 1996 Paper 2*

7 'Networked computer systems (e.g. Internet) will revolutionise the way in which we shop.' With the aid of specific examples, discuss this statement. Include in your discussion:

   • The types of organisation likely to advertise on such systems.

   • The capabilities and limitations of such systems for this activity.

   • The potential security risks for the customers in using such systems.

   • The organisational impact of such systems.

   • The social impact of such systems. (20) *NEAB 1997 Paper 1*

8 'The growth of communication systems may result in an increasing number of people working from home, often referred to as telecommuting.'

   Discuss, with the aid of specific examples, the advantages and disadvantages to individuals, organisations and society of such methods of working. (20) *NEAB Specimen Paper 1*

9 a) Give **two** differences between a Local Area Network (LAN) and a Wide Area Network (WAN). (2)

   b) Discuss the relative merits of server based networks and peer to peer networks. (6) *NEAB 1997 Paper 2*

10 Recent changes in communications technology have resulted in a blurring of the distinction between telecommunications and computing. Information services are starting to be provided on what is becoming known as the Information Super Highway (ISH).

   a) State the minimum facilities needed to gain access to these services. (3)

   b) Identify and briefly describe **three** types of information service you would expect to find when linked to the ISH. (6) *NEAB 1998 Paper 1*

11 Explain the function of a gateway when used with Local and Wide Area Networks. (2) *NEAB 1998 Paper 2*

12 At the central office of a landscape gardening company there are six employees. Each employee has a stand-alone computer system and printer. The company director has commissioned a business survey which indicated that it would be more efficient if the six PCs were formed into a peer to peer network.

   a) State **three** benefits that the company would gain from networking their computer systems as a peer to peer system rather than a server based system. (3)

   b) What additional hardware would be needed to connect the six stand-alone computer systems as a peer to peer network system? State why each item is required. (4) *NEAB 1998 Paper 2*

13 A company specialises in organising international conferences for doctors. The company has decided to make use of the Internet for advertising and organising the conferences.

a) State, with reasons, the hardware that the company would need, in addition to their PC and printer, in order to connect to the Internet. (4)

b) State the purpose of the following software when used for the Internet:

i) Browser

ii) Editor

iii) E-mail software (3)

c) Explain **three** potential advantages for this company of using the Internet as opposed to conventional mail/telephone systems. (6) *NEAB 1999 Paper 1*

14 Ring and Star are two common network topologies.

a) Explain what is meant by 'network topology'. (2)

b) Give **two** advantages for each of the ring and star topologies that are not held by the other. (4)

c) State **two** factors that affect the rate of transfer between the computers in a network. (2) *NEAB 1999 Paper 2*

15 Most banks and building societies now offer cash withdrawal facilities through the use of Automatic Teller Machines (ATMs). The data that needs to be entered before a transaction can take place will include the customer's account number and Personal Identification Number (PIN).

a) i) State **three** validation checks that should be made on the customer's account number. (3)

ii) State **one** validation check that should be made on the PIN. (1)

b) These transactions will normally involve transmission of data from and to an ATM via a communications link. State **one** security precaution that should always be taken, giving a reason why it is needed. (2) *NEAB 1999 Paper 2*

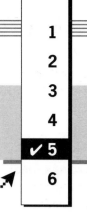

Computer crime

What is computer crime?

Computer crime is any criminal act that has been committed using a computer as the principal tool. As the role of computers in society has increased, opportunities for crime have been created that never existed before.

Computer crime can take the form of the theft of money (for example, the transfer of payments to the wrong accounts), the theft of information (for example, from files or databases), the theft of goods (by their diversion to the wrong destination) or malicious vandalism (for example, destruction of data or introducing viruses).

Why is computer crime on the increase?

The rapid spread of personal computers and particularly, distributed processing and WANs, has made information held on computer more vulnerable.

Every single one of the top 100 companies in the FTSE index has been targeted or actually burgled by the new computer criminals. The British police have evidence of 70 000 cases where systems have been penetrated and information has been extracted. One inquiry revealed three hackers had been involved in making 15 000 extractions from systems. The criminals are after information worth billions of pounds.

The arrival of Automated Teller Machines provides a good example of how a new technological device creates new opportunities for fraudulent activity. In the 'phantom withdrawals' scandal, British banks and building societies were sued by hundreds of customers in 1992 who claimed that they had been wrongly debited throughout the 1980s for withdrawals they did not make. The banks claimed that the customers must have withdrawn the money and that phantom withdrawals from their machines were 'impossible'.

In another case a criminal gang rented a shop, made it look like a bank and installed a fake cash-machine. The machine did not issue any money (saying it was out of order) but copied the magnetic strip on the back of the card and stored the card's PIN-number. The gang then made duplicate cards and used real cash machines to steal money.

The advent of the mobile phone has led to another computer crime – cloning the chip inside the phone so that you can use your phone but the charge appears on someone else's bill.

Banking security experts in the USA estimate that an average bank robbery nets $1900 and the perpetrator gets prosecuted 82 per cent of the time. With a computer fraud, the proceeds are nearer to $250 000 and less than two per cent of the offenders get prosecuted.

Software Copyright

The Computer Misuse Act 1990 which makes hacking illegal and **The Data Protection Act 1984** which says how personal data on computer must be stored were introduced in Chapter 2. Computers have led to new laws as well as new ways of breaking the old laws. Computer users must also obey old laws such as **The Copyright Act**. When you buy software you do not buy the program, only the right to use it. You do not have the right to give a copy to a friend.

If you have more than one computer, you should check with the licence agreement, which comes with the software, to see if you have the right to run the software on both machines. The most common licence agreements are:

◢ Single user licence where the software can only to be used on one machine.

➤ Network licence. This may be for up to 15 or 20 stations on a network, depending on the licence. This licence is obviously much more expensive than a single user licence.

➤ Site licence which enables the software to be used on any computer on the site.

The Federation Against Software Theft (FAST) aims to prevent illegal use of software and has a policy of prosecuting anyone found to be breaching copyright law. Software companies are getting more and more sophisticated in their attempts to prevent breaches of copyright. The program may only operate if a code is typed in. This code changes each time the program is run and can be found by looking in the manual or using a special code-wheel which comes with the software. Some programs will only run if the CD-ROM is in the CD drive or if a special piece of hardware called a **dongle** is plugged into the back of the computer.

Preventing computer crime

Good computer security is vital to protect information, but the Audit Commission has repeatedly reported that computer security is far too lax in most companies and government departments.

Raymond Cheng, President of a computer company in Malaysia, boasted in a magazine interview in 1995 that his system was impenetrable and promised £12 000 to anyone who could prove otherwise. Two Malaysian computer experts broke into an Internet-based system, saying it took only minutes to hack into the electronic mail-box. Cheng paid up the £12 000.

As more and more PCs are networked, security problems will only get worse. Companies must not be complacent. Computer security undoubtedly is being improved quite rapidly, but it seems that every time one security loophole is plugged, the computer criminals discover another one.

Detecting computer crime

Crimes may go on for some time before anyone suspects anything. The give-away is often the two Porsches on the drive and the trips to Hawaii. The neighbours get jealous and suspect something. Even

when the fraud is found, some companies will not prosecute. It would be too embarrassing for a bank to admit in court that its security was poor.

Stories abound about computer crime. There is a story of an American bank employee who rounded every interest calculation down to the nearest cent. All the odd fractions of a cent left over went into his account. It added up to millions but no-one missed the odd half cent. No-one even noticed. He would have got away with it, if he hadn't started spending the money. (Well you would, wouldn't you?) His jealous colleagues checked his bank account.

Computer crime is now so big that a new special police unit has been set up at New Scotland Yard to deal with it. Companies are vulnerable because they often underestimate the need for communication security. They are at risk because they need their networks to keep in touch with personnel in the field and other offices. The police unit points out that computers can be as much a force for good as for evil. Like many human inventions it just depends who is using them.

■ USING COMPUTERS TO SOLVE CRIME

Computers are used to help solve all types of crimes. For example, they

➤ store details of all known criminals

➤ can identify the owner of a suspect vehicle very quickly

➤ can create identikit pictures

➤ can trace telephone calls

➤ link crimes in different parts of the country

■ PROBLEMS WITH STORING CRIMINAL DETAILS ON COMPUTER

Some people are concerned about the police use of computers because

➤ they are exempt from most of the Data Protection Act

➤ you cannot inspect your own records

➤ information is stored on suspects, not just those proved guilty

➤ incorrect information could be stored

➤ the information is very sensitive and may not be secure from hackers

Case Study 1

A hacker seizes control of a US battleship

In a case frighteningly similar to the film 'War Games,' a young hacker gained unauthorised control of the US Navy's Atlantic Fleet via the Internet in 1995. The hacker was a young US air force captain, who showed the US Navy how easy it was to hack into their control systems.

With the sceptical 'top brass' from the Pentagon watching, the young officer used a PC and a modem to control the US Navy's warships. He had no special insider knowledge but was a computer whizz-kid – just the sort of person the Pentagon most want to keep out.

Within a few minutes of logging on to the Net, the computer screen announced 'Control is complete'. Meanwhile, out at sea, the ship's captain had no idea that the hacker had command of his multi-million-dollar warship. Other ships' controls were also taken over.

The US Navy realised that its security was not good enough. 'This shows we have a long way to go in protecting our information systems,' said a senior executive. The exact method of entry remains a classified secret. ☐

Case Study 2

Plot to steal millions from cash points

In 1996 seven men were jailed for conspiring to defraud banks. The prosecuting counsel said that 'the banking system of Britain would have been put at risk' if the plot had succeeded.

The judge said that, with the help of corrupt BT employees and ex-employees, the plotters intended to tap into lines that run between cash dispensers and main banking computers. Confidential information from customers would have been downloaded on computers to make false cards. A large number of plastic cards were recovered by detectives.

The gang planned to use the counterfeited cash cards to obtain hundreds of millions of pounds from cash machines. The conspiracy was only foiled when the computer expert recruited by the gang became an informer. ☐

Case Study 3

The blonde 'divorcee'

An attractive blonde opened a bank account using false information, saying that she expected to receive her divorce settlement shortly. Later she returned to the bank and surreptitiously removed all the paying-in slips (used by customers to pay money into their accounts) and replaced them with paying-in slips that she had had specially printed. These paying in slips were exactly the same except they had her account number printed at the bottom in MICR characters – just like the paying-in slips at the back of a cheque book.

When reading paying-in slips, the computer looks for the MICR numbers. If there are none, the operator has to type in the bank account number given. If there are MICR numbers, the information is automatically read and not checked. Money paid in with the fake paying-in slips was paid directly into the blonde's account.

Customers did not notice any errors until they checked their bank statements. By this time the blonde had withdrawn over $150 000 in cash from her 'divorce settlement,' disappeared and was never seen again. This demonstrates the dangers when data is fed automatically into a computer. ☐

Using computers to catch fraudsters

Barclays have been catching fraudsters trying to use stolen Barclaycards by using their *Fraud 2000* computer system.

The system is linked to card authorisation terminals used in shops for approving card transactions. It carries out the checking process so quickly that the shop has time to call the police to arrest people trying to use stolen cards. Barclays said the system was designed to carry out checks in one fifth of a second.

In the vast majority of cases the criminal runs off and leaves the card behind, allowing the bank to return it to its rightful owner and stop any fraudulent transactions. The bank said: 'We are trying to make it so difficult that at least some of the fraudsters will give up'.

As well as spotting stolen cards, the *Fraud 2000* system alerts staff to unusual spending patterns such as high spending on stereo equipment, drink or jewellery. This triggers suspicion even if the card is not listed as lost or stolen. The computer compares the purchases with the cardholder's normal spending habits. If the two do not match up, retailers are told to keep the card and the goods while staff try to check the identity of the card user. Genuine customers who come under suspicion because of high spending are normally grateful that Barclays is taking steps to fight fraud. Fraud has fallen by 15 per cent at Barclays, partly due to *Fraud 2000*. ☐

Activity 1 – Fraud at the bank

In 1994, the security of *Citibank*, the fifth-largest bank in the world, was breached. $200 000 disappeared overnight from the bank account of an Argentinian investment company. Four electronic cash transfers were made without authorisation, sending money to four unknown destinations.

Investigations at *Citibank* headquarters in Wall Street showed money transfers were taking place without the knowledge of customers, and during the rest of the year clients' money flowed illegally into accounts in California, Latin America, Finland, Israel and the Netherlands.

Citibank and the FBI began a joint investigation. Previously hackers who have defrauded computerised cash transfer systems were insiders. This was an outside job. Suspicion focused on Vladimir Levin, a Russian computer programmer from St. Petersburg, who allegedly hacked into the bank's computers and carried out the attempted multi-million-dollar fraud from a laptop at the offices of a company in St. Petersburg.

Another hacker – known only as Megazoid – was the first Russian to break into *Citibank*'s computers. Megazoid claimed he entered the *Citibank* computer network undetected for months, using a computer and modem he bought for $10 and a bottle of vodka. One of Megazoid's fellow hackers got drunk one night and sold the secrets of how to break into *Citibank* for $100 and two bottles of vodka.

'Transfers can take place from anywhere,' a New York attorney told one of the many US court hearings on the Levin case. 'You have the right user ID and the right password and the right hardware, and so long as you know what you're doing on the computer, you can get into the system from literally anywhere.'

Citibank says that it lost less than $400 000 in the scam. The bank has since installed a new security system which uses passwords which are changed each time they are used. *Citibank* now claims to be a leader in fraud prevention in the finance industry.

1 It is much more difficult to spot computer crime and catch those responsible than with conventional crime. Why do you think this is?

2 What would you say to reassure investors that their money is safe in a bank account and not at risk from 'phantom' withdrawals?

Activity 2 – The teenage hacker

In 1995, Richard Pryce, 18, a music student from north London, appeared in court on 12 charges of unlawfully gaining access to US Air Force computers. Following a 13-month investigation by US security agents, Pryce, who used the name *Datastream Cowboy,* was arrested by the British police. It was alleged that he gained access on at least 69 occasions.

In 1997 Pryce's fellow-hacker Matthew Bevan, known as *Kuji,* walked free from court even though he had been accused of almost starting World War III from an old 486 PC in his bedroom.

Bevan had been arrested in 1996 after allegedly accessing US Air Force and South Korean military computer systems. After 14 court appearances, the case was dropped when the prosecution declared it would not be in the public interest to pursue the matter.

US Air Force investigator Jim Christy described how

Datastream Cowboy (then aged 16) hacked into the Korean Atomic Research Institute's database. When the Americans discovered the problem, they were not sure whether the database belonged to North or South Korea. If the North Koreans had been penetrated, they would have suspected the US military of the data theft, bringing East and West to the brink of war.

The case collapsed as it was simply too expensive to prosecute Bevan, who may only have had a small fine if found guilty. After the Bevan case lawyers agreed that there were loopholes in the law and that the legal system was finding it hard to keep pace with technological change.

The US Air Force believes that more than 500 attacks are made on its computer networks in one year. The Pentagon has more than 12 000 computer systems. The Americans spend $1 billion a year on improving security on networks alone.

3 What would you say to someone who said that hackers who break into files for fun are relatively harmless and there is little to worry about?

Summary

- [ ] The increase in distributed networks carrying details of financial transactions or valuable information has increased the opportunity for criminal activity.

- [ ] It may be possible for unauthorised users to access these networks with everyday hardware and knowledge of a few passwords.

- [ ] All large UK companies have been targeted by computer criminals.

- [ ] Computer security is vital for any company.

- [ ] Scotland Yard has a special unit to deal with computer crime.

- [ ] 'Lending' or 'borrowing' software is theft of copyright.

Computer crime questions

1 Give three different examples of computer-related crime. (3)

2 Describe how the bank might have prevented the theft in Case Study 3 on page 54. (3)

3 The illegal use of computer systems is sometimes known as computer-related crime.

a) Give **three** distinct examples of computer-related crime. (3)

b) Give **three** steps that can be taken to help prevent computer-related crime. (3) *NEAB Specimen Paper 1*

4 Describe **three** different major ways in which computers may assist the police in their task of combating crime. (3) *AEB Computing Specimen Paper 2*

5 Give **two** reasons why the storing of information on a police computer constitutes a potential danger for the individual in society. (2) *AEB Specimen Paper 2*

6 Give **three** steps that should be taken to help prevent computer-related crime. (6)

7 Describe the dangers of crime that the Internet presents. (3)

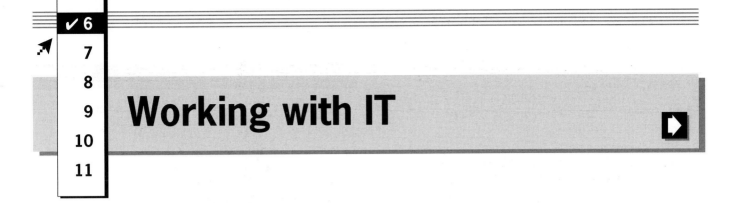

Working with IT

There are many different jobs associated with IT. Some jobs will involve using IT only some of the time. Other jobs will involve using IT most, if not all of the time. These include **development personnel** such as systems analysts and programmers who produce new systems and **operational personnel** such as operators and support staff who keep the system running.

There has been a blurring of definitions as some staff are responsible for development and support, as can be seen from the following advertisements. The Database Administrator post, for example, is involved in investigation, implementation and support.

Database Administrator   Up to £27 500

Responsible for identifying operational needs and configuring the software accordingly. During implementation, you will be working on-site and managing change in a high pressure environment. After implementation, you will be the first line support for users.

Systems Analyst Package
£36 000 to £45 000

Responsible for working directly with clients and end-users to gather requirements. Investigate, design and develop new systems. Work with end users in User Acceptance Testing.

Network Administrator
£19 500 to £22 000

Responsible for the day-to-day management and maintenance of the network, including providing operational and problem solving support and introducing new applications.

Programmer/Analyst Package
£25 000 to £35 000

Responsible for program development and modification, program and data changes, testing and documentation.

E-mail Administrator   £23 000

To monitor and maintain the company's e-mail systems; installing, supporting and administering all services

IT Support Officer
Package £18 000 to £23 000 + Car

You will be responsible for IT within our Midlands offices, providing first-line software and hardware maintenance support. Day-to-day duties will include installation, support and maintenance of PC hardware, software and netware.

Web Administrator   Up to £20 000

Responsible for developing, managing and co-ordinating the posting of material from all departments of the company to web sites.

The qualities required

- Possessing good inter-personal skills, you will have the ability to communicate with all people at all levels of the company.

- You will have a strong technical competence. You will also be a committed, flexible team player.

- You are self-motivated, a fast learner and a good negotiator. You'll need plenty of patience, have your own ideas of the office of the future and a sense of humour.

- A flexible, user-focused, hands-on approach is essential to this role.

As can be seen from these cuttings from job advertisements, although good technical knowledge is essential for IT personnel, there are other important qualities that are necessary.

- A lot of the work of IT personnel involves communicating with other employees; for example, investigating how the work is done at present, dealing with users who need support. **Good inter-personal and verbal communications skills** (together with the ability to avoid jargon) are essential.

- **The ability to write clearly and concisely** is important for writing reports recommendations, systems specifications and documentation.

- IT personnel must be able to **work well in a team**. Projects like introducing a new system are very large and will involve a team of workers, each with different roles. It is important to be able to work well together with co-operation and teamwork.

- A **flexible approach** to a problem is essential. The solution should not be based on the needs of the IT manager or a programmer, but on the requirements of the user.

- **Thoroughness and reliability** are vital as well as the organisation will probably not be able to operate without its computer systems. Failure to test a system properly or not backing-up files, for example, could have disastrous consequences. Deadlines must be met in introducing new systems.

There are many different jobs involving Information Technology, requiring different levels of skill and different training. Important qualities necessary for IT personnel include:

- good technical knowledge
- good inter-personal and verbal communications skills
- the ability to work well in a team
- a flexible approach
- thoroughness and reliability.

Health and Safety

Thousands of computer users have blamed computers for various problems with their health. Trade unions representing these workers have claimed compensation from their employers.

Among the problems suffered are:

Repetitive Strain Injury (RSI)

It is widely accepted that prolonged work on a computer can cause repetitive strain injury. This can be a serious and very painful condition that is far easier to prevent than to cure once contracted, and can occur even in young physically fit individuals. It affects the shoulders, wrists and fingers of those typing all the working day. The symptoms are pain, numbness, stiffness, swelling and in severe cases paralysis.

Possible solutions: ergonomic keyboards, wrist supports, correct positioning on chair

Eye strain

Looking at a screen for a long time can lead to eye strain; particularly in glare or poor light, where screens flicker or where light from other sources is reflected in the screen.

Possible solutions: Non-flickering screens and screen filters to prevent glare and reflection are now required by law. Appropriate practices, for example, taking a break from the screen every few minutes. Appropriate spectacles should be worn when using a computer – employers are now required to pay for employees' eye-tests.

Back problems

Uncomfortable or awkward posture sitting at a computer can lead to serious back problems. Awkward foot positions may cause ankle problems.

Possible solutions: An adjustable chair is now required by law. Employees have the right by law to be provided with foot supports which can reduce problems with ankles. Screens can tilt and turn to a suitable position.

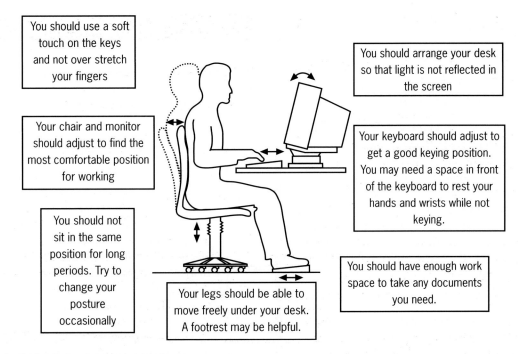

You should use a soft touch on the keys and not over stretch your fingers

Your chair and monitor should adjust to find the most comfortable position for working

You should not sit in the same position for long periods. Try to change your posture occasionally

Your legs should be able to move freely under your desk. A footrest may be helpful.

You should arrange your desk so that light is not reflected in the screen

Your keyboard should adjust to get a good keying position. You may need a space in front of the keyboard to rest your hands and wrists while not keying.

You should have enough work space to take any documents you need.

Figure 6.1 Adjustable chairs can prevent injuries

Ozone irritation

Laser printers emit ozone and so can become a health risk. Experts believe that ozone acts as an irritant, but it is harmless in the quantities present in printers. This has become more of a problem since the advent of affordable, small laser printers. Bubble-jet or dot-matrix printers do not emit ozone.

Possible solutions: Some ergonomists recommend that personal laser printers should be located at least one metre away from where someone is sitting to avoid the ozone emissions. Good ventilation is essential.

Other problems

In the USA research has revealed that an unusually high percentage of pregnant computer users have suffered miscarriages. This may be pure coincidence and there is no evidence to link these problems to the computer. Computers have also been blamed for the incidence of epilepsy in users.

The responsibility for the cause of health problems is very hard to prove, as they may have been the result of other activities. Eye strain may result from reading in poor light or watching too much TV, back and foot problems may result from wearing

unsuitable shoes. However RSI occurs from repeated physical movements in occupations ranging from meat packers to musicians resulting from the typical tasks they perform. If RSI occurs in these occupations it is logical to assume that it occurs in IT workers as well.

How serious is the problem?

As computer use has increased the incidence of RSI has risen. In 1995 the US Occupational Safety and Health Administration said that of all the illnesses reported to them, 56 per cent were RSI cases, compared with 28 per cent in 1984 and 18 per cent in 1981. Among the generally accepted causes of these cases were poorly designed workstations, ill-fitting chairs, stressful conditions and extended hours of typing. In the early 1990s computer games were blamed for RSI and attacks of epilepsy, giddiness and headaches among British youngsters.

In 1993 it was reported that Blue Cross of California estimated that every case of repetitive strain injury it treated cost an average $20 000 in medical expenses and lost productivity. One single case of carpal tunnel syndrome, a common form of RSI in the wrists, could cost as much as $100 000 in medical and administrative expenses and lost productivity.

Daily Mail, 24 January 1992

HEALTH HELP FOR SCREEN WORKERS

Millions of workers who use computers are to get a better deal under new regulations announced yesterday. They will be entitled to regular screen breaks, free eye tests and special glasses.

The new rules – put forward by the European Commission – will come into force at the end of the year and should combat repetitive strain injuries (RSI), eye problems and stress. Last year computer-related problems affected 100 000 victims.

Sickness

Unveiling the proposals in London, the Health & Safety Commission said they will cost employers £40 million a year, £42 per worker. But the Commission's Chairman, Sir John Cullen, said the money should be recouped by reducing days lost through sickness.

The regulations – which suggest breaks of five to ten minutes after 50 to 60 minutes work – are still to be finalised.

The Guardian, 24 January 1992

New controls on VDU work 'inadequate'

Plans to control the growing tide of injuries caused by working with computer screens and keyboards were attacked yesterday by trades unions as **hopelessly inadequate** and falling far short of the European Community directive they are supposed to implement.

John Edmonds, leader of the 930 000 GMB union, said the proposals published yesterday by the Health and Safety Commission let British employers off the hook. 'The HSC is doing everything it can to stop the new laws that would make life safer for people who use VDUs (visual display units).'

Under the regulations employers would have to assess work on VDU work stations and reduce risks found, plan work to allow for regular breaks away from screens and pay for employees' eye tests and special glasses if required.

Preventing health problems

In 1992 the European Union introduced regulations to try to prevent health problems associated with computer equipment.

The regulations called the Health and Safety (Display Screen Equipment) Regulations 1992 *(Health and Safety Executive)* said as follows

Employers must

1 assess the risks due to display screen equipment workstations and take steps to reduce risks which are discovered;

2 conform to minimum standards for workstations so that the working environment is suitable for the work involved. This means providing tiltable screens, anti-glare screen filters, adjustable chairs and foot supports. Room lighting must be suitable and workstations must not be cramped;

3 plan work at a computer so that there are breaks or changes of activity;

4 provide information and training for display screen equipment users;

5 pay for appropriate eye and eyesight tests by an optician or doctor for display screen equipment users and to provide special spectacles if they are needed and normal ones cannot be used.

As well as covering the working environment (such as lighting, noise, and sitting position) and hardware (such as the screen and keyboard), health and safety guidelines aim to ensure that new software should be designed to reduce health risks. For example using screen layouts that are clear, colour schemes that have good contrast and text that is a suitable size, minimising the chance of eye-strain.

Note: These regulations apply only to offices – not to students or pupils in schools or colleges.

Case Study

A sufferer from RSI

Paul is a university lecturer who works a lot on computers and suffered from RSI.

■ WHAT SYMPTOMS OCCURRED?

- Tightness, discomfort, stiffness, or pain in the hands, wrists, fingers, forearms, and elbows

- Tingling, coldness, or numbness in the hands

- Clumsiness or loss of strength and co-ordination in the hands

- Pain that kept him awake at night

- Feeling a need to massage his hands, wrists, and arms

■ WHAT DID HE DO ABOUT IT?

- Made sure his seating position was comfortable

- Took lots of breaks to stretch and relax

- Avoided pounding on the keys

- Held the mouse lightly

- Kept his arms and hands warm

- Eliminated unnecessary computer usage

■ DID IT WORK?

- There are no quick fixes but pain and discomfort were reduced

- Many RSI victims do regain the ability to work but find that they remain vulnerable to flare-ups

- Prevention is still the best prescription ☐

It is always difficult to develop regulations to cover an area that is evolving so quickly that by the time you have researched the problems and made recommendations on a given technology, it may be out of date.

I use a computer a lot. Should I be worried?

This depends on how you use it. You can get muscle soreness, tired eyes and repetitive stress injuries from using PCs, just as you can from any number of other activities like jogging or working-out. Common sense and simple precautions should reduce the risk.

Close, intricate work on PCs could lead to eye strain. You should have adequate, glare-free lighting in the room. If you need glasses, wear them.

You should use a good, adjustable office chair. Your chair should support the lower back and you should sit with your thighs horizontal and your feet flat on the floor. Your keyboard should be in a position so that your shoulders are relaxed and your upper arm and forearm form a 90-degree angle to the desk. Your wrist and hand should be in a straight line and a wrist support, which sits in front of the keyboard may help.

Use a light touch on the keyboard or mouse. Joysticks and games control pads put an awful lot of strain on your hand, so don't play them aggressively for long periods. A tracker ball may lead to fewer problems than a mouse.

Modern screens meet stringent radiation limits so you shouldn't have to worry about them, but there may possibly be cumulative effects over a period of years.

Take short, frequent breaks from the screen. A couple of minutes every half hour having a cup of tea is recommended. Playing games for five hours non-stop is not.

Summary

There are many different jobs involving Information Technology, requiring different levels of skill and different training. Important qualities necessary for IT personnel include:

☐ good technical knowledge

☐ good inter-personal and verbal communications skills

☐ the ability to work well in a team

☐ a flexible approach

☐ throughness and reliability

Regular use of IT equipment over a long period of time may lead to health problems, particularly:

☐ RSI (Repetitive Strain Injury) mainly in the wrists and hands

☐ Eye problems

☐ Back problems

☐ Ankle problems

☐ Irritation due to ozone

These problems can be reduced by taking sensible precautions, not using equipment for too long and introducing adjustable chairs and screens, wrist supports and screen filters.

New regulations have been introduced throughout the European Union setting standards for using IT equipment in offices. Employers have duties to:

☐ provide adjustable chairs

☐ provide foot rests

☐ provide adjustable screens

☐ provide anti-glare screen filters

☐ pay for appropriate eye and eyesight tests

☐ plan work so that there are breaks from the screen

☐ provide information and training for IT users.

These regulations apply only to offices – not to students or pupils in schools or colleges.

Display screen equipment users are entitled to regular checks by an optician or doctor, and to special spectacles if they are needed and normal ones cannot be used. It is the employer's responsibility to provide tests, and special spectacles if needed.

Working with IT questions

1 What is meant by RSI? List the parts of the body it affects, and describe the symptoms.

2 List **six** factors which could give rise to health or safety problems associated with the use of information technology equipment. (6) *NEAB 1996 Paper 1*

3 A trades union is concerned about its members who use computers all day. Produce a leaflet describing the dangers and the rights of computer users.

4 The trades union leader (Mr. Edmonds) thought the new regulations weren't strong enough. The employers complained that the new regulations would cost £42 per worker. Imagine you are either

a) the boss of a company using computers or

b) a trades unionist using a computer at work.

Write a letter to Sir John Cullen of the Health and Safety Commission telling him whether you think the new regulations are a good idea and whether they go far enough or too far, explaining why.

5 Describe **three** health hazards associated with computer use. (6) *NEAB Specimen Paper 1*

6 The introduction of computer terminals and personal computers has been associated with a number of physical health hazards.

a) State **three** health hazards which have been associated with prolonged use of computers. (3)

b) Describe **five** preventative actions which may be taken to avoid computer related health hazards, explaining clearly how each action will assist in preventing one or more of the hazards you have described in part (a). (10) *NEAB 1998 Paper 1*

7 The use of Information Technology equipment has brought Health and Safety risks for employees. Describe **four** such risks, and the measures that an employer should take to protect their staff from them. (12) *NEAB 1999 Paper 1*

8 A firm is recruiting staff to work within its Information Technology department. In addition to technical skills what other qualities should the firm be looking for, and why are these important to the effective working of an Information Technology department. (4) *NEAB Specimen Paper 1*

9 A software house is advertising for an analyst, programmer to join one of their development teams. State **four** personal qualities that the company should be looking for in the applicants. (4) *NEAB 1999 Paper 1*

Data capture and hardware

Automatic data capture

Data capture means collecting data for computer processing. There are alternative methods of collecting data other than typing it in at a computer keyboard.

Multiple choice exam papers use a special **Optical Mark Reading (OMR)** answer sheet. The user puts a mark with a pencil into spaces on a grid. The computer can be connected to an input device that can 'read' documents by shining a light at them and detecting reflected patterns. The National Lottery and football pools coupons use a similar system. These systems avoid keyboard input which can lead to mistakes.

A questionnaire may be printed on a sheet that can be read by **Optical Character Recognition (OCR).** This recognises letters and numbers written into squares on the paper by again detecting patterns of reflected light. This is not as reliable as using OMR as it depends on people's handwriting being clear, but is still a very useful data capture method.

Bar-code readers in supermarkets read the stripes in the bar code, again by detecting patterns of reflected light. Each number in the bar code is represented by four stripes (two black and two white). The bar code is read very quickly with very few mistakes. Most products have a numeric code printed underneath the bar code. This is used if, for some reason, the bar code reader fails to read the code: the operator can key in the number of the product.

Numbers on the bottom of cheques are printed in magnetic ink. The banks' clearing house computers can read these numbers using **Magnetic Ink Character Recognition (MICR).** This is another fast and reliable method of entering data. It is used on cheques instead of OCR to help minimise fraud.

Data stored on the **magnetic strip** on the back of credit and debit cards is automatically read by computer when the card is 'swiped' across a reader. This is much quicker than typing in the card number when a product is bought.

Other methods of data capture

Another example of data capture in use is to be found in the electricity companies who provide meter readers with **hand-held data loggers**. These are small computers little bigger than a pocket calculator, with a key-pad for input. Meter readings are entered into the data logger as the meter is read. Later on, the data logger is connected to a computer and the data down-loaded.

Data can be collected in such a way as to be automatically entered into the computer. For example, computers can be connected to **sensors** which monitor temperature, humidity, light or wind speed.

Other examples of uses of computer sensors for data capture include:

- a traffic count
- an automatic weather station, particularly in remote and inhospitable places
- timing devices in swimming or athletics races
- temperature sensors to allow computers to control central heating
- PIR sensors for burglar alarms

Key-to-disk data entry

Where large volumes of data have to be entered, off-line key-to-disk systems may be used. In these systems the data entry clerks type in the data, which is stored straight away on disk. This method would be used in batch processing, where large batches of data have to be entered before being processed. (As the entry is off-line, it does not affect the performance of the main computer.)

If the data is processed immediately, as in a real-time system, on-line interactive data entry must be used.

Data capture in schools

Many schools no longer use the old method of registering pupils by putting a mark in a book.

St Joseph's High School uses an OMR system. Pupil names are automatically printed on an OMR sheet. The teacher taking the register puts a mark in the right place. All the OMR sheets are batched together and read by an OMR reader, which automatically inputs the data into the computer. The computer records lateness and absences, and can calculate statistics.

This is a form of batch processing. The data is processed after registration and so may not be up-to-date. There can be no other input, for example if a pupil arrives late.

Kentland Sixth Form College uses a swipe card system. Every student has a swipe card and there is a swipe card reader in every classroom. Each student swipes their card through the reader at the start of each lesson and the data is transmitted to the computer. This method is much more expensive and requires a lot of cabling in a large school. It keeps a record of attendance at all lessons including the time of arrival, so the teacher does not need to take a register and a traditional registration period is not required.

Woodland Brook High School uses the EARS (Electronic Attendance Registration System) system. Each teacher carries a small hand-held computer. In each block of classrooms is a radio antenna which sends and receives messages to and from the teachers' computers. The teacher completes an electronic register on the computer for every lesson. This system is also expensive to install but keeps very good records on attendance. It also can be used for other administrative tasks, such as sending a message to a teacher and storing pupil marks.

The systems used by Kentland Sixth Form College and Woodland Brook High School are both on-line systems. The data is always up-to-date and the systems can be searched to find out where a pupil or a teacher is. □

Is the information correct?

All the above methods of data capture are reliable and automatic. For example, by using bar code readers at a supermarket check-out, no data needs to be typed in, thus saving time and leading to fewer errors. However no method of data entry can guarantee to be error-free. Examples of errors might be:

- Transcription errors where details are copied incorrectly. For example, data may be entered in the wrong row of a table or the decimal point may be missed out.

- Transmission errors where data is corrupted when sent from one computer to another. This may be due to a poor connection or 'noise' on the line.

- Errors in data capture, possibly because the system has not been set up correctly, for example a timing device may not start immediately at the start of a race or a traffic count may be set up at the wrong place.

- Processing errors for example where a calculation has not produced the right result. This may be due to a mistake in the programming or may be a truncation error when the computer cannot cope with very large numbers.

Reducing errors

We know that Garbage In leads to Garbage Out. As businesses and other organisations are dependent on information from computers, it is essential to try to reduce these errors, to ensure that the information is correct.

Processing errors should be spotted at the system's testing stage. Testing should not just include simple test data but should include extreme data (such as large numbers) and incorrect data. Results should be compared with expected results. Incorrect data should be rejected. Testing may need to be performed on different platforms, for example computers with different memory sizes to make sure it works on all machines.

Transmission errors can be avoided by the recipient computer sending the same message back to the original sender. If the two messages are the same, the data will be correct.

Human errors such as transcription can be detected using two techniques, **validation** and **verification**.

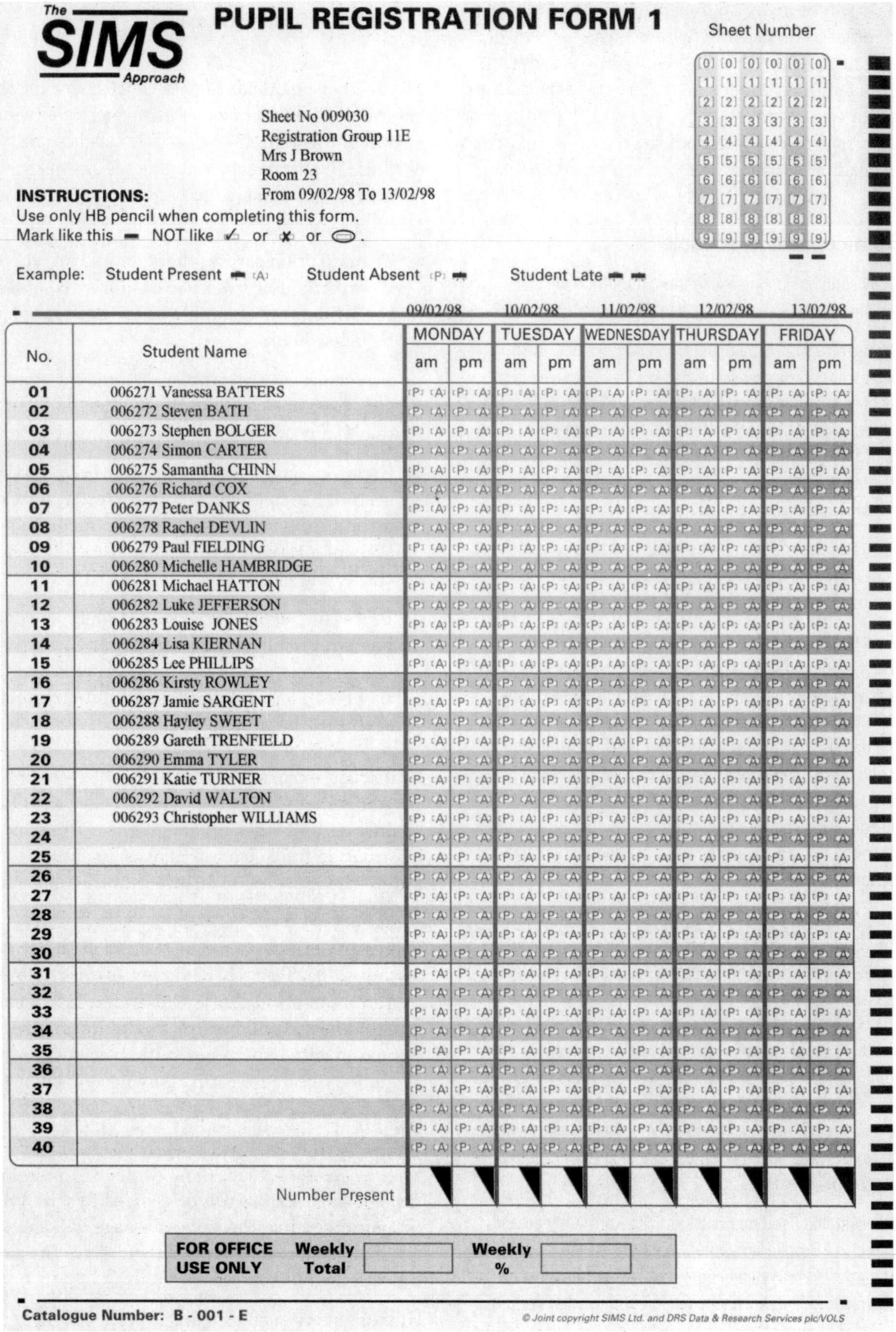

The SIMS Approach

PUPIL REGISTRATION FORM 1

Sheet Number

[0] [0] [0] [0] [0] [0]
[1] [1] [1] [1] [1] [1]
[2] [2] [2] [2] [2] [2]
[3] [3] [3] [3] [3] [3]
[4] [4] [4] [4] [4] [4]
[5] [5] [5] [5] [5] [5]
[6] [6] [6] [6] [6] [6]
[7] [7] [7] [7] [7] [7]
[8] [8] [8] [8] [8] [8]
[9] [9] [9] [9] [9] [9]

Sheet No 009030
Registration Group 11E
Mrs J Brown
Room 23
From 09/02/98 To 13/02/98

INSTRUCTIONS:
Use only HB pencil when completing this form.
Mark like this ▬ NOT like ✓ or ✗ or ⬭.

Example: Student Present ▬ [A] Student Absent [P] ▬ Student Late ▬ ▬

No.	Student Name	MONDAY 09/02/98 am	MONDAY 09/02/98 pm	TUESDAY 10/02/98 am	TUESDAY 10/02/98 pm	WEDNESDAY 11/02/98 am	WEDNESDAY 11/02/98 pm	THURSDAY 12/02/98 am	THURSDAY 12/02/98 pm	FRIDAY 13/02/98 am	FRIDAY 13/02/98 pm
01	006271 Vanessa BATTERS	[P] [A]	[P] [A]	[P] [A]	[P] [A]	[P] [A]	[P] [A]	[P] [A]	[P] [A]	[P] [A]	[P] [A]
02	006272 Steven BATH	[P] [A]	[P] [A]	[P] [A]	[P] [A]	[P] [A]	[P] [A]	[P] [A]	[P] [A]	[P] [A]	[P] [A]
03	006273 Stephen BOLGER	[P] [A]	[P] [A]	[P] [A]	[P] [A]	[P] [A]	[P] [A]	[P] [A]	[P] [A]	[P] [A]	[P] [A]
04	006274 Simon CARTER	[P] [A]	[P] [A]	[P] [A]	[P] [A]	[P] [A]	[P] [A]	[P] [A]	[P] [A]	[P] [A]	[P] [A]
05	006275 Samantha CHINN	[P] [A]	[P] [A]	[P] [A]	[P] [A]	[P] [A]	[P] [A]	[P] [A]	[P] [A]	[P] [A]	[P] [A]
06	006276 Richard COX	[P] [A]	[P] [A]	[P] [A]	[P] [A]	[P] [A]	[P] [A]	[P] [A]	[P] [A]	[P] [A]	[P] [A]
07	006277 Peter DANKS	[P] [A]	[P] [A]	[P] [A]	[P] [A]	[P] [A]	[P] [A]	[P] [A]	[P] [A]	[P] [A]	[P] [A]
08	006278 Rachel DEVLIN	[P] [A]	[P] [A]	[P] [A]	[P] [A]	[P] [A]	[P] [A]	[P] [A]	[P] [A]	[P] [A]	[P] [A]
09	006279 Paul FIELDING	[P] [A]	[P] [A]	[P] [A]	[P] [A]	[P] [A]	[P] [A]	[P] [A]	[P] [A]	[P] [A]	[P] [A]
10	006280 Michelle HAMBRIDGE	[P] [A]	[P] [A]	[P] [A]	[P] [A]	[P] [A]	[P] [A]	[P] [A]	[P] [A]	[P] [A]	[P] [A]
11	006281 Michael HATTON	[P] [A]	[P] [A]	[P] [A]	[P] [A]	[P] [A]	[P] [A]	[P] [A]	[P] [A]	[P] [A]	[P] [A]
12	006282 Luke JEFFERSON	[P] [A]	[P] [A]	[P] [A]	[P] [A]	[P] [A]	[P] [A]	[P] [A]	[P] [A]	[P] [A]	[P] [A]
13	006283 Louise JONES	[P] [A]	[P] [A]	[P] [A]	[P] [A]	[P] [A]	[P] [A]	[P] [A]	[P] [A]	[P] [A]	[P] [A]
14	006284 Lisa KIERNAN	[P] [A]	[P] [A]	[P] [A]	[P] [A]	[P] [A]	[P] [A]	[P] [A]	[P] [A]	[P] [A]	[P] [A]
15	006285 Lee PHILLIPS	[P] [A]	[P] [A]	[P] [A]	[P] [A]	[P] [A]	[P] [A]	[P] [A]	[P] [A]	[P] [A]	[P] [A]
16	006286 Kirsty ROWLEY	[P] [A]	[P] [A]	[P] [A]	[P] [A]	[P] [A]	[P] [A]	[P] [A]	[P] [A]	[P] [A]	[P] [A]
17	006287 Jamie SARGENT	[P] [A]	[P] [A]	[P] [A]	[P] [A]	[P] [A]	[P] [A]	[P] [A]	[P] [A]	[P] [A]	[P] [A]
18	006288 Hayley SWEET	[P] [A]	[P] [A]	[P] [A]	[P] [A]	[P] [A]	[P] [A]	[P] [A]	[P] [A]	[P] [A]	[P] [A]
19	006289 Gareth TRENFIELD	[P] [A]	[P] [A]	[P] [A]	[P] [A]	[P] [A]	[P] [A]	[P] [A]	[P] [A]	[P] [A]	[P] [A]
20	006290 Emma TYLER	[P] [A]	[P] [A]	[P] [A]	[P] [A]	[P] [A]	[P] [A]	[P] [A]	[P] [A]	[P] [A]	[P] [A]
21	006291 Katie TURNER	[P] [A]	[P] [A]	[P] [A]	[P] [A]	[P] [A]	[P] [A]	[P] [A]	[P] [A]	[P] [A]	[P] [A]
22	006292 David WALTON	[P] [A]	[P] [A]	[P] [A]	[P] [A]	[P] [A]	[P] [A]	[P] [A]	[P] [A]	[P] [A]	[P] [A]
23	006293 Christopher WILLIAMS	[P] [A]	[P] [A]	[P] [A]	[P] [A]	[P] [A]	[P] [A]	[P] [A]	[P] [A]	[P] [A]	[P] [A]
24		[P] [A]	[P] [A]	[P] [A]	[P] [A]	[P] [A]	[P] [A]	[P] [A]	[P] [A]	[P] [A]	[P] [A]
25		[P] [A]	[P] [A]	[P] [A]	[P] [A]	[P] [A]	[P] [A]	[P] [A]	[P] [A]	[P] [A]	[P] [A]
26		[P] [A]	[P] [A]	[P] [A]	[P] [A]	[P] [A]	[P] [A]	[P] [A]	[P] [A]	[P] [A]	[P] [A]
27		[P] [A]	[P] [A]	[P] [A]	[P] [A]	[P] [A]	[P] [A]	[P] [A]	[P] [A]	[P] [A]	[P] [A]
28		[P] [A]	[P] [A]	[P] [A]	[P] [A]	[P] [A]	[P] [A]	[P] [A]	[P] [A]	[P] [A]	[P] [A]
29		[P] [A]	[P] [A]	[P] [A]	[P] [A]	[P] [A]	[P] [A]	[P] [A]	[P] [A]	[P] [A]	[P] [A]
30		[P] [A]	[P] [A]	[P] [A]	[P] [A]	[P] [A]	[P] [A]	[P] [A]	[P] [A]	[P] [A]	[P] [A]
31		[P] [A]	[P] [A]	[P] [A]	[P] [A]	[P] [A]	[P] [A]	[P] [A]	[P] [A]	[P] [A]	[P] [A]
32		[P] [A]	[P] [A]	[P] [A]	[P] [A]	[P] [A]	[P] [A]	[P] [A]	[P] [A]	[P] [A]	[P] [A]
33		[P] [A]	[P] [A]	[P] [A]	[P] [A]	[P] [A]	[P] [A]	[P] [A]	[P] [A]	[P] [A]	[P] [A]
34		[P] [A]	[P] [A]	[P] [A]	[P] [A]	[P] [A]	[P] [A]	[P] [A]	[P] [A]	[P] [A]	[P] [A]
35		[P] [A]	[P] [A]	[P] [A]	[P] [A]	[P] [A]	[P] [A]	[P] [A]	[P] [A]	[P] [A]	[P] [A]
36		[P] [A]	[P] [A]	[P] [A]	[P] [A]	[P] [A]	[P] [A]	[P] [A]	[P] [A]	[P] [A]	[P] [A]
37		[P] [A]	[P] [A]	[P] [A]	[P] [A]	[P] [A]	[P] [A]	[P] [A]	[P] [A]	[P] [A]	[P] [A]
38		[P] [A]	[P] [A]	[P] [A]	[P] [A]	[P] [A]	[P] [A]	[P] [A]	[P] [A]	[P] [A]	[P] [A]
39		[P] [A]	[P] [A]	[P] [A]	[P] [A]	[P] [A]	[P] [A]	[P] [A]	[P] [A]	[P] [A]	[P] [A]
40		[P] [A]	[P] [A]	[P] [A]	[P] [A]	[P] [A]	[P] [A]	[P] [A]	[P] [A]	[P] [A]	[P] [A]
	Number Present										

FOR OFFICE USE ONLY	**Weekly Total**	**Weekly %**	

Catalogue Number: B - 001 - E

© Joint copyright SIMS Ltd. and DRS Data & Research Services plc/VOLS

Figure 7.1 School registration sheet

Verification and validation

Verification is used to check that data is entered correctly. One method of verification is typing data into the computer twice. The computer automatically compares the two versions and tells the user if they're not the same. When network users change their password, they have to type in the new password twice to verify it.

Validation is a method that checks that data is sensible. A validation program checks that data typed in obeys a number of constraints. There are many different validation techniques.

1 **Range check.** Checks a value to see if it is in a certain range. For example the month of a year must be between 1 and 12. Numbers less than 1 or greater than 12 are rejected.

2 **Field check** (or **picture check**). This checks that data is of the right format. Is it alphabetic? Is it numeric? A National Insurance number must be of the form XX 99 99 99 X. The first two and the last characters must be letters. The other six characters are numbers. The total length is nine characters. Any other format is rejected.

3 **Cross-field check.** This checks that data in two fields corresponds. For example, if someone's sex is stored as Female, their title cannot be Mr. If the month in a date is 04, the day cannot be 31 as there are only 30 days in April.

4 **Look up table.** This checks that the data is one of the entries in the list of acceptable values. For example, the day of the week must be from the list Monday, Tuesday, and so on.

5 **Hash total or control total.** This is used when a collection of numbers is to be typed in. The operator types in the sum of the figures. The computer adds them up and checks it, for example 5,6,3,4,1,19. If the first five numbers do not add up to the last, the figures are rejected. A control total may be used to check how many records have been entered. This is particularly used in batch processing.

A hash total has no meaning in itself. It could be the total of all the account numbers added up.

6 **Check digit.** This is used to check the validity of code numbers, for example supermarket bar codes or bank account numbers. These numbers are long and prone to data entry errors. There are a number of ways of calculating the check digit. One way is as follows:

A short identity code number may consist of five digits plus one check digits at the end.

For example ID code = 69247

Multiplying each digit in the code by another number called the weight. Then add them, for example

Formula is multiply	first number by 6	(36)
	second number by 5	(45)
	third number by 4	(8)
	fourth number by 3	(12)
	fifth number by 2	(14)
Add them up	36+45+8+12+14 = **115**	
Divide by 11	10 remainder **5**	

Take the remainder away from 11 gives **6**. The check digit is 6.

(Note: if the remainder is 0, the check digit is 0. If the remainder is 1, the check digit is X.)

ID code number = 692476

Try changing one number or swapping two numbers around and working out the check digit. Would the computer spot the mistake?

Remember: Check digits can only be used to valid a code such as a bar code, bank account number, credit card number.

Accurate data, wrong information

However, it is possible that information entered could be both sensible and incorrect. A temperature sensor recording the temperature in a furnace may give valid data, numbers in a certain range, but the data may be wrong if the sensor is not set up correctly.

It is possible for data to be accurate but the information to be wrong. A survey of a sample of the population may lead to storing a lot of data. The information from the survey can be incorrect even if the data is correct. For example, the sample may not be representative of the population as a whole. The information is correct for this sample but not for all the population.

Burton Advertiser, 28 May 1997

Council blunder sends Annie a demand for 3p

PENSIONER Annie Duignan thought the world had gone mad when she received a council tax demand for just THREE pence.

And she was frightened by the threat along with it that if she didn't pay she would have to cough up £351. Now East Staffordshire Borough Council chiefs have admitted they blundered, and telephoned Mrs Duignan to say sorry.

But they also confessed that 24 other people got similar baffling bills – and similar threats – over amounts of less than £1 this month.

Mrs Duignan, 71, accidentally forgot to pay the 3p when she paid her bill.

But she says: 'To get this demand was a complete waste of money. The stamp to send it alone is worth six or seven times the value of the demand. Then there's the time it takes council workers to send it – it's crazy.'

The council's director of finance, Mr Ken Birch, said the problem arose during the computerised printing of demands for people who had not paid.

He said normally those who owed amounts less than £1 were 'hand pulled' from the pile of bills, and the amounts were added to next month's bill. But pressure of work had meant the hand-sorting was missed this time, and 25 bills for less than £1 – including Mrs Duignan's – had gone out by mistake.

Mistake

He said, 'We put our hands up and admit we made a mistake – there was a rush getting the bills out.' Mr Birch said one of his staff would be telephoning Mrs Duignan this week to apologise and explain that she did not have to pay up the 3p.

Hardware

Computers work on the basis of input, process and output. The various parts of the computer (the hardware) can be defined as input devices, the processor, backing store and output devices.

Input devices are used to enter information. The processor is in the 'box' part of the computer and is

Case Study 2

▶ Validation of meter readings

An electricity company reads the meters in individual houses. These readings will be validated by a range check to make sure that they are within a sensible range. The records are fed into the computer.

The computer can check that the details for every house have been entered by checking that a control total – the number of houses – is correct. Of course, one house may have been omitted and another entered twice. This can be checked by making sure that no customer number is repeated.

When the bill is calculated it should be validated by another range check to make sure that it is not ridiculously high or low. ☐

made up of printed circuits and microchips. The processor includes the computer's memory. Backing store is where information is stored when the computer is turned off. Output devices present the information to the user.

Input devices

The keyboard and mouse are not the only input devices. OMR readers, OCR readers and MICR readers are used by commercial businesses. Games computers use a joystick. Hand held computers use keypads. Computers controlling manufacturing processes may use sensors.

The ergonomic keyboard is an alternative keyboard that is reputed to reduce the risk of repetitive strain injury. (See Chapter 6.) A concept keyboard is a simple keyboard for people with learning difficulties. Fast food chains use special keyboards where there is a key for each menu item.

Touch screens allow users to make selections by actually touching the screen. In fact there is a grid of infra red beams in front of the screen. Pointing at the screen breaks the beams so giving the position of the finger. Touch screens are input and output devices combined.

Scanners can be used to scan in pictures or text. Pictures can be stored in a number of formats like **jpg** or **bmp**. Text can be stored as a text file and imported into other applications. Hand held scanners

Figure 7.2 The typical computer consists of the processor, input and output devices

are cheap and suitable for home use but the quality is not as good as flat-bed scanners which have reduced in price to around £50. Flat-bed scanners often come with OCR software, enabling text to be scanned in and stored so that it can be loaded into a word-processing program. The accuracy of this software depends on the quality of the original text.

Digital cameras have reduced in price considerably and now cost little more than conventional cameras. Some digital cameras store pictures in jpg format directly on to a conventional floppy. Other digital cameras connect to a computer's serial port so that pictures can be down-loaded. It is possible to imagine that conventional film will be obsolete in the near future.

Graphics tablets have a stylus pen which is used to draw on a special flat surface. The drawing is automatically read by the computer. These are ideal for graphic designers, artists and technical illustrators.

Speech recognition is a growing area of computer input. As computers get faster and memory increases, reliable speech recognition has been cheaper. Uses include for people with disabilities or to control access to an area.

IBM VoiceType

IBM produce a speech recognition system called VoiceType. This system claims to recognise 32 000 words at approximately 70 to 100 words per minute, with 97 per cent accuracy. The system is compatible with many existing applications such as Lotus Notes, AmiPro, Microsoft Excel, Word and Quicken.

There is a Windows version and a version for notebook and laptop computers. They are available in American English, British English, French, German, Italian and Spanish.

Although the market for speech recognition is growing, deterrents to growth include the lack of appropriate applications, the limited accuracy of the system, the fact that background noise increases errors, users disturbing other people in their office and users going hoarse or losing their voice.

Output devices

Output devices include the monitor, printers, plotters and loud speakers.

The most common forms of printers all form their images out of dots. The smaller the dots, the better the quality of the print. Dot matrix printers print by hammering pins against a ribbon on to the paper to print the dots, so they can be noisy. They are cheap and although quality of print and noise levels have improved, dot matrix printers have lost popularity as the price of ink jet and laser printers has fallen.

Ink-jet printers squirt ink on to the paper and form letters from tiny dots. There are quiet, quality is good and colour versions are available for as little as £100. Laser printers are the fastest and produce the best quality print but are the most expensive type of printer. Prices have dropped and they are now on sale for little more than £250.

LCD projectors which project computer output on to a large screen are expensive but are now common in business demonstrations projecting the output of a laptop computer. Prices continue to fall and quality is improving.

Speech synthesis is an expanding area of output, common in computer games. If you phone up directory enquiries or the speaking clock, you will be told the number you require or the time by the computer in synthesised speech. It sounds almost human.

Backing store

Backing store is a permanent storage medium on which data can be stored for later retrieval. Normal backing store is a **magnetic disk** although in certain circumstances **magnetic tape** is used.

Tape is much slower as it only offers serial access – it has to be read in the order in which it is stored on the tape. Access times for data held at the wrong end of the tape are very slow. Today tape is mainly used for back-up where the whole disk is copied on to tape.

Personal computers use two sorts of magnetic disk.

- The **hard disk** inside the computer can hold as much as 8 Gb. It is used to store the operating system, software and datafiles. These files are vital and some form of back-up is necessary in case of disk failure. It is possible to have a removable hard disk for better security but these are only rarely used.

- The small removable **floppy disk** normally stores only 1.44 Mb and is much slower than a hard disk. It is mainly used for keeping back-up copies of small files or for transferring small files between two or more computers.

New floppy disk drives can store up to 120 Mb on one special floppy disk. This means that much more can be backed up on to one floppy but several floppies will still be needed to back up the contents of a hard drive. As might be expected, these drives and their disks cost more than the standard drives.

CD-ROMs (Compact Disk – Read Only Memory) are small plastic discs, similar to audio CDs. They store large quantities of data (650 Mb) permanently with short access times. CD-ROM drives are now common in PCs. New DVD CDs have a much greater capacity than conventional CD-ROMs.

The high capacity of a CD-ROM means they are more convenient than using floppy disks. Whereas a few years ago newly purchased software was stored on several floppy disks, today software is normally purchased on CD-ROM. CD-ROMs are also ideal for storing large reference files for example encyclopaedias or files of clip-art.

CD-Rs (Compact Disk – Recordable) are writable compact disks. Using a special writable CD drive, up to 650 Mb of data can be recorded on a blank disk which costs as little as £1. This data cannot be altered but it can then be read by any other PC with a standard CD-ROM drive. This means that CD-R is suitable for back-up as it holds much more than a floppy disk, yet can easily be taken off site.

CD-RWs (Compact Disk – ReWritable) are re-writable compact disks. Using the same writable CD drive, up to 650 Mb of data can be recorded, deleted and re-recorded on these disks which cost around £15. This means that a CD-RW can be used in the same way as a floppy disk, but obviously storing far more data.

Writable CD drives can be purchased for PCs from around £250.

Summary

Computers consist of input devices, the processor, output devices and backing store.

Computers can capture data automatically. Examples of automatic data capture include:

☐ OCR

☐ OMR

☐ MICR

☐ Bar code readers

☐ Sensors

Validation and verification are used to avoid mistakes when data is input. Validation checks to see that data is sensible. Examples of validation include range checks and field checks.

Hardware and validation questions

1 Describe **three** methods of data capture and give applications for which each would be appropriate.

2 Write a brief paragraph on each of the following stating when it is likely to be used and why.

   a) Ergonomic keyboard

   b) Concept keyboard

   c) Touchscreen

   d) Voice recognition

   e) Graphics tablets

   f) Laser Printers

   g) Dot-matrix Printers

   h) Ink-jet Printers

   i) Speech synthesis

3 Use of voice recognition is increasing steadily. Give three reasons why this is so.

4 Why are loud speakers now common as computer output?

5 The Driver and Vehicle Licensing Centre want to validate car registration numbers. These are all valid:

   R5ERP   R55ERP   R555ERP

   a) Suggest five different checks which accept all the above which together would reject the following invalid codes, for example The first character must be a capital letter.

   R555P   R055ERP   r555erp   R5555ERP
   R400R7P

   b) Why might a valid registration number still be incorrect?

6 The date of birth of applicants for jobs is stored on file as a six-figure digit e.g. 120167. Suggest three ways you could validate this data.

7 How could the council in the newspaper cutting on page 70 use validation to prevent mistakes in the future?

8 An insurance company wishes to enter the data from hundreds of proposal forms filled in by customers each day.

   a) Describe a suitable method of data input.

   b) Justify your choice

   c) Explain a method used to reduce the number of errors made at this stage. (NEAB Computing Specimen Paper 2)

9 A company sells a range of health foods at five different shops. It also sells directly to the home from a number of vehicles. There are hundreds of different items of stock and many items are seasonal, so items in stock are constantly changing. Customers purchase goods and pay by cash, cheque or credit card.

   The company is considering a computerised system to help manage sales and stock control. Discuss the capabilities and limitations of current

   • communications devices

   • input devices

   • output devices and

   • storage devices

   appropriate for establishing a computerised system for this company. (20) NEAB 1997 Paper 2

10 A school uses an information system to store details of students' examination entries and results. As part of this system a program is used to check the validity of data such as candidate number, candidate names and the subjects entered.

   a) Suggest **three** possible validation tests which might be carried out on this data. (3)

   b) Explain, with an example, why data which is found to be valid by this program may still need to be verified. (2) NEAB 1996 Paper 2

11 An international chain of department stores uses an information system to assist in marketing their produce by forecasting public demand for certain products.

   a) Suggest **two** factors which might be considered when planning a sampling strategy for forecasting public demand. (2)

b) Describe a situation where an inappropriate sampling rate could lead to accurate data but inaccurate marketing information. (2) *NEAB 1997 Paper 2*

12 A well designed information system should be able to check that input data is valid, but it can never ensure that information is accurate.

a) Explain the distinction between accuracy of information and validity of data. Illustrate this distinction with a suitable example. (4)

b) Describe **two** ways in which data capture errors may arise, together with techniques for preventing or reducing these errors. (4)

13 SupaGoods is a home sales company. Catalogues are left at people's homes. A local agent calls two days later to take orders and collect the catalogues. The agent sends the details of the goods ordered to the Head Office where they are processed. The completed order is returned to the agent who distributes the goods and collects payment.

a) Describe **two** distinct methods of data capture for the agent. State **one** advantage and **one** disadvantage of each method. (6)

b) The orders are validated at Head Office.

i) Explain what is meant by validation. *(2)*

ii) Describe briefly **two** validation checks that might be carried out on an agent's order. (4) *NEAB 1998 Paper 2*

14 Speech recognition systems for Personal Computers are now becoming more affordable and useable.

a) State **two** advantages to a PC user of a speech recognition system. (2)

b) Give **two** different tasks for which a PC user could take advantage of speech recognition. (2)

c) Speech recognition systems sometimes fail to be 100% effective in practice. Give **three** reasons why this is so. (3) *NEAB 1999 Paper 2*

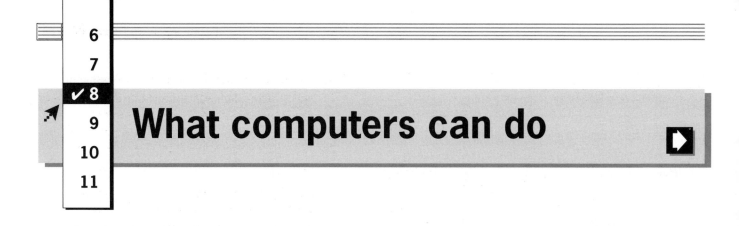

What computers can do

What is a computer?

A computer is a very fast information processing machine. Computers are very good at repetitive processing and have a vast storage capability. They can **search**, **sort** and **combine** data. Their very fast response speed makes them very suitable for applications where large files of data have to be searched to find information.

For example when you use a bank automatic teller machine (ATM) the system can quickly find out how much money you are allowed to withdraw. Supermarket systems read the bar code from an item and quickly find the name of the item and its price. Stock market computers store details of share prices which are adjusted as brokers buy and sell shares.

Internal computer memory is often called RAM (Random Access Memory). This is where computers store programs and data. This is measured in Megabytes (MB). One Megabyte of memory stores roughly one million characters. Programs and data needed for later use are stored on an internal hard disk. The size of the disk is measured in Gigabytes (GB). One Gigabyte is roughly one billion characters.

There are limitations in the use of computers. Old hardware will have smaller memory and will run much more slowly. Old computers, modems and networks may not be fast enough to cope with graphics, sound or video. You may spend a long time waiting for data to load or be processed, for example on the Internet. Poor design of the program or the data files will also lead to problems.

Methods of processing

Batch Processing

All computers work on a basis of

Input	Process	Output

For example, a computer controlled greenhouse uses sensors to input the temperature and humidity directly into the computer. The computer's processor processes this information, controlling outputs which turn on heaters and sprinklers when needed. This is an example of **real-time processing**. The computer is operating all the time, receiving data, processing it and outputting information in time to influence events.

Batch processing is another common method of computer processing used when there are large numbers of similar transactions. An example of its use is in electricity billing. All the input information is prepared away from the main computer, put into **batches** and input. (After this there can be no more input.)

The computer already has much of the information required to calculate the bill, stored on disk in a file called the **master file**. This includes:

- customer number
- name
- address
- last meter reading (to calculate units used)
- amount of electricity used in last four quarters (for checking that this quarter's usage is sensible)
- special instructions for meter reader (for example meter round back)

A computer program prints the card for the meter reader, which includes the address. After the meter is

MidWest Electricity	Customer number
MRS A CLIFTON 37 FARM CLOSE BANKTOWN	0040 419 67392

Notes OUTSIDE	DATE	METER READING
	0 0 0 0	0 0 0 0 0 0
	1 1 1 1	1 1 1 1 1 1
Re read if not between	2 2 2	2 2 2 2 2 2
316500 and 334200	3 3 3	3 3 3 3 3 3
	4 4	4 4 4 4 4 4
Special	5 5	5 5 5 5 5 5
0 NEW CUSTOMER	6 6	6 6 6 6 6 6
1 RE READ CORRECT	7 7	7 7 7 7 7 7
2 METER CHANGED	8 8	8 8 8 8 8 8
3 PREMISES VOID	9 9	9 9 9 9 9 9

Figure 8.1 An electricity meter reading card

read, the data is entered into the computer. Only two items of data are required:

- customer number
- meter reading

The processing involves matching a record from the transaction file with the corresponding record in the master file. The number of units used and the total bill can then be calculated. The batch processing takes place as follows:

1 Customer numbers and meter readings for about 100 houses are made up into batches.

2 A control total (the sum of all the meter readings) is calculated manually.

3 The batches of data are typed into the transaction file, followed by the control total.

4 At this stage the data may be typed in a second time and compared with the original data as an extra check.

5 The transaction file is read by a data validation program which detects errors such as a customer number that has the wrong number of digits.

6 At the end of the batch, the computer calculates the control total and compares it with the input control total to check for errors.

7 The transaction file is **sorted** into the same order as the master file. This is probably the customer number order.

8 The computer reads the first record from the **sorted transaction file** and the first record from the **master file**. (They should be the same customer.) The bill is calculated and **printed**. A machine will fold the bill and place it in the envelope.

9 The up-dated master file details are stored in a third file called the **new master file**. The changes will include the latest meter reading which will be needed to calculate the next bill in three months' time.

10 The new master file is called the son file. The old **master file** is the **father file**. The previous master file is called the **grandfather file**. These are generations of files. Companies usually keep three generations of files as back-up in case of problems.

Using this method of processing means that:

- Details of only one customer are processed at one time – the whole file does not have to be loaded in, so the computer does not need a large amount of memory.

- This method of processing can store files on **magnetic tape** or magnetic disk. The computer would normally have three tape drives – one for the transaction file, one for the father file and one for the son file.

- Magnetic tape is cheap, robust and reliable with very high storage. Today it is being replaced by magnetic disks. Data is stored along the tape in order. This is important as the computer can only read the tape in order. *(This is called serial access).* It can't jump back and forward as the tape is too long and this would take a long time. Therefore, tape is not suitable for a real-time retrieval system.

- The master file is up-dated in batches, daily, for example (not immediately). This would be ideal for a payroll program, which is run once a week or once a month. Batch processing is not suitable for a police database where there is a need to search quickly for a record in the file to produce information immediately.

- Huge volumes of data can be processed by this method.

- Once the program is running, there is no human intervention.

Data flow

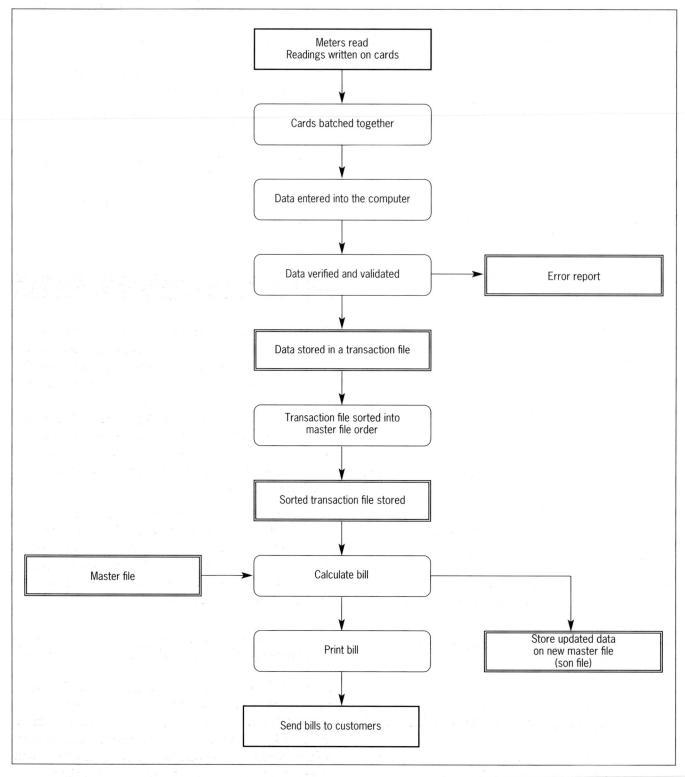

Figure 8.2 Diagram for electricity meter reading showing flow of data

Real-time (interactive) processing

When a computer system uses batch processing, all the information has to entered at the start. No more data can be entered once the system is running. This means:

- information is out-of-date and new data can only be entered with the next batch.

- response time is slow. You cannot perform an immediate search for information.

As we have seen, batch processing is suitable for payroll or billing systems, which are run once a week, once a month or once a quarter. It is not suitable for a system that needs an immediate response.

It is much more common today for businesses to use real-time processing. This is where the computer is used interactively; that is where there is communication between the user and computer. PCs are usually used interactively. If a quick response time is essential, real-time processing must be used. Examples of real-time processing systems are

- a computer controlled greenhouse

- an hotel booking system

- computer controlled robots

As computer hardware has developed, more and more processing has become real-time. The advantages over batch processing are:

- information is up-to-date at all times

- you can interrogate the computer

- the computer can respond to events: for example, it can close the windows in a greenhouse if it is too cold

- real-time processing can be used for many more applications

Multi-access (pseudo real-time) processing

Multi-access processing allows several terminals to communicate interactively with the main computer. A piece of hardware called a multiplexer allows each terminal to be in contact with the main computer for only a fraction of a second in turn. As the main computer processes the data so quickly, it appears to the user that the terminal is in contact with the main computer all the time.

As it appears to be real-time processing, some people call this '**pseudo real-time.**'

Multi-access processing is used by:

- bank cash point machines which are connected to the bank's central computer

- supermarket tills which are connected to the store's main computer

- travel agents' computers which are connected to travel companies' computers

- police station computers connected to the Police National Computer

Methods of back-up

The value of data can far exceed the value of their computer system itself. Loss of this data could lead to the collapse of the business. A study by the University of Texas Centre for Research on Information Systems showed that businesses never recover from a loss of computers which lasts for ten days or more. The same study found that 90 per cent of data losses were due to accidents such as power failures, water leaks, loose cables, user mistakes, and other hardware, software and human errors.

It is essential to have back-up copies of data in case of accidental or deliberate damage. Regular back-ups are essential and back-up tapes should be removed from the site of the business, otherwise the back-up information could be destroyed in the same disaster that destroyed the computers.

Back-up tapes

Floppy disks are not suitable for backing-up files as they do not store enough data. Even new high capacity floppies are not large enough to back up hard disks easily. Data files should be backed-up on to high volume back-up tapes automatically at the end of each working day. Many computer users have a **tape streamer** fitted to their computer. This is a special sort of magnetic tape drive, used to copy a whole disk or selected files on the disk on to tape. A program can be set up to copy the disk at a chosen time.

Multi-access (pseudo real-time) processing

The **Visa credit card company** used to use batch-processing to keep a record of sales to bill card holders. All sales were recorded in triplicate. One copy was for the customer, one copy was for the shop and the third copy was posted to Visa at the end of the day. Every day thousands of sales records arrived at Visa and the data from them was typed into a transaction file. These transactions were then processed over night using batch-processing.

There were problems associated with this method:

- The information was out of date as it was only as accurate as the last processing

- The information was slow to arrive at Visa HQ

- Errors caused when the data was mistyped were not spotted

- The system was slow to spot customers going over their credit limit or the use of stolen cards

The alternative method

Visa has now installed a real-time processing system. Data can be entered directly into the computer and processed straight away. As data is entered, the master file is updated immediately.

Tills in large stores are connected to the Visa HQ computer by a computer network. The data is input automatically:

details of the goods sold using a bar-code reader and the card details from the magnetic strip on the back of the card. Entering data directly like this reduces errors.

The information is sent immediately to the Visa computer, which can check that the card hasn't been reported as stolen and that the customer has not exceeded their credit limit. The master file is updated with details of the transaction. Two copies of a receipt are printed, one for the customer and one for the shop to keep.

This system requires the many terminals located in the stores to be connected to the central computer at once. Such a system requires a powerful computer with a fast response time and multi-access processing.

Note: Batch processing may still be used by Visa, if for some reason the real-time system isn't working or for entering data from small shops who cannot afford an on-line facility.

The advantages of Visa's new system:

- it is very fast and can access the database in a fraction of a second

- the information is up-to-date as the master file is updated as a transaction takes place

- the system is interactive; users enter data and get an immediate response, unlike batch processing where all the data has to be input before the processing starts

- the system can be accessed from many different points, e.g. several shops

It is a good idea to do this in the middle of the night, as the computer is unlikely to be in use at the time. While the back-up program is running, any other applications are likely to run more slowly. Data from the tape can be restored to the disk in an emergency. However it is necessary to restore data in the right order by following agreed recovery procedures.

If a user backs-up the whole of the disk every day, they will often keep six tapes. One is the original tape with all the software on but no data files. The other five are one for every day of the working week to make sure that a back-up exists even if today's tape is damaged.

Other media

It is possible to buy special disk drives for backing-up. A **Zip** drive can store up to 100 MB on a small removable disk. A **Jaz** drive is more expensive but can store one GB on one disk. Its transfer rate is much faster. It claims to be able to store a two-hour movie on one disk.

It is essential to back-up the right files. Some files don't ever change. These only have to be backed-up and verified once. Some files change regularly, and should be backed-up often.

Figure 8.3 A network server with internal tape back-up drive

Types of back-up

Some programs automatically produce back-up files. Word-processing programs can be set to save the last two copies of a file. The current version is called something like LETTER.DOC; the older one is the back-up copy called LETTER.BAK saved in case the other version is corrupted. The program can also be set to auto-save. Work is saved automatically every few minutes which avoids embarrassment in the event of a power cut. Users should save their work regularly, of course.

A **global back-up** copies every file on the hard drive onto back-up medium. As the back-up involves large volumes of data, the back-up medium is usually magnetic tape.

This back-up saves all files even if they have not been changed since the last back-up and serves as the starting point in restoring your system in the event of a disk failure.

An **incremental back-up** saves any files that have been modified since the last back-up. This type of back-up is best performed on the entire disk every day. This ensures that you back-up all changed files from day to day and enhances your ability to restore your computer to the same working condition as it was before.

If computers are networked, you may even back-up to a disk or tape in a different computer, a different building or even a different country (see *Case Study 1*).

Case Study 1

A remote back-up service

Mercer DataSafe is a company that offers businesses an automated off-site back-up service. Data from the company's computer is automatically backed-up after business hours to Mercer DataSafe's computer in Lawrenceville, New Jersey, USA.

Using Mercer DataSafe's back-ups, businesses can quickly recover any lost data, usually within 24 hours of a catastrophic loss (such as a flood, fire, theft or operator error), or within hours of a minor loss (such as a deleted file).

To use Mercer DataSafe's system, you only need a modem. At the preset time, the data to be backed-up is compressed in a securely encrypted format. Mercer DataSafe's computer in Lawrenceville is contacted using the modem. After identification checks, the data is transmitted to the Lawrenceville office.

When finished, the data is verified. If everything is correct, a confirmation message is sent back to the company's computer and the computer shut down. The back-up company won't have access to the user company's records as all files are securely encrypted and only the user has the password.

Users who already have a tape back-up system may still find it better to use a remote service. Many users forget to perform a back-up even though they have a tape drive. Other users think they may be doing a back-up correctly, but when they need to restore a file they find out that their tapes are useless. Few users take their tapes off site so if they have a fire or other disaster, they lose all their data. □

Case Study 2

The National Lottery

The National Lottery, run by Camelot, is the biggest on-line lottery system in the world, selling around 100 million tickets a week. In a high-profile application, with large prizes, any loss of data could be disastrous.

The data is stored in three separate data centres in case of any breakdown. Each of the three central computers is backed-up three times at the end of every day. For increased security, the day's records are then sent to the computers of Oflot, the lottery regulator. □

Summary

Computers are very fast information processors. One method of processing is batch processing using magnetic tapes.

☐ All the input data is batched into a transaction file.

☐ The transaction file is then sorted into master file order.

☐ The transaction file and the master file (father file) are merged.

☐ A new master file (son file) is created.

Batch processing is suitable if a large number of similar transactions take place. It offers good security as older generations of files are kept as back-up. It is ideal for payroll files where the data needs only to be up-dated every week. However it is not interactive. It does not allow you to search the files, all data has to be input before you start and the information will not be up-to-date.

The other method is real-time processing when the user can enter information as it happens and keep files up-to-date. This system is interactive and you can search through the files. It has to store data on disk.

Companies are more and more dependent on Information Technology. It is hard to imagine them working without IT. Loss of data would be catastrophic, as it would bring business to a halt. It is vital to have alternative hardware in case of problems and to back-up data held on computer to avoid embarrassing and expensive loss of data. This can be done with back-up copies and back-up tapes.

Batch and real-time processing questions

1 Using an example, not given in the text, describe what is meant by real-time processing. (2)

2 Name a method of processing each of the following:

a) Details of payments to a firm selling plumbing supplies are sent weekly to the head office computer to up-date records.

b) An aeroplane needs to fly at a steady height. The altimeter sends data directly to the computer which controls the automatic pilot. (2)

3 Batch processing and multi-user pseudo real-time are two forms of computer processing.

a) Distinguish between a batch processing system and a multi user pseudo real-time system, identifying those features in a multi user pseudo real-time system not needed in a batch processing system.

b) Give an example of each type of system.

c) Give two advantages of each type of processing.

d) State how security of data can be maintained by each method. (8)

4 A bank's computer supports real-time enquiries in the day but at night works in batch processing mode. Suggest three tasks that the computer would perform in batch processing mode. (3)

5 A chain of estate agents has eighty branches. Daily transactions relating to house sales, purchases and enquiries are processed using a batch system based on a mainframe computer at head-office.

a) Outline the flow of data through such a batch processing system. (4)

b) The company is considering changing from the batch system to an interactive system. Describe the advantages and disadvantages of moving to an interactive system. (4) *NEAB 1996 Paper 2*

6 A company is about to replace its old batch processing system for the preparation of customer accounts, by a real-time system.

a) Give **two** distinct advantages, which could be expected as a result of this change over.

b) Suggest **two** problems that are likely to be encountered during the changeover period. (4)

7 A computer system that is normally in use 24 hours a day holds large volumes of different types of data on disk packs. The main types of data stored are:

- applications software that changes only occasionally during maintenance;

- data master files that are up-dated regularly every week;

- transaction files which are created daily;

- database files which are changing constantly.

It is vital that these different types of files can be quickly recovered in the event of file corruption. Outline a suitable back-up strategy for each of these types of files explaining what data is backed-up and when, the procedures to be followed and the media and hardware needed. (8) *NEAB Specimen Paper 2*

8 Suggest six reasons why a business might wish to use an Automatic Remote Back-Up system. (See Case Study 1.) (3)

9 List the circumstances when batch processing is more appropriate than real-time processing. (4)

10 A computer system can be described as being a 'pseudo real-time system'.

a) State clearly what is meant by pseudo real-time. (2)

b) Give a situation where pseudo real-time is essential, stating a reason why it is needed. (2) *NEAB 1998 Paper 2*

11 A nation-wide chain of retail clothing stores processes its daily sales transactions using a batch system based on a mainframe computer at a central location.

a) Outline the flow of data through such a batch processing system. (6)

b) The company is considering a change from a batch system to an interactive system. Describe the advantages and disadvantages of moving to an interactive system. (4) *NEAB 1999 Paper 2*

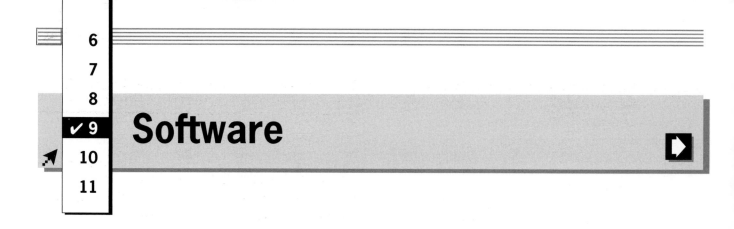

Software

Software means the programs that are written to run on the **hardware.** The hardware is the physical components of the computer, such as printer, processor, or disk drive. There are two main types of software:

- systems software

- applications software

Systems software

Systems software is the name given to the programs which help the computer run more smoothly. Systems software helps the user to control and make best use of the computer hardware, rather than carrying out a specific task such as producing a newsletter, for the user.

Utility programs which perform regular housekeeping functions are a form of systems software.

Perhaps the most important type of system software is the operating system.

Operating systems

An operating system is needed to run a computer. It controls and supervises the computer's hardware and

		Windows		_ □ ✕
File	Edit View Help			
Open	Size	Type	Modified	
Quick View	1KB	Configuration Se...	08/04/97 08:40	
	11KB	Application Exte...	22/01/96 01:43	
Send To ▶	35KB	Application Exte...	22/01/96 01:43	
	1KB	Setup Information	09/04/97 11:15	
New ▶	31KB	Wave Sound	31/03/95 16:03	
	1KB	Bitmap Image	24/08/96 11:11	
Create Shortcut	1KB	Bitmap Image	24/08/96 11:11	
Delete	3KB	Bitmap Image	24/08/96 11:11	
Rename	58KB	Application	24/08/96 11:11	
Properties	1KB	Bitmap Image	24/08/96 11:11	
	1KB	Configuration Se...	18/07/97 14:49	
Close	125KB	Application	05/04/94 00:00	
Cdplayer.exe	86KB	Application	24/08/96 11:11	
Charmap.exe	15KB	Application	24/08/96 11:11	
Cinklow.wav	31KB	Wave Sound	31/03/95 15:59	
Circles.bmp	1KB	Bitmap Image	24/08/96 11:11	
Clipbrd.exe	17KB	Application	24/08/96 11:11	
Clouds.bmp	301KB	Bitmap Image	24/08/96 11:11	
Cmd640x.sys	25KB	System file	24/08/96 11:11	
Cmd640x2.sys	21KB	System file	24/08/96 11:11	

Contains commands for working with the selected items.

Figure 9.1 A program that allows users to see the names and size of files in the directory is a utility program

supports all other software. It also provides an interface between the user and applications, and the computer's hardware.

Some specific tasks carried out by an operating system are:

- Allocating internal memory (RAM): this is particularly important if the computer is running more than one program and the memory available may be relatively small

- Scheduling of programs (if more than one is running) according to preset priorities

- Transferring programs and data between disk and RAM

- Controlling input and output devices

- 'Booting up' the computer: carrying out the initial set up when the computer is switched on

- Checking and controlling user access to prevent unauthorised access

- Logging of errors

An operating system may allow more than one program to run at the same time. With a microcomputer this is called **multi-tasking**. The programs are not actually running at the same time: the operating system decides which program to run and switches from one to the other. This often happens so fast that the user *thinks* that the programs are running at the same time.

An operating system may allow more than one input device. The operating system checks which input devices are present and decides which should have priority.

The operating system manages the memory. It loads the program into RAM from disk and decides how much RAM to allocate to the program, how much to allocate to data and how much is needed for its own use. If there is not enough memory available, the operating system will give the user an error message.

The user interface of an operating system is the section of the program that interacts with the user. It can be text based (for example DOS) or graphics based (for example Windows) (see Chapter 8).

Software companies that are developing new software must write the package for a specific operating system. A user must be aware of which operating system they use in order to establish whether a particular piece of software will run on it.

More packages have been written for MS-DOS than any other operating system. The most popular off-the-shelf software is usually available for a range of different operating systems.

Network operating systems are very complex. They will check usernames and passwords, restrict access to forbidden areas of the disk, manage the computer's memory and manage the printer queue.

Case Study

Operating systems – Linux

Nearly all PCs use Microsoft's DOS operating system together with a Microsoft Windows graphical user interface. As well as giving Microsoft a virtual monopoly in this market, many users have commented that as PC hardware has developed, DOS has become inefficient, particularly for network servers.

Linux is an alternative operating system for the PC written by Linus Torvalds, available for free over the Internet. At present, Linux is not yet suitable for desktop PCs, as there is little application software available. However some users report that Linux is better for servers than Windows.

In 1998, SBC Communications Inc. replaced 36 Windows 95 and Windows NT workstations at its Kansas City operations centre with Linux workstations because they handled the results of a giant network monitoring system better. The graphics-intensive system caused the Windows 95 workstations 'to lock up on average every 4.2 minutes. The Windows NT workstations locked up every 2.58 minutes,' said Randy Kessell, a manager at the centre. The Linux workstations haven't had a problem.

Gary Nichols, manager of network administration at WaveTop runs Linux on 30 of WaveTop's 45 servers for such tasks as e-mail, Web servers and the firewall. He worked out that he saved $30,000 in licence costs of Windows NT. 'I bought $100 worth of Linux CDs and books and got the same functionality,' he said.

Not surprisingly Linux has received financial support from Microsoft's rivals like Netscape and Microsoft has acknowledged that Linux is one of the few threats on its horizon. However Linux will only threaten Windows when there is a large amount of application software available. At present, there are no Linux equivalents for word processing, presentation graphics and spreadsheets. However in October 1998, Oracle released its Oracle 8 database for the Linux platform. Other software companies were also working on applications software for Linux such as SUSE's Linux Office Suite '99. Only time will tell if Microsoft has a serious rival. □

Utility programs

Utility programs are programs that are tools to help you use your computer more effectively. Although utility programs are not part of an operating system, they are often provided with operating system software. Examples of utilities include:

- file conversion
- file copying
- file compression
- comparing the contents of two files or disks
- deleting files
- renaming files
- sorting data

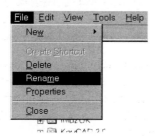

Figure 9.2 Programs to copy, delete or rename files are utility programs

Applications software

Applications software is a term used to describe programs that have been written to help the user carry out a specific task. These tasks can be very specific – for example a booking system for hairdressers – or more general – for example a spreadsheet.

Applications software is usually broken down into two sections: general purpose or generic applications software and specialised applications software.

Generic application software

The most common form of software is generic or general purpose application software. Many such programs are pre-installed on a computer when it is sold and can be used in many ways. You will use this type of software for your coursework. The following are examples of generic applications software:

Word-processing packages are used to produce documents such as letters and reports. Faster processing speeds and increased computer memory have allowed modern packages to include many extra features such as mail-merge, macros, text wrap around imported graphics, e-mail, spelling and grammar checks and text in tables and columns.

Database packages are used for information storage and retrieval (see Chapter 11). These packages allow users to set up tables, link them together and set up queries and reports. More sophisticated packages will include wizards to help the user set up the database, import data from other packages and customise the database to hide the workings of the package from an inexperienced user.

Spreadsheet packages are used to store and manipulate tables of numerical data; automatically recalculating as the data is altered. They are ideal for storing, calculating and displaying financial information such as cash-flow forecasts, balance sheets and accounts. Today's spreadsheet users can set up several different types of graphs, perform many different mathematical functions and use macros, customising and wizards.

Simple graphics software like *Paint*, which comes free with *Windows*, store graphics files in bitmap form. This stores the colour of each pixel (dot) in the picture. As a result, bitmaps are large files and when they are resized, the image tends to be distorted.

Sophisticated graphics packages, like *Corel Draw*, store graphics files in vector graphic form. This means it stores pictures as a series of lines, arc, text,

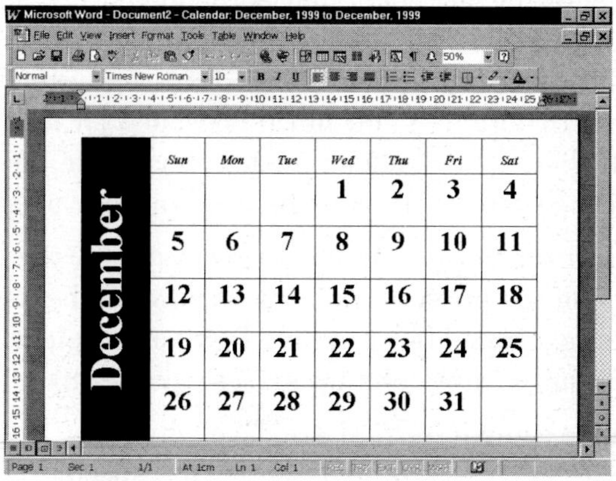

Figure 9.3 Word processing packages can be automated to produce calendars, agendas, invoices and many other types of document

etc. When storing a line, it will store the co-ordinates of the start and finish of the line, its width and colour. Not only does this save space, it means that the line can be moved, deleted or changed in colour or width without affecting the rest of the picture.

Gifs and jpg images are used on the Internet. They use compression techniques to store graphics files. This means that they too are a fraction of the same picture in bitmap format. Utility programs like Paint Shop Pro can convert between different types of file.

CAD (Computer-aided design) programs are used for designing, for example in engineering or architecture. Users can draw accurate straight lines and arcs of different types and thicknesses. By zooming in, designs can be produced more accurately. Designs can be produced in layers to show different information, for example one layer might show electrical wiring, another gas pipes and so on.

Data communication software allows users to access the Internet. E-mail features will include the ability to store a mailing list of recipients. Other documents such as graphics files can be attached to an e-mail. A Web browser lets the user view Web pages, search the Internet for key words, store the addresses of 'favourite' pages and store details of previous pages in RAM to reduce loading time if the user decides to go back.

Presentation software such as Microsoft PowerPoint has become increasingly common with the development of the Liquid Crystal Display (LCD) Projector. These projectors can project a computer display on to a large screen and are ideal for a presentation to an audience, usually replacing the old OHP (Overhead Projector).

Presentations in PowerPoint can include text and graphics, displays can be animated to attract the audience's attention and only display part of a page at a time. Sound files can be added for extra effect. Presentations can be stored on disk and edited for later use.

Specialised applications software

Specialised applications software is designed to carry out a specific task, usually for a particular industry. It is of little use in other situations. The hairdressing booking software mentioned above would be an example of specialised system software. A payroll application program would be another example. It is designed to be used exclusively for payroll activities and could not be used for any other tasks.

Specialised applications software is available for wide ranging areas such as engineering and scientific work, which include a range of specialist design tools including specialist CAD packages and sophisticated mathematical software.

■ BUSINESS APPLICATION PACKAGES

This includes programs for accounting functions such as discounted cash flow, banking and investment software.

■ EDUCATIONAL APPLICATION SOFTWARE

This is a fast developing area that includes administrative tasks such as timetabling and attendance monitoring as well as classroom applications. There are a number of examples of the use of software in the classroom: computer-assisted instruction (CAI) packages guide the user through a course of study. Simulation software such as 'Managing the Economy' allows pupils to model real life situations. There are many drill type programs available that provide exercises for revision.

■ HOME COMPUTING APPLICATION SOFTWARE

More and more households now own computers and there is a huge range of software available for use. Examples include family tree programs, menu selectors, flight simulators and games.

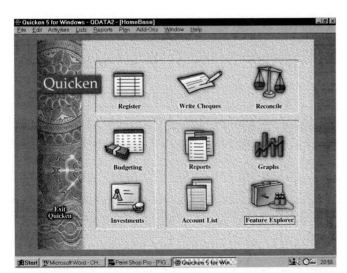

Figure 9.4 An accounts program

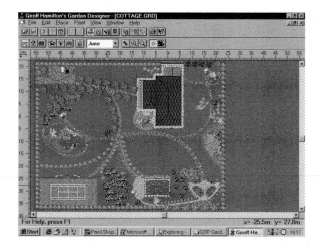

Figure 9.5 A home computer program

Integrated packages

Application packages such as Word for Windows, Excel and Access are each designed to perform a distinct function: process text, perform calculations and manage data. In the past few years many integrated packages, often called 'suites' or 'works', have come on to the market.

Integrated software packages can perform a variety of tasks combined in one easy to use package. An integrated package contains several generic programs, probably a word-processor, a graphics package, database, communications (e-mail) software and spreadsheet. You buy it all together and data can be transferred from one program to another. Examples include Microsoft Office, Lotus Smart Suite, Microsoft Works and Claris Works. All the component parts have the same 'look and feel' and it is very easy to transfer data from one function to another. For example, data produced in the spreadsheet can easily be transferred to a word-processed document.

Integrated software packages are suitable for users on a tight budget (as they are usually very much cheaper than separate packages) and for use on notebook computers.

Beginners and occasional users may like to choose an integrated package as they are generally simple to use and have a limited range of functions without confusing extras.

There are a number of advantages of buying an integrated package rather than separate packages:

- the total cost is likely to be less
- integrated packages provide a cheaper way of buying a variety of functions
- integrated software is often 'bundled' with hardware, that is it is given away 'free' by manufacturers to encourage purchase
- integrated packages need less RAM and less disk space
- they are easier to use as all applications share a common interface and command structure and it is easy to pass data from one part to another

■ DISADVANTAGES OF AN INTEGRATED PACKAGE

Integrated software can have more functions than are needed and therefore costs more than a single application. To reduce the amount of RAM required, some of the features of an application may not be available.

Bespoke Software

Application software can be bought 'off the shelf' or can be written especially for a user. The latter is often called **bespoke** or **tailor-made** software. Bespoke software can either be developed in-house, by the programmers employed by the user's company, or by using an outside agency.

Case Study

▶ Bespoke software – Strudwick & Hill Auctioneers

The company is an auction house holding numerous auctions per week throughout Britain. They required a system to do the following:

- an automatic invoicing system to reduce queuing time for buyers wanting to claim their goods
- creation of catalogues
- compilation of a selective mailing list.
- up-to-the-minute reports of the day's progress for Head Office
- post-auction analysis reports, for example profit/loss. As there was no software on the market for such a specialised activity, Strudwick & Hill hired computer consultants to write the software for them. □

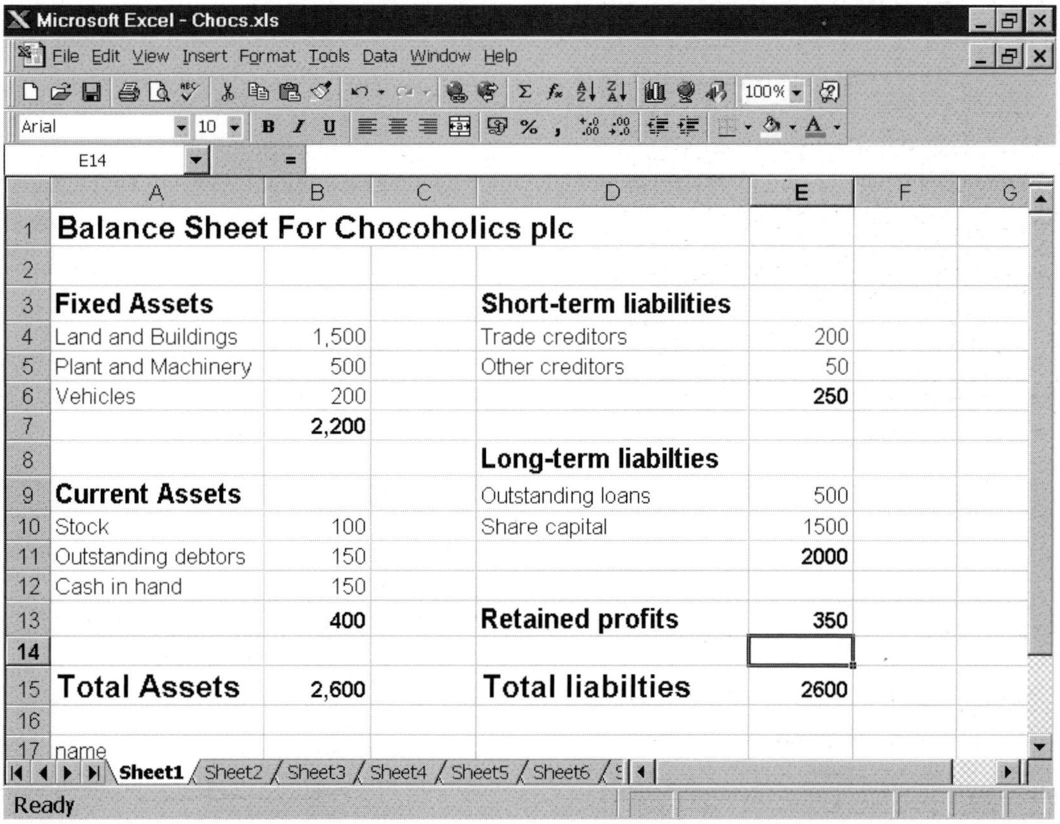

Figure 9.6 A spreadsheet, typically part of an integrated package

Generally, pre-written software can be bought directly from the author or publisher, from a shop or by mail order through specialist computer magazines.

In-house software is designed to meet the exact needs of the user by personnel employed within the firm. If expertise is not available in-house it is possible to contract a software house to develop software to meet specific user needs.

■ ADVANTAGES OF USING PRE-WRITTEN SOFTWARE

Pre-written software is ready to use immediately and less expensive than bespoke software. The user can be confident that it has been fully tested. Some pre-written software can be customised to fit the individual user's need.

However bespoke software can precisely fit user requirements and may be the only way of solving the problem (see below).

Capabilities of software

There are a range of features which are desirable in software and should be looked for when choosing a suitable package.

Links to other packages

This allows for the embedding or importing of data from other packages. For example, Microsoft Word offers links to a range of software.

Search facilities

Such facilities allow the user to move quickly to the desired data. Search facilities are of course a feature of databases, but the ability to search for words or phrases is valuable in other packages such as word-processors or spreadsheets.

Macro capabilities

A sequence of key strokes and menu choices can be saved as a macro and then repeated simply by invoking the macro. Macros are used to customise a package for a particular use. They can speed up operations, and also allow a complex operation to be set up by an experienced user and used by less experienced ones.

Applications generators

This facility allows users to customise software without specialist programming knowledge, by describing the input and output required. For example, it will typically allow menu systems to be produced without the need for an extensive programming code to be undertaken.

Editing capabilities

Modern software packages allow the user to move around in a document or sheet using mouse or cursor keys, delete and insert data in the appropriate place with ease. Facilities to 'cut and paste' are usually included.

Ability to change or extend data and record structures

Flexible database packages allow for fields to be added or deleted when data is already recorded. Suitable safeguards are usually included to ensure that data is not lost unintentionally.

Short access times

Packages need to be able to retrieve data as fast as possible. Speed of access can be a major factor in software selection.

Report generators

Report generators can take data from a database and turn it into a text report. The format of the report and its wording can be defined by the user and customised to suit the application.

The report may be customised to include graphs of data. Data in fields can be aggregated to present totals and sub-totals as required. The advantage of this sort of software is that once the report has been specified with this year's figures (say), when next year's figures are available, the report can be prepared automatically.

Data portability

The ability to transfer data to or from another package or hardware platform is a feature that is increasingly becoming an important requirement for users. This means data does not have to be typed in again, which would waste time and could lead to errors.

Data is said to be **portable** if it can be transferred from one application to another in electronic form. (Portability has a specialist meaning here – it doesn't mean 'you can put a floppy disk in your pocket and carry it around'!)

It is very common to need to transfer data produced in one package to another package. Different packages have different functions, many of which are used by businesses today.

A user may have different packages, or versions of packages, available at home and at work and will need to be able to transfer data. The growth in the use of networks has increased the need for portable data files. Portability ensures that documents produced on one package can be accessed by other similar packages, or by the same package on different hardware platforms.

For example, a freelance journalist carries out much of his work at home using Microsoft Word on a PC. One of the magazines he works for has a network of Apple Macs on which they run Microsoft Word as their word-processor. The second magazine uses WordPerfect on PCs. The journalist needs to be able to transfer documents produced at home to either place of work.

Consider a sales manager who is writing a report on the performance of her sales representatives during the past year. She is producing the report using a word-processing package. Details of sales throughout the year are maintained on a spreadsheet package that has graphing capabilities. The sales manager would like to include graphics and tables into her report. Ideally this data could be **imported** from the spreadsheet into the word-processor.

Users buying a new package will want to be able to import files from their previous package. Portability is an important marketing feature for companies producing new software. Early microcomputers had no common standard for storing data. As a result it was very difficult to transfer data between computers made by different manufacturers. Now manufacturers have standardised PC format, making it easy to transfer for data between different

computers – for example data can be quickly transferred from a pocket organiser to a PC.

Portability and Windows

Windows offers portability – the ability to take data from one program to another. By using **Edit** and **Copy**, you can copy data from one program into the Clipboard. You can then use **Edit, Paste** in another program.

Windows programs usually allow you to import and export data from and to another program. Many Windows programs are published by the same company Microsoft, for example Word for Windows, Publisher, Excel and Access. You can import into Word or Access from Excel into Publisher from Word and into Excel or Word from Access.

Programs must also be able to import from earlier versions of the same program, e.g. Word 6 and Word 95 can import from Word 1 and Word 2. Word 97 can import from all previous versions.

Many software suppliers provide **filters** that allow documents developed in other companies' packages to be converted into a suitable format for input. Word 97 can import from WordPerfect 5.*, or WordPerfect 6.*. A WordPerfect user could buy Word 97 and not lose previous work.

Windows Draw for example can import or export to several different types of picture format, for example PCX, BMP and TIF. Word 97 can import from at least 12 different formats of picture. Excel can import from Lotus 1-2-3.

Setting the standard

Most programs store data in its own particular way. Special codes are used in storing the data. It is important that programs and computers have access to common standards for data storage. Without such standards, portability would be impossible. A number of standard file formats have developed over the years.

All text can be transferred in ASCII format. ASCII is an internationally agreed standard coding system for representing characters and certain special codes. It is recognised by virtually all small computer systems.

Case Study

▶ Portability – Cranmer Estates

Cranmer Estates is a firm of estate agents with around ten branches in the north of England. Until recently they have been using old PCs, which did not run Windows, and a DOS version of the word-processing program WordPerfect 4.1 to produce details of houses for sale and to type letters.

Owner Derek Cranmer decided to invest in new equipment because he wanted to store details of houses for sale in a database, use a spreadsheet to keep accounts and to be able to send house details from one branch to another by e-mail. He decided to buy new Windows PCs and an integrated package including a word-processing package, a database package and a spreadsheet package, together with e-mail facilities.

Derek considered many packages such as Microsoft Office Professional, Corel Office Professional (which includes the latest version of WordPerfect) and Lotus Smart Suite. His priorities in choosing were:

1 **Portability.** It was essential to be able to import old WordPerfect files as he did not want his staff to have to type data in again.
2 **Links to other packages.** He also wanted to be able to take data about houses from the database into the word-processor.
3 **Training.** He wanted to use a package where training was available. Derek found out that his local college ran courses in Microsoft Word, Excel and Access – all part of Microsoft Office.

The need for training persuaded Derek to choose Microsoft Office, which could also import from WordPerfect. After a pilot project at one branch was successful, Derek introduced the new system into all his branches. ☐

CSV (Comma Separated Variable) format is used for transferring data to and from databases and spreadsheets. Each field or cell is separated from the next by a comma.

Graphics can be stored in a number of formats and most programs can import in a number of forms, for example **BMP** Bitmap picture, **PCX** Paintbrush picture, **JPG** a compressed image used for storing Internet pictures.

One of the most widely used protocols is TCP/IP (transmission control protocol/internet protocol). TCP/IP is the protocol used to establish a connection with the Internet. TCP and IP map on to levels 4 and 3 of the ISO OSI model.

■ HOW THE IT INDUSTRY HAS HELPED PORTABILITY

- introduced standards for formatting disks so that different manufacturers' equipment can share disks

- adopted standard methods of storing pictures, for example in Bitmap (BMP) format

- adopted standard methods of storing files (for example CSV files – Comma Separated Variable) which can be accessed from many programs, for example Excel

- adopted standard methods of storing text (for example TXT files that just store the text, WRI – Windows Write format, a simple word-processor that came free with Windows 3.1 so it is very common, RTF – Rich Text Format that also stores font sizes as well as text)

- sold utility programs which convert from one version of a program to another

Up-gradability of software

Every few years a new version of the same program is produced. Each version has a number such as Excel 97, PageMaker 6, or Word 95. New versions take advantage of increased speed, memory and processing power of later computers to offer new facilities. Excel 2 has no graph wizard. Excel 3 does. Excel 3 has no spell check. Excel 4 does. Excel 5 offers two sorts of macros. Word 97 and Excel 97 offers links to the Internet and files can be saved in HTML format. Word 6 includes an AutoCorrect to correct some spellings automatically. Word 95 has an automatic spell check, underlining a mistake as you type.

The life of a successful software package usually goes through several stages. The '.0' versions (4.0, 5.0) are usually major up-grades from the previous versions. Minor changes are then incorporated in later .1 or .2 versions.

Work produced on an earlier version should be useable on later ones, although it is rarely true the other way round. For example, all data produced in WordPerfect 5.1 can be used in WordPerfect 6.1, but not vice versa. It will also be an advantage to be able to import data from other software used by the organisation.

Software up-grades should be sufficiently similar to the previous version so that users are immediately comfortable with the new version. The increased processing power of recent computers means that up-graded software can be even more user-friendly. Microsoft say that Office 97 has resulted in 52 per cent fewer helpdesk calls. Even so up-grading software may still result in the need for re-training.

In 1995 Microsoft brought out Windows 95 to replace Windows 3.1. Within a short time, it was not possible to buy Windows 3.1 although many computers were still using the older version. Windows 98 was developed as an improvement to Windows 95.

When a new program is brought out it is important that:

- a user can load work done in the previous version of the program (backwards compatibility)

- it looks similar enough to the old version so that users can start straight away

Reliability of software

The production of software that will not fail and is bug-free is obviously a desirable aim. For commercial software producers, customer satisfaction and product credibility will depend upon the reliability of the software.

However, to produce absolutely bug-free software is a near impossibility. A reasonably sized package might contain hundreds of thousands of lines of program code with millions of different pathways through. It would be unrealistic to contemplate testing every possible route. Programming staff are not always as rigorous as they should be in designing and testing programs.

Software producers need to impose very high standards for the writing and testing of programs to minimise the likelihood of errors. A program should be tested by colleagues who have not been involved with the actual production themselves. Then they will be given to selected users to test out and report bugs before the official release date. This means that they are tested on different hardware platforms that may operate differently. Examples of such differences could be in memory size, specifications or clock speeds.

However, there are often strong commercial demands to keep software development costs within defined limits and development time as brief as possible. Publishing before the competition might be vital to the success of the project. These demands mean that it is not feasible to test all software fully. Producers therefore usually provide help lines for users where they can record new bugs and receive help, often in the form of program 'patches', which will repair the error. Up-graded, .2, versions of software are often distributed free or at considerable discount to existing users of the .0 version.

Configuring hardware

Different computers, even ones made by the same manufacturer, will have different peripherals: input devices, output devices and backing storage. For example, the keyboard is different in different countries for different languages and currencies. The computer will need to be set up, or configured, for the appropriate keyboard.

A user may wish to add a second internal hard disk to a PC or add a CD-ROM drive. The configuration settings of the computer will need to be altered for any new peripherals.

Different printers operate at different speeds, using slightly different sizes for lines and numbers of lines on a page. A computer needs to be configured for the printer by installing the appropriate **printer drivers**: software that converts the output information into a form that the particular printer can understand and print accurately.

Even so, the software does not necessarily use all the facilities of the peripherals. Many Windows programs do not use the right hand mouse button. Some mice have a third button in the middle, which is rarely used by Windows software. A scanner may be capable of scanning text (OCR) but the software may only scan images.

■ CONFIGURING SOFTWARE

Software also needs to be configured to make best use of the peripherals available. For example communications software must be configured for the speed of the modem, a games program will have to be set up for the appropriate joystick and software that scans images must be set up for the right sort of scanner.

Output format

WYSIWYG programs allow users to see how the output will appear when it is printed and make alterations if necessary. Most word-processing programs today enable the user to present information in different formats, for example text, tables, charts and graphs. Different formats will be used depending on the target audience.

- A director may want to see sales figures at a glance in a **graph**.

- An accountant may want to see all the information in a **table**.

- A manager may want a full written **report**.

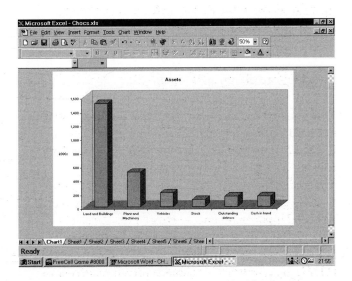

Figure 9.7 Presenting information in graphical form

Even if a WYSIWYG (What You See Is What You Get) program is used, the printed output may not be exactly how it appears on the screen. Most printers only print on A4 size paper, whereas the software may include different paper sizes. A screen may display a page in colour but the printer only print in black and white. A printer cannot print close to the edge of the paper whatever appears on the screen.

Summary

Hardware means the physical parts of the computer. **Software** means the computer programs. There are two main types of software:

☐ Systems software – programs which help the computer run more smoothly. Utility programs and operating systems are types of systems software.

☐ Applications software – programs that carry out a specific task, such as calculating staff wages. Word-processing, spreadsheets, databases, graphics packages, web browsers and e-mail are examples of applications software.

The tasks of an operating system include allocating internal memory (RAM), transferring programs and data between disk and RAM, checking user access and logging of errors.

Examples of utilities include programs to delete, rename or copy files.

Integrated packages include several applications programs combined in one easy to use package. An integrated package probably contains a word-processor, a graphics package, database, communications (e-mail) software and spreadsheet.

Bespoke software is application software written especially for a user.

Features of packages:

☐ Today most packages have links to other packages allowing them to share the same data. Data is said to be **portable** if it can be transferred from one application to another. *Windows* software offers portability.

☐ New versions of software are similar to the previous version so that users can use the new version straight away. The new version should be able to load work saved in the previous version.

☐ Output may be in many forms, for example, a graph, a table or a report.

Software questions

1 By using an example of each, explain the difference between systems software and applications software. (4)

2 What is a utility program? Give an example. (3)

3 Users may encounter problems when software manufacturers up-grade a software package. With reference to specific **examples** describe **two** such problems. (4)

4 A particular word-processing package is described as having a WYSIWYG (What You See Is What You Get) output capability. Give the advantages of using such a package rather than one which does not possess this capability. (3) *NEAB 1996 Paper 2*

5 A microcomputer with high-resolution monitor, scanner and a laser printer are used in the development of a school magazine. Suggest a suitable software package to help produce the magazine. State **four** features that make this package useful. (4)

6 By giving examples, explain the difference between applications software and systems software. (4) *NEAB 1996 Paper 2*

7 When purchasing software, it is often possible to buy either an 'integrated package' or separate applications that run under a common operating environment.

   a) What is meant by the term 'integrated package'? (2)

   b) What applications would you normally expect an integrated package to offer? (4)

   c) What are the relative advantages and disadvantages of an integrated package over a collection of separate applications packages running under a common operating environment? (4) *NEAB Specimen Paper 2*

8 Two users are using two different versions of the same word-processing package. When user A sends a file on disk to user B there are no problems in reading the file. However, when files are transferred the other way the transfer is not successful. Explain why this may happen and how to overcome this problem. (4) *NEAB Specimen Paper 2*

9 From the user's point of view, give three functions of an operating system. (3) *NEAB Specimen Paper 2*

10 A freelance reporter who regularly contributes articles to various newspapers and magazines is considering which word-processing package she should purchase. A friend has said that 'most modern application packages enable users to produce files which are portable'.

   With the aid of specific examples discuss this statement. Include in your discussion:

   • an explanation of what portability means in this context;

   • why portability is important;

   • how the Information Technology industry can encourage this portability. (16) *NEAB 1996 Paper 2*

11 Articles in the media referring to computer software which fails to work properly are commonplace. Discuss the difficulties facing software companies when testing and implementing complex software, and the measures that software providers could take to minimise these problems. (6) *NEAB 1996 Paper 2*

12 Given an existing hardware platform, selecting the most appropriate software package for a specific application can be a difficult process.

   Describe the criteria and methods you would use to select the most appropriate software package for a specific application. (16)

   Users may encounter problems when software manufacturers upgrade a software package. With reference to specific examples describe **two** such problems. (4) *NEAB Specimen Paper 2*

13 A spreadsheet package is described as having a macro facility. Describe what is meant by the term 'macro' and suggest a situation in which the use of a macro would be appropriate. (4) *NEAB 1997 Paper 2*

14 When installing or configuring a particular word-processing package, the documentation states that the correct printer driver must also be installed. What is a printer driver, and why is it necessary? (4) *NEAB 1997 Paper 2*

15 When using any applications software package on a network, the user is often unaware that an operating system is working 'behind the scenes', managing system resources. Give **three** of these resources and in each case briefly explain the role of the operating system in its management. (6) *NEAB 1997 Paper 2*

16 You have installed a new piece of applications software onto a stand-alone PC. You then find that the printer attached to the PC fails to produce what can be seen on screen in that package.

Explain clearly why this might happen. (2) *NEAB 1998 Paper 2*

17 Why does commercially available software not always function correctly when installed onto a computer system? (2) *NEAB 1998 Paper 2*

18 There is now a wide range of software tools available to increase the productivity of the end-user. Two such software tools are Application Generators and Report Generators.

a) Explain what is meant by an Application Generator. (2)

b) Explain what is meant by a Report Generator. (2)

c) Give an example of when it might be sensible to use each one. (2) *NEAB 1998 Paper 2*

19 Graphic Designers are making increasing use of hardware and software systems to assist in the representation of pictorial and textual information on paper or on screen.

a) State **two** benefits to a graphic designer of using these hardware and software tools (2)

b) While graphic design software can be used on a PC based system, additional or specialised hardware can be used to assist the designer. Specify **four** devices that might be used, stating clearly why they would be needed. (4)

c) The completed image may need to be transferred to another package. One method is to store the image as a bitmap.

   i) What is meant by the term bitmap in this context? (1)

   ii) State **three** problems that could occur when a bitmapped image is used. (3) *NEAB 1998 Paper 2*

20. An office worker has created a macro which imports data from one spreadsheet file to another and then performs some calculations. However, the macro fails to work as expected when used.

a) Explain the term 'macro' as used in the above description. (2)

b) What could the office worker have done to reduce the chance of the macro failing when it was used? (3) *NEAB 1999 Paper 2*

21 The head of a sales team has developed a presentation. It is planned for members of the sales team to deliver this presentation as part of a sales talk to large audiences at various locations.

a) State **three** advantages to be gained by using presentation software as opposed to the use of traditional methods, e.g. OHP. (3)

b) State **three** design considerations that should be taken into account when the head of the sales team is developing the presentation. (3) *NEAB 1999 Paper 2*

22 Floppy disk was once the usual method for distributing software adopted by software houses and developers.

a) Describe **one** occasion where it would still be sensible to use a floppy disk for software distribution. (2)

b) State **two** different ways, other than by floppy disk, in which software can be distributed. Give with reasons, an example of when each one might be used. (6) *NEAB 1999 Paper 2*

23 'The rise of de facto standards due to commercial sales success can only benefit organisations and individuals.' Discuss this statement. Particular attention should be given to:

- Operating systems

- Portability of data between applications

- Portability of data between different computer systems

Illustrate your answer with specific examples. (16) *NEAB 1999 Paper 2*

User interfaces

Humans and computers

Computers are very good at repetitive calculations and manipulation of data. Unlike humans, computers never get tired, lose concentration or feel ill. Computers have a much larger information storage capacity than humans and can process data much more quickly than humans do.

However humans can respond to a wide variety of situations. Computers are only useful in a limited number of situations for which they have been programmed. Although a computer can perform logical operations, it cannot really think for itself but executes instructions given by the user. Computers are still rather stupid assistants to the human user.

The user interface

It is important that the computer is easy to use so that the user can operate it at all times. A user interface is any hardware or software designed to make it easier for someone to use a computer. It may involve:

- special input devices

- special output devices, for example a loud speaker
- special screen layouts

Dialogue between humans and computers

Dialogue is the exchange of instructions and information between the user and computer. The instructions may be typed in at a keyboard, spoken using speech recognition and speech synthesis or chosen using a mouse or another pointing device.

The user interface should make it easy for the user to operate the computer. The mouse, for example, has been designed so that it can be used while looking at the screen. The user does not need to look at the mouse itself.

The user interface can also make it easy for the user to specify instructions and the computer should provide feedback so that the user knows that the right instruction is being carried out. For example, if a user copies or deletes a file in Windows 95, a message appears on the screen.

Error messages should appear if the instruction cannot be executed, for example if there is no disk in the floppy disk drive or the printer is not turned on.

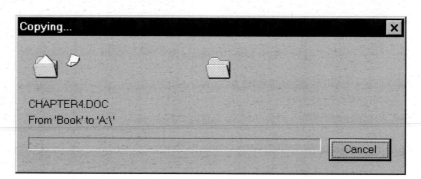

Figure 10.1 Feedback from the computer tells the user that the instruction to copy a file is being executed

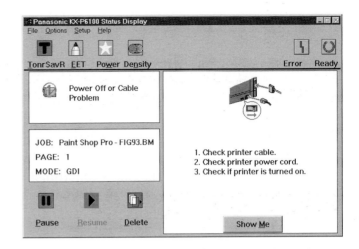

Figure 10.2 and 10.3 Error messages telling the user why instructions cannot be executed

Types of dialogue

Command based systems, for example MS-DOS

If a computer was first turned on and loads the operating system MS-DOS, a **C:>** (called the C prompt) appears on the screen, representing drive C – the computer's hard drive. The user then types in the command. The user has to remember the command and optional choices at the end of the command called switches. For example in MS-DOS
DIR gives a list of all files in the directory.
DIR/P gives the same list but pauses at the bottom of the screen. /P is the switch.
DIR/W gives the same list but lists the files in columns.
DIR A: gives a list of files on floppy drive A.

The user will need to memorise many commands and switches. Command based systems were common in early computers but are still a good idea for IT experts such as programmers.

■ MENU SYSTEMS

MS-DOS systems can be programmed to include a menu. The user then only has to recognise the required command from a list. There is much less chance of the user forgetting the command or typing in the wrong command. A hierarchy of menus will probably be needed to include all the options available in a menu. This may involve many levels, which can be tedious for the experienced user. Even so, it is possible that some commands will not be included in the menu.

```
C:\>
C:\>
C:\>
C:\>BACKUP C:\WORK\HOME\*.DOC A: /D:09-01-97
```

Figure 10.4 A command in MS/DOS, backing-up all files that have a .DOC extension that have been modified since 1 September 1997 to a disk in drive A. The syntax in this command is critical and will not operate if typed in incorrectly

Figure 10.5 A simple to use menu system

Graphical User Interfaces

A graphical user interface (GUI) uses graphics, icons and pointers to make the computer more user friendly. The command required can be specified by clicking or double-clicking with a mouse. The menu screen or desk-top can easily be edited to include extra options.

A GUI was first developed for the **Apple Macintosh** using WIMPs (Windows, Icons, Mice and Pointers.) **Windows 3.1** and **Windows 95** are GUIs. GUIs make the computer much more user friendly and more suited to the casual IT user.

Features of GUIs

GUIs such as Windows include many features that make dialogue between the user and the computer easier. These include:

■ PULL DOWN MENUS

All Windows programs have pull-down menus, making it easy to choose options. Many of the commands on the menus are the same in all programs, for example File Open..., Edit Copy... Users become very familiar with the techniques required.

This ease and familiarity means it builds confidence in novice users and it is easier for users to learn new packages.

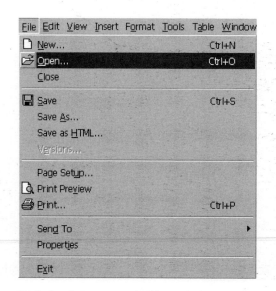

Figure 10.6 A pull down menu in Microsoft Word

Figure 10.7 Common icons seen in many Windows programs

■ ICONS

An icon is a small picture on the screen, used as a short cut to a function such as open a file, save, or print. Nearly all Windows software uses icons, which often can be customised to suit the user. Different programs often user the same icons for common functions.

■ DIALOGUE BOXES

Many Windows programs include dialogue boxes on the screen and asking the user certain questions. This is very easy for the user and makes sure that all the questions that need to be answered are answered. A dialogue box may be part of a wizard – a quick way of automatically performing a task in a program.

Figure 10.8 A dialogue box in Microsoft Access

■ FORMS

An on-screen form looks like a form on a piece of paper and has boxes to be filled in. It is easy for a novice user to enter information. List boxes can be used to restrict information entered to a list of alternatives.

■ PORTABILITY

As we saw in the section entitled Portability and Windows in Chapter 9, data from one Windows application can easier be taken into another application. This increases the versatility of Windows

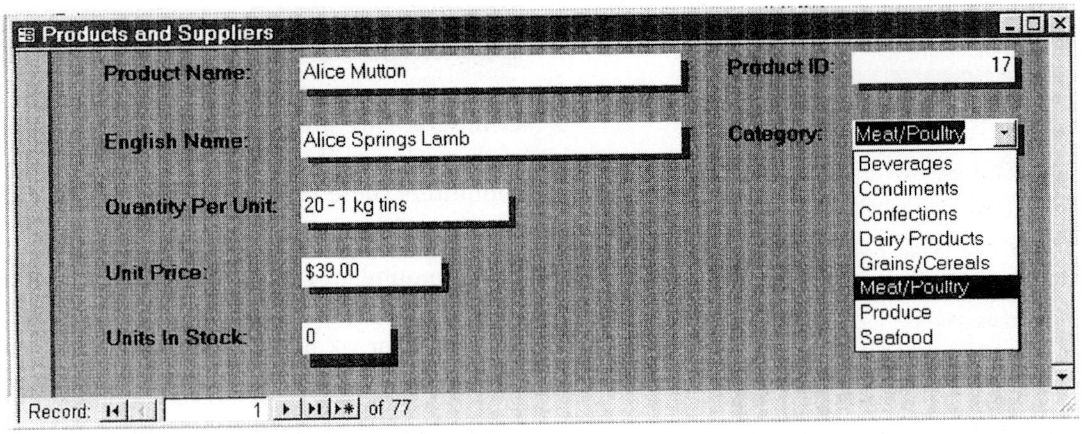

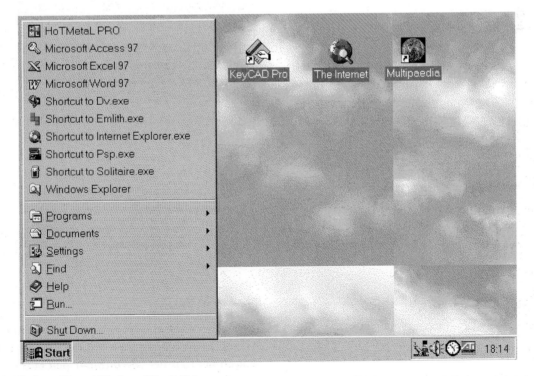

Figure 10.9 An on-screen form for entering data in Microsoft Access showing a list box on the right for choosing the category of the product

and by using more than one package users can solve an even greater range of problems.

■ WYSIWYG

Most Windows programs, particularly word-processing and DTP programs are **WYSIWYG** (What You See Is What You Get). What you see on the screen is what is printed on the printer. WYSIWYG programs such as Word for Windows, Excel and PageMaker are much easier to use. You should not waste paper, as you know how it will print. You can see how different fonts and graphics will look. You can also see what sizes are available and get a preview of the whole page before printing.

Reasons why GUIs are popular

GUIs

▸ have programs that tend to have a common interface and common commands

▸ are user friendly as they are mouse driven

Figure 10.10 Windows 95 and Windows 98 are GUIs

File	Edit	View	Create	Text
New Document...				
Open...		Ctrl+O		
Close				
Save		Ctrl+S		
Save As...				
Import/Export...				
Import Picture...				
TeamMail...				
TeamReview...				
TeamConsolidate...				
TeamSecurity...				
Internet		▶		
Versions...				
Print...		Ctrl+P		
Choose Another SmartMaster...				
Document Properties		▶		
User Setup		▶		
Exit Word Pro				

Figure 10.11 Menus are similar in different programs. This menu is from Lotus Word Pro. See how similar it is to the menu from Microsoft Word in Figure 10.6. Both programs use the same short cut keys, for example Ctrl and P to print

- have icons that are easier to understand
- can exchange data between applications
- lots of application software is available

Problems with GUIs

The extra software required for GUIs like Windows takes up extra disk space. The software is graphics intensive and so needs more computer memory, for example Windows 95 needs at least 80 MB hard disk space and 16 MB RAM to run smoothly. For the latest PCs, this is not a problem but older machines with smaller hard disks may not be able to run Windows 95 successfully. If they can, running programs may be easier but probably will be slower.

GUIs

- are slower to load
- take up more RAM
- take up more disk space
- need a powerful computer
- do not enable some software to work
- need a better quality display on the monitor

Natural language interface

Computers are programmed in computer languages such as C++ or PASCAL. It would be easier if computers could be programmed in a 'natural language' like English.

Programming languages are precise with a limited number of commands. English is less precise with a much greater vocabulary. Sentences can be ambiguous.

It is not yet possible to develop systems that accept a natural language but some systems, mainly expert systems, have been developed using a limited sub-set of a natural language, that is by restricting instructions to a number of acceptable commands.

The user still has to learn how to use these commands and how to construct sentences that can be understood by the system. This may defeat the object of trying to develop an easy-to-use natural language.

Commands are usually typed in at a keyboard, but if such a system is combined with speech recognition, there is potential for humans and computers to have a limited dialogue.

Natural language interfaces may still be the stuff of science fiction but there are many research projects exploring the possibilities of talking to computers. The Spoken Language Systems (SLS) Group at MIT's (Massachusetts Institute of Technology) Laboratory of Computer Science have been trying to get computers to communicate the same way as humans since 1989.

Progress has been made as can be seen from the following excerpt from a conversation between a user and JUPITER, an SLS-based weather forecast system:

User: Yes, I would like the weather forecast for London, England, please.

JUPITER: In London in England Wednesday, partly cloudy skies with periods of sunshine. High 82 and low 63. Is there something else?

User: What is that in degrees Celsius, please?

JUPITER: In London in England Wednesday, high 28 Celsius and low 17 Celsius. What else?

Check out their progress at their web site: http://www.sls.lcs.mit.edu/sls/whatwedo/index.html

Some banks are starting to use natural language computers to deal with telephone customers. By asking a limited number of questions, and asking the customers to press the number pad of their phone, they can handle many queries.

In 1997 the world chess champion, Gary Kasparov, was beaten in a series of matches by an IBM computer, Deep Blue. The computer could consider millions of different combinations of moves very quickly to consider the right option.

Although the computer can play a game of chess, the complexity and the large vocabulary of the English language mean it still cannot have much of a conversation with a human. The human/computer conversations of science fiction films are still a long way away, if they will be possible at all.

Ways of improving human/computer interaction

There are many ways in which the design of dialogue between human and computer can be improved. The interface can be varied depending on the user and what they are using the computer for. Different interfaces can be used for different projects. Examples of how human/computer interfaces can be set up:

Dedicated keys	Using a key for a particular purpose, for example F1 is used for Help. You can customise keys in many programs using macros. See Hot key.
Dialogue Boxes	A box on the screen asking the user certain questions that need to be answered.
Forms	Similar to a form on a piece of paper where data has to be entered.
Hot key or soft key	A key, usually a function key such as F12, on the keyboard which the user can change within a program so that when pressed it performs a command.
Hot spots	An area on the screen where the user may click to find more information. For example, on a database it may lead to more detailed information; on a web page it may provide a link to another page.
Icons	Illustrated buttons used in a program, usually in a toolbar at the top of the screen. The toolbars can be customised by adding, deleting or editing icons.
List box	A box on the screen that displays a list of options to be chosen by user

Pointing devices	These are not just mice but could be tracker balls, joysticks, paddle – the device used will depend on the situation.
Screen colours	You can vary the colours in an application, for example in a word-processor, some people prefer white letters on a blue background.
Sound	Warning sounds can be used to tell you of errors
Speech recognition	This is now getting very sophisticated and is expected to be very common in word-processors in the future.
Touch screens	Users make selections by pointing at a required menu item on the screen.

Talking PCs improve bit by bit

Natural conversation with our PCs is a step nearer. David Boothroyd looks at the latest developments
Computing, 26 November 1997

Over the past two decades, effective speech recognition has been a technology that was always just around the corner, according to its developers. Yet far more was promised than vendors could realistically deliver. Finally it has arrived, and over the next 10 years the impact on the industry is destined to be enormous.

The advent of continuous speech recognition will increase the pressure to develop much more sophisticated natural language processing.

How long might it be before people start to prefer chatting to their PC rather than to each other? Decades ago, artificial intelligence researchers demonstrated that people can be fooled, at least for a short time, into thinking that they are conversing with someone (or something) intelligent, when in fact the responses are being generated by relatively simple software that makes use of linguistic tricks to imitate a conversation.

Ambiguities in language

Languages like English can be ambiguous. We only know the meaning from the context of the sentence. For example, the word *lead* has several meanings. It can mean the leash for a dog. It can mean the person in front in a race. Pronounced differently it can mean the writing part of a pencil. The written sentence:

'I want the lead.' could mean
a) I want the leash for my dog
b) I want the lead to put in my pencil
c) I want to be in front

There are probably many other meanings too. We only know which one by the context. Programming computers to recognise as well as words is not easy.

Sometimes when humans communicate they mean exactly the opposite of what they say. When someone says, *'That was brilliant,'* we might recognise it as sarcasm by the tone and intonation. Could a computer recognise this as well?

Case Study 1

Railway ticket machines

There are automatic ticket machines at many mainline stations. These machines allow customers to purchase a ticket quickly without having to queue in a ticket office. Customers use a touch screen to choose their destination from a list and the type of ticket, such as single or return. Payment is by credit card: the ticket machine can automatically read the card details from the magnetic strip on the back.

A touch screen is used because

- it is more durable than other pointing devices such as a mouse

- it is easy to operate, even for the inexperienced computer user ☐

Case Study 2

Novell network manager

Carol Price is the network manager for a Novell local area network of around 50 PCs in an office building. Her work involves running several network management utilities on the network server. Most of her network maintenance is carried out using a menu system. For example, one option on one menu is for connection information. This leads to another menu listing all logged on users. Choosing a particular user from the list gives information on this user such as station address, log on time and open files.

However not all network management utilities appear on the menu system. If Carol wishes to run another program, she has to use a command line interface. For example, if a print server has crashed in room 2, she has to type in:

>LOAD PSERVER PSERVER_RM2

to restart the print server. ☐

Summary

- [ ] Computers are very good at repetitive calculations and manipulation of data but they are only useful in situations for which they have been programmed.

- [ ] Using a command line interface means the user must type in commands to run programs.

- [ ] Graphical user interfaces involve WIMPs (Windows, Icons, Mice and Pointers) to make a computer more user friendly.

- [ ] GUIs need a powerful computer as they are slower to load, take up more RAM, take up more disk space than command line interfaces.

- [ ] It is not yet possible to program computers in 'natural languages' like English as computer commands need to be very precise and English has a large vocabulary.

- [ ] Program designers must design a human/computer interface to suit the user's demands, using such techniques as colour, hot keys and pointing devices.

Interface questions

1 A manufacturing company intends to use an information system to store details of its products and sales. The information system must be capable of presenting the stored information in a variety of ways. Explain, using **three** distinct examples, why this capability is needed. (6) *NEAB 1996 Paper 2*

2 A travel agent uses an information system to help customers chose their holidays. The system is used by different types of user. Justify different user-interface features which would be appropriate for each of the following:

   a) customers, who can interrogate a local off-line system to find details of all the holidays on offer;

   b) travel agents, who use the system to make bookings;

   c) staff who set up the system and maintain the accuracy of the database. (10) *NEAB 1996 Paper 2*

3 A different human-computer interface would be needed for each of the following users:

   a) a young child in a primary school,

   b) a blind person,

   c) a graphic artist,

   For each user describe and justify an appropriate human-machine interface. (9) *NEAB Specimen Paper 2*

4 User interfaces have gradually become more and more oriented to the needs of users over recent years.

   a) Briefly describe *three* features of user interfaces which have been developed and explain how each has benefited the user.

   b) Describe two ways in which user interfaces could be further developed to make computers more accessible and friendly to untrained users. *London Board Computing Specimen Paper 1 q9*

5 Give three reasons why you think speech recognition is likely to expand in use.

6 Many machines now offer a *graphical user interface* such as Windows.

   a) Describe two features of such interfaces that are likely to be helpful to a non-technically minded user.

   b) Give **three** disadvantages of this type of interface. *AEB Computing Specimen Paper 2*

7 A college uses a range of software packages from different suppliers. Each package has a different user interface. The college is considering changing its software to one supplier and to a common user interface.

   a) Give **four** advantages of having a common user interface. (4)

   b) Describe **four** specific features of a user interface which would benefit from being common between packages. (4)

   c) Discuss the issues involved, apart from user interfaces, in the college changing or upgrading software packages. (8) *NEAB 1997 Paper 2*

8 Describe and justify a suitable human-computer interface for:

   a) a user of a bank ATM (automatic teller machine)

   b) a games computer

   c) a teletext user

   d) a computer programmer

9 A large entertainment and leisure complex has a wide range of facilities available including a cinema, live entertainment, indoor sports and exhibition facilities. They have made use of computer systems since it opened, but due to the popularity of the complex, and the wide range of activities available, they are considering introducing a computer based information system. This system will be used by the general public who visit the complex.

Discuss how such a system would operate in practice. Particular attention should be paid to the following issues:

   • the dialogue between the user and the computer;

   • the types of interface suitable for this system;

   • the hardware required;

   • the timeliness and accuracy of the information displayed.

Quality of language will be assessed in this question. (20) *NEAB 1998 Paper 2*

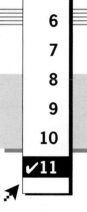

Relational databases

What is a database?

All organisations need to store and retrieve data. The data stored and structure is referred to as a database. A software package, called a database management system (DBMS) organises the database, allowing it to be created and accessed. Microsoft Access is an example of a DBMS. Many different applications can be built to access the same database.

Traditional filing

In the early days of computing, individual applications were developed independently within an organisation. Computers were used to process data and in many organisations the payroll was the first application to be computerised, with the benefit of increasing the speed of production of payslips compared with manual methods.

At your school or college, a number of applications will have been developed, most probably including:

- Examination entry system
- Attendance system
- Student record system
- Timetabling

The **examination entry system** would need a file of data relating to students and the subjects they are taking.

The **attendance system**, perhaps using OMR data entry from class registers, would need to access a file of student and class details.

The **student record system** would use a file that held details of a student's previous educational history and basic personal data.

Each separate application had its own data files which were accessed by a number of programs. Although these computer systems brought many benefits over non-computerised methods, there were still a number of problems.

There was a duplication of data: fields such as name, address and data of birth were stored in different files for each of the systems. Such duplication often led to data inconsistency. Inconsistency occurs when the same data item, held in two separate files, is different. This could perhaps occur when data is originally entered into the systems (date of birth could be typed correctly into the record system but incorrectly into the examination entry system) or if a data item were changed in one system but not another (a student changes home address – the amendment is made in the student records system but not the examination entry system).

The data files used in each system were linked very closely to the programs. Any changes to the structure of the file – for example, if an extra field, such as previous school, needed to be added to a student's record in the student record system then every program using the file had to be modified.

Data in one system was not available to another system and if information was needed in a different format then a new program would have to be written.

The term **flat file** is often used to refer to these single files which are like 2-dimensional tables. A flat-file is similar in structure to a manual card file: there is one card (or record) stored for each entry in the file.

Development of databases

More and more complex filing methods were developed and from these developments database systems were evolved. In a database system data is kept separate from the applications programs that use it: the DBMS works between the two. Ideally, each data item is stored only once in the database. Thus inconsistencies of data do not arise.

Your school or college will probably now hold all the required data in one database which will support a wide range of applications. Basic student data need only be captured once when a student joins the college and other data added throughout his/her stay. Timetabling, examination entry and attendance tracking can all be integrated into one system. The DBMS used will allow ad hoc, occasional reports to be produced easily as required without the need for complex, time consuming programming.

There are several ways of organising data in a database but the model used most often these days is the relational database.

A relational database consists of tables of data. Each table refers to an object, person or **entity** – a table is made up of a number of columns or **attributes** that represent individual data items.

Before we examine in detail how a relational database is used, it is worthwhile studying entities and attributes further.

Entity relationships

Consider a video hire shop that has a computerised system to keep track of loans. Data will need to be kept on each member (such as name, address, phone number, date of birth) and each video stocked (such as title, category, charge).

Member and video are called **entities** and name, address, phone number, date of birth and title, category and charge are their respective **attributes**.

There is a link made between *member* and *video* when a loan is made and this is called a **relationship**. A member can borrow many videos and a video can be borrowed by a number of members at different times, so we call this a **many-to-many** relationship.

Not all relationships are many-to-many. In a hospital, *patient* and *medical record* are two entities which have a **one-to-one** relationship as each patient will have just one medical record and each medical record will refer to only one patient. In a veterinary practice, *owner* and *pet* are entities that have a **one-to-many** relationship as an owner can have more than one pet, but a pet only has one owner.

Entities, attributes and relationships need to be worked out for the data before a relational database is set up. This process is called data modelling.

Figure 11.1 shows how relationships are represented diagrammatically.

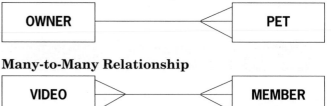

One-to-One Relationship

One-to-Many Relationship

Many-to-Many Relationship

Figure 11.1 Examples of entity relationships

Primary Key

Records need to be able to be identified individually. A key is a field used to identifying a record. A **primary key** is a unique field or group of fields chosen to identify a record. For a given table there may be more than one possible field that could perform this function and these are called **candidate keys**.

An example of this can be seen in an employee entity that could include Surname, Forename, Address, Employee_Number and National_Insurance_Number as attributes. Employee_Number and National_Insurance_Number are both unique for an employee and are thus candidate keys. It is likely that Employee_Number would be used as the primary key as it is shorter. The Surname is not a candidate key – there could very well be more than one employee with the same surname – and if it were chosen as the primary key all sorts of mix-ups could occur: for example the wrong employee called Jones could be allocated a pay rise! Another problem would be caused by the fact that many women change their surname on marriage. If the surname had been used as the primary key it would have to change – causing all sorts of problems.

Flat File Databases

It is possible to store data using just one table. This is often referred to as a **flat file** database and can be thought of as the electronic equivalent of a card file where rectangular cards storing information are held, usually in a specified order, in a box.

If you wished to store the details of all of the CDs in your collection, you could create a card file with a card for each CD holding relevant details. In a database this could be stored as a table called CD, made up of columns each of which represents an attribute relating to the CD. Each row in the table refers to one specific CD.

For a file holding details of video shop members the following fields may be stored:

Surname, Initial, Member_Number, Phone_Number, Date_of_Birth

The Member_Number is likely to be the primary key as it is unique for each member and identifies which member the record is referring to. Name could be a secondary key.

Date_of_Birth could be a sort key. The list produced after such a sort would be in order of age.

Surname and Initial could be a composite key. If the data were sorted using this key all the Motts would be put together, in alphabetic order of Initial.

Entity – the whole table – details of all CDs in collection

CD

CD Code	Title	Artist
CD027	Joshua Tree	U2
CD033	Mutations	Beck
CD045	You've Come a Long Way, Baby	Fat Boy Slim
CD046	Ray of Light	Madonna
CD055	Urban Hymns	The Verve

One occurrence of entity details of CD045 – a row

Attribute – **title:** one column

Figure 11.2 Tables, rows and columns

Relational database

A relational database is the most common form of database in use today. There are many database management packages using relational methods on the market, from sophisticated systems such as Oracle and SQL server that are used by large organisations, to systems running on stand-alone computers or over local area networks, such as Microsoft Access, Paradox for Windows or Foxpro.

Moving on from one table

Consider the application discussed earlier for a veterinary practice. Figure 11.3 shows some of the data items that might be held in a flat file database.

PET

Pet Code	Name	Type	Date of Birth	Owner Code	Owner Name	Telephone Number
P0123	Misty	Cat	23/1/92	2234	Mary Preston	889976
P0345	Rover	Dog	12/12/90	1995	Julian Giles	765095
P0887	Foggy	Cat	23/1/92	2234	Mary Preston	889976
P1559	Gladys	Gerbil	16/4/98	1942	Amelia Alderson	565643
P1985	Slinky	Tortoise		1995	Julia Giles	765095
P2233	Speedy	Tortoise		1772	Sally Ann Taylor	876875

Figure 11.3 Single table file for Pet details at a veterinary practice

The single table shown is not the best way of storing the data as certain data items are repeated. For example, there are instances when one owner owns two pets. Owner 2234 (Mary Preston) owns both Misty and Foggy. Repeating the name and Phone number of this owner for every pet that she owns, has led to unnecessary data duplication.

Owner 1995 owns Rover the dog and Slinky the tortoise. But is this owner **Julian** or **Julia** Giles? Unnecessary data duplication can lead to **inconsistency** where two or more entries for the same data item differ. In this case it is likely that the discrepancy occurred when the data was entered into the database.

The use of a more sophisticated data structure would allow the data to be stored in a way that avoided such unnecessary duplication. Such a structure would store the data in more than one table with links between the tables.

It can be seen that we need two tables to store the veterinary data without unnecessary duplication. See Figure 11.4.

PET

Pet Code	Name	Type	Date of Birth
P0123	Misty	Cat	23/1/92
P0345	Rover	Dog	12/12/90
P0887	Foggy	Cat	23/1/92
P1559	Gladys	Gerbil	16/4/98
P1985	Slinky	Tortoise	
P2233	Speedy	Tortoise	

OWNER

Owner Code	Owner Name	Telephone Number
2234	Mary Preston	889976
1995	Julian Giles	765095
1942	Amelia Alderson	565643
1772	Sally Ann Taylor	876875

Figure 11.4 Two tables

The two tables in Figure 11.4 equate to the entities PET and OWNER. Note how the details of Owners 2234 and 1995 now only appear once in the database. The problem of unnecessary duplication has been resolved – but the link between owner and pet has been lost.

An extra field needs to be added to one of the tables to provide a link.

We could try adding Pet Code to the OWNER table. We could fill in P1559 for Owner 1942 (Amelia Alderson owns Gladys the gerbil) but which code should we enter for owner 2234? She owns both P0123 and P0887 (Misty and Foggy) and there is only space for one code.

We do better if we add the Owner Code to the PET table. Here, as every pet has only one owner, we do not have any problem. Owner Code in the PET table provides the link that is needed. The data item Owner Code is duplicated in the database, but the duplication is necessary. Figure 11.5 shows the two tables linked.

In a relational database the link between tables is achieved by duplicating the primary key of one table with an identical field in the other table.

In the veterinary example, an extra field Owner Code was added to the PET table to provide the link with the OWNER table. This field is known as a **foreign key** in the PET table. The format of the foreign key should be identical to that of the primary key in the linked table.

PET

Pet Code	Name	Type	Date of Birth	Owner Code
P0123	Misty	Cat	23/1/92	2234
P0345	Rover	Dog	12/12/90	1995
P0887	Foggy	Cat	23/1/92	2234
P1559	Gladys	Gerbil	16/4/98	1942
P1985	Slinky	Tortoise		1995
P2233	Speedy	Tortoise		1772

Same field – provides link

OWNER

Owner Code	Owner Name	Telephone Number
2234	Mary Preston	889976
1995	Julian Giles	765095
1942	Amelia Alderson	565643
1772	Sally Ann Taylor	876875

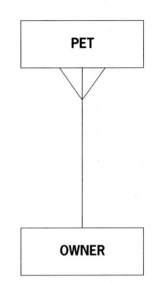

Figure 11.5 Two linked tables

- A **foreign key** is a field that is not a key in its own table but is a primary key in another table.
- In relational databases foreign keys are used to enable the links between tables.
- The use of foreign keys brings some, necessary, duplication of data.
- The use of foreign keys allows for the use of one-to-many or many-to-one relationships in a relational database.

In the example given, the attribute, Owner–Code, that was used as the foreign key, was already present in the original, single, table. (see Figure 11.3). Very often this would not be the case and an extra field, a primary key for the new table, would need to be created.

The veterinary example shows how the use of a foreign key allows for the use of one-to-many or many-to-one relationships. Implementing many-to-many relationships in a relational database is more complicated.

For example in a video hire application where a member can have borrowed many videos and a video can have been borrowed by many different members,

the use of a foreign key would not provide the required solution (see Figure 11.6). If Mary O'Sullivan has borrowed three videos

128	The Jungle Book
981	Babe
1769	The Lion King

which of these three codes should be inserted in the foreign key field for Video Code in the MEMBER record? Adding a Member Code as a foreign key to the VIDEO table cannot solve the problem. If the video had been loaned on three successive days to Mary O'Sullivan, Frazer McKensie and Paola Price, then which Member Code should be entered as the foreign key in the VIDEO table?

Creating an extra table that provides the link between the VIDEO and MEMBER tables in this many-to-many situation solves the problem. This table could be called LOAN and might contain attributes, Member Code, Video Code, Date Of Loan.

The two attributes Members Code and Video Code would together form the primary key of the LOAN table. See Figure 11.7.

VIDEO

Video Code	Title	Category	Loan Price
0981	Babe	PG	2.50
1197	Shakespeare in Love	15	3.50
1280	The Jungle Book	PG	2.00
1543	Antz	PG	2.50
1694	Silence of the Lambs	18	2.00
1769	The Lion King	PG	2.50

MEMBER

Member Code	Surname	Forename	DOB	Phone Number
A7652	Nayyar	Mukesh	12/3/84	832514
A9856	O'Sullivan	Mary	11/4/56	897543
B1100	Price	Paola	5/11/76	675643
B6993	McKenzie	Frazer	23/8/38	877665

Figure 11.6 Two tables with a many-to-many relationship

Database design

When you start designing a database system, you must:

1 decide on the tables in a proposed database system

2 identify the fields that belong to each of the tables

3 identify the relationship between these tables

4 define the structure of each individual table in the database

5 decide upon the properties of each field (see below)

The first three factors may well be arrived at by going through the process of normalisation (see Chapter 25). Attention will have to be given to the choice of primary keys, some of which may already exist while other tables may require new fields to be established specifically for this purpose to ensure that each record can be uniquely identified. It will then be necessary to determine the properties of each field.

When storing information about people, surname, forename and title are usually stored as separate fields. This:

- Enables the table to be sorted into alphabetical order by surname (it is unusual to sort by forename)
- Allows the name to be put together in different ways for different uses

<Title>	<Forename>	<Surname> on an envelope
Mr	Patrick	McGowan
Dear	<Forename> etc.	
Dear	Patrick	

LOAN

Member Code	Video Code	Date Out	Date In
A9856	1280	22/1/99	23/1/99
B1100	1769	24/1/99	1/2/99
B6993	1694	4/2/99	5/2/99
A9856	0981	12/3/99	14/3/99
A7652	1543	23/3/99	24/3/99
A9856	1769	6/4/99	7/4/99

The many-to-many link

Has been replaced with:

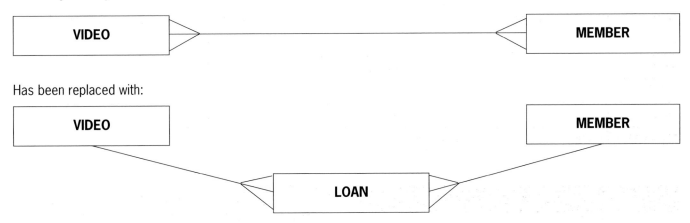

VIDEO

Video Code	Title	Category	Loan Price
0981	Babe	PG	2.50
1197	Shakespeare in Love	15	3.50
1280	The Jungle Book	PG	2.00
1543	Antz	PG	2.50
1694	Silence of the Lambs	18	2.00
1769	The Lion King	PG	2.50

MEMBER

Member Code	Surname	Forename	DOB	Phone Number
A7652	Nayyar	Mukesh	12/3/84	832514
A9856	O'Sullivan	Mary	11/4/56	897543
B1100	Price	Paola	5/11/76	675643
B6993	McKenzie	Frazer	23/8/38	877665

Figure 11.7 Three tables needed for video hire application

Properties of fields

For each field you must decide:

- **the name**

- **the data type**, for example integer, currency value with 2 decimals, text, date or a Yes/No field. Software such as Microsoft Access allows the user to choose from several different types

- for a text field, the maximum **length of the field**. By specifying the length, disk space can be saved. It is not easy to choose the appropriate length. For example, how many characters should be in a surname field? Ten? Fifteen? What about double-barrelled names? Twenty should be sufficient

- the **default value** (optional) is a value that will be inserted into a field at data entry if the user does not enter one for himself. An example for a default value could be today's date: in most situations the user will not to change this value, thus saving time at data entry

- any **validation** checks for a field; for example a month number must be between 1 and 12. The data type also can validate data for example a number field will not accept any other characters than digits of the decimal point

- an **input mask** can be applied to a field to ensure that it satisfies a pre-determined format. For example, in a particular application a membership number might always have to be a letter followed by four digits; a post code must consist of letters and digits in the correct combination.

A database design will depend on what needs to be stored and how it is to be accessed. Planning is essential. It is difficult to alter a design at a later date. Changes to the design can be difficult once a system has been set up.

Choosing data types

- In the VIDEO table, *Video code* is a 4 character primary key, for example 1197. This is a *text* field maximum length 4.

- It would not be a good idea to store this code as a number. Firstly, this code is not used for any calculations. Secondly, numbers cannot begin with 0. This field should be defined as a text field.

- The Loan Price field would be currency type set to 2 decimal places. As currency is a number, an application would be able to perform calculations on that field, for example to add up the cost of hiring more than one video.

Sometimes there are several possible data types which are acceptable. A field which shows whether or not a customer has paid their subscription, could either

- use a text field of length 3, and enter 'Yes' or 'No' into it

- use a single character field and enter 'Y' or 'N' into it

- use a Boolean field which will only accept 'Yes' or 'No'

The advantage of using a Boolean field here is that the entry is automatically validated since only one of two values can be entered, and it saves space in the database.

Manipulating the data

Once the structure of the database has been set up, the DMBS software allows the user to access the data without needing knowledge of how to set up the database.

■ EDIT THE DATA

The data will not always stay the same. Users can

- input new data: for example enter details of new members or new videos

- alter the data: for example a change of member's address

- delete data: for example when a member leaves (see Archiving data).

■ PRINT INFORMATION

Tables, results of queries, reports and so on can be printed. For example a list of all members, a list of videos borrowed today, a report of all videos not presently borrowed can be printed out.

Archiving data

In most systems, records are not deleted from a file or database. They are either stored immediately in an archive file or table, or marked in some way as deleted. Periodically, a utility program will be run to remove all records marked for deletion to an archive file.

This is done to ensure that records are not deleted in error and lost as well as for audit purposes (see Chapter 16).

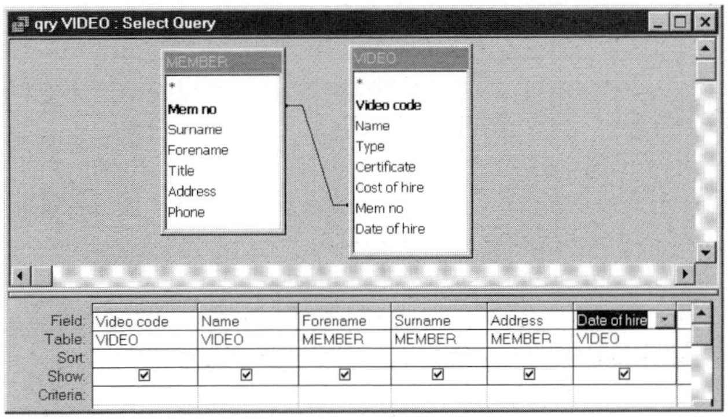

■ QUERY THE DATABASE

Users can search the database using a query to find certain data. Users can:

- search for information in one table
- search for information in related tables
- combine the information from related tables into a new table
- choose which fields are to be shown in the new table
- specify criteria for searching on; for example find the names of all videos borrowed today
- save the query so that it can be executed when needed
- save the new table so that it can be displayed, printed or used in a report

Programs like Microsoft Access allow users to set up queries in two ways. They can write the query themselves in SQL (Structured Query Language), a special language for querying the database, or they can get the software to write it for them using a query by example (QBE). There may even be a wizard to help write the query using QBE.

Using QBE involves choosing the fields to be output, establishing the necessary relationships and specifying the criteria to select the appropriate records.

Using the query by example method is much easier but some queries cannot be done by QBE and SQL has to be used. Queries can be amended by changing the SQL script.

If a query is set up in Microsoft Access using QBE it looks like this:

Figure 11.8 A query in Microsoft Access

This query searches information from the VIDEO table and the MEMBER table. It will print these fields: Video code, Name, Forename, Surname, Address, Date of hire.

Queries can be set up to ask what information they are searching for. This is a *parameter query*. When you run the query you will be asked for the video code (see Figure 11.9):

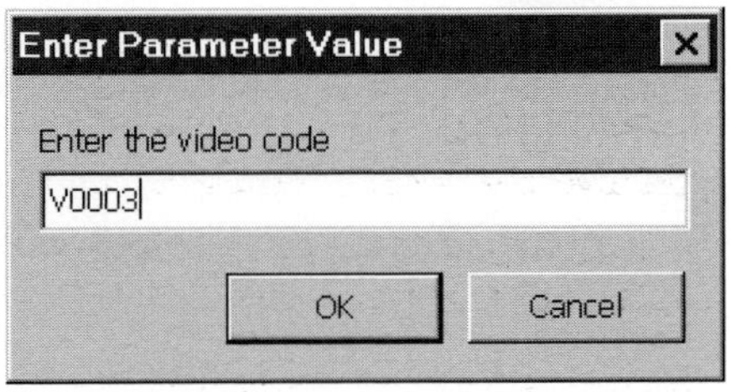

Figure 11.9 A *parameter query* in Microsoft Access

When the video code is entered, the computer will search for the details on that video (the result is shown in Figure 11.10).

Producing information is the prime purpose of a database, so query facilities are fundamental to an efficient DBMS. Structured Query Language is a data manipulation language developed for databases. The main constructs of SQL are **select**, which specifies which fields are required and **join** which sets up the links between tables.

Select	(fields required)
From	(name of table)
Where	(criteria for selection eg Certificate = '18')

SQL can be used to write very complex queries.

SQL is a language that is supported by many DBMS.

Many DBMS, such as Mircrosoft ACCESS provide an easier, graphical way for use to produce simple queries. This method of QBE (Query by Example) is illustrated in Figure 11.8. Access stores queries produced by QBE in SQL form.

■ SORT DATA

The information in a database will probably be stored in primary key order. The DBMS can sort the information into any sequence. For example records may be needed in alphabetical order of surname for one purpose but in chronological order for another.

■ INDEX DATA

Sorting a large database can take a long time. Sorting can be made much faster by creating an index for a field. This keeps a record of the information in the order of this field. If the records of a table are likely to be frequently needed in order of a particular field, then it is worthwhile indexing that field.

It is also a good idea to index fields that are often used in a query criteria; for example to find all videos hired yesterday we need to query on *Date of hire*. This field should be indexed.

Note: Indexing a particular field will speed up access to data, but will slow down record up-dating since every time an indexed field is changed, the index entry will have to be changed as well.

Indexed fields other than the primary key field are known as **secondary keys**.

■ RUN MACROS

Macros can be stored which automate important actions. Buttons can be set up on the screen to run these macros. A macro might be used to run a query, to add a new member to the MEMBER table, to delete an old video from the VIDEO table or to print a report.

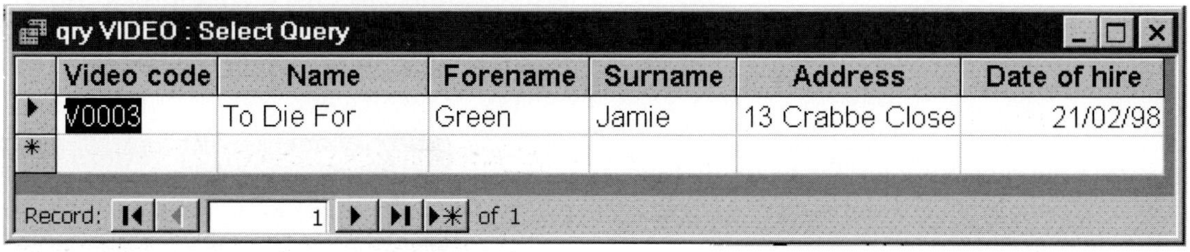

Video code	Name	Forename	Surname	Address	Date of hire
V0003	To Die For	Green	Jamie	13 Crabbe Close	21/02/98

Record: |◄ ◄ | 1 | ► ►| ►* | of 1

Figure 11.10 The result of a search in Microsoft Access

Summary

Files and databases are used for storing data. Storing data in a flat file often leads to duplication of data, wasting disk space. Storing data in a relational database is more efficient.

A database usually has more than one file. A relational database has links between its files. A relational database can avoid redundant data. It is easier to update the information and the data is consistent.

The DBMS (Database Management System) includes the stored data and programs to access, search and edit the data.

In a database:

☐ a **table** is another name for a file

☐ an **entity** is a subject about which information is stored in a table

☐ a **relationship** is a link between two entities

☐ an **attribute** is a property of an entity, for example Surname, Forename

☐ a **primary key** is an attribute used to ensure no two records are the same

☐ a **index** or **secondary key** is used when a table is often sorted into another order

It is vital to design the database structure before starting work. This design includes:

entities, attributes, primary keys, data types, validation and relationships.

Queries search through the database for required conditions. They can be written using

☐ SQL (Structured Query Language)

☐ QBE (Query By Example)

The Database Administrator (DBA) manages the database.

Relational databases Questions

1 A council has a database to keep track of council tax payments for houses in the district. One table in the database is **HOUSE.** Name 8 attributes of **HOUSE.** (8)

2 A motoring magazine store details of new cars. Suggest four attributes of the table **CAR MANUFACTURER.** (4)

CUSTOMER FILE

Surname	Forename	Street	Town	City	PostCode
Smith	James	11 The Avenue	Bemersley	Ruston	RS12 5VF
Penfold	Jayne	67 Bathpool Road	Outclough	Wignall	WG5 6TY

ORDERS FILE

Surname	Forename	Post Code	Order Date	Item Ordered	Quantity Bought	Price	Total Cost	Paid
Smith	James	RS12 5VF	6/5/98	Magic Duster	2	£10.99	£21.98	Yes
Penfold	Jayne	WG5 6TY	1/6/98	Banana Rack	1	£12.50	£12.50	No
Smith	James	RS12 5VF	12/5/98	Winsor Doormat	1	£29.95	£29.95	Yes
Smith	James	RS12 5VF	12/5/98	Easee Food Grater	1	£11.99	£11.99	Yes
Penfold	Jayne	WG5 6TY	1/6/98	Winsor Doormat	1	£29.95	£29.95	No

STOCK FILE

Item Name	Price	Quantity in Stock
Winsor Doormat	£29.95	11
Magic Duster	£10.99	34
Electric Potato Peeler	£39.00	0
Easee Food Grater	£11.99	9
Banana Rack	£12.50	1

NEAB 1998 Paper 2

3 A company makes use of a computerised flat file information storage and retrieval system. The company is experiencing problems due to the use of this flat file system.

a) Describe three benefits that the company would gain by using a relational database as opposed to a flat file system. (6)

b) The company currently has three files in use: computer, stock and orders. During conversion to a relational database system these files would need to be normalised. Explain clearly what you understand by the term normalisation. (2)

c) Examples from the three files are shown above. Normalise these files explaining any assumptions or additions you make to these files. (5)

4 Draw an entity relationship diagram for the following relationships:

a) Council Tax Payer and Payment Details

b) Pupil and Teacher

c) Supplier and Item in a Shop

d) Customer and Newspaper (on a delivery round)

e) Vehicle Registration Number and Vehicle Chassis Number

f) Computer and Repair details

g) Racehorse owner and Racehorse (7)

5 In database modelling, explain what is meant by the terms attribute and relationship. (2)

6 A hospital is organised into a number of wards. Each ward has a number and a name recorded, along with the number of beds in that ward. Each ward is staffed by nurses. Nurses have their staff number and name recorded, and are assigned to a single ward. Each patient in the hospital has a patient ID number and their name, address and date of birth recorded. Each patient is under the care of a single consultant and is assigned to a single ward. Each consultant is responsible for a number of patients. Consultants have their staff number, name and specialism recorded.

a) State four entities for the hospital in-patient system and suggest an identifier for each of these entities. (6)

b) Describe three relationships that can be inferred from these data requirements. (5)

c) A relational database is to be used. Describe tables for two of the entities for which you have described a relationship. (7) *AEB Computing Specimen Paper 3*

7 A medium sized electrical shop is to be computerised using a relational database. Two tables are called **PRODUCT** and **MANUFACTURER**.

a) State eight fields associated with the **PRODUCT**. Draw a diagram showing the relationship between the **PRODUCT** and the **MANUFACTURER**. (6)

b) State two other tables which could be related to either or both of the original tables. Describe their relationships. Suggest an identifier for each of these two tables. (6)

Activity 1: Exercise to practise the use of a foreign key to create one to many links between tables

For each of the scenarios described there are 2 entities. Name the entities and list appropriate attributes for each entity. Show how tables can be set up in a relational database for these entities using a foreign key as the link. Choose the attribute that is to be the primary key in each table with care, and justify your choices.

1 A conference has been organised for students of IT A level in a large city and students are coming from a number of schools. The organisers wish to store data about individual students and the schools involved.

2 The organisers of an arts festival wish to store data about venues where events are held the artists who are taking part (assume that an artist only performs at one venue).

3 A holiday company works with hotels in a number of resorts.

4 An art dealer sells the paintings and other works of art produced by a number of artists.

Activity 2

Databases or a manual system?

Why should a video shop use a database? Could a manual system be better than a database system? Both systems record details of members and who has hired which video. (This could easily be done in a paper based system by using a list of all videos and the name of the hirer written next to it.)
A manual system is cheaper, unlikely to break down and requires little training. However the computerised system will probably be better for the following reasons:

▸ **Management information** is automatically gathered, for example details of each hiring, financial details, how many times a customer has hired a video and how many times a video has been hired. These figures can be used in preparation of accounts or to analyse which videos are most popular.

▸ **Better service to customers.** Using a bar code reader to enter the video code and the member code is very quick. Queues at the counter will be shorter.

▸ Details of members and videos can be found and printed quickly.

▸ The names and addresses of members can be used for advertising purposes in a **direct-mail** shot. The database can be queried to come up with a list of people who haven't hired a video for six months and a letter written offering them a discount if they hire a film this week. The letter can be personalised using the mail-merge from a word-processing package.

▸ Similarly automatic reminders can be sent out to members who have not returned a video by the due date.

▸ The database can be extended to include the member's date of birth. The computer can be used to ensure that a member is old enough to hire, say, an 18 video.

Describe in a similar form to the example given above, the advantages of implementing a database system for:

a) a veterinary practice
b) a private leisure centre
c) a knitware business

Revision questions

1 Give an example of when computer data may have a commercial value. (3)

2 Jim gets paid on the last day of the month. In the afternoon of pay-day Jim goes to his bank's ATM (Cashpoint machine) to get some money out. The ATM says that Jim's account is over-drawn and won't let him have any money. Suggest three reasons for this. (3)

3 Using examples, explain the difference between information and data? (4)

4 A national supermarket chain uses batch processing to up-date its stock records every night. It is considering moving to an on-line system that would be expensive. By listing additional facilities available, explain why this may still be good value. (5)

5 Suggest some sensible back-up procedures for a college network to make sure no work can be lost. Include in your answer: protecting student files, virus prevention, protecting network software and applications software, avoiding malicious damage. (4)

6 A department store introduced a computerised till system where the code number has to be typed in by the shop assistant. The store subsequently discovered that transposition errors (swapping two numbers round) were common. Describe the steps the store should have taken to avoid these errors. (4)

7 A school issues passes to its students. These are checked at the door of the classroom against a computer file. Describe two types of pass the school might use and the equipment needed and comment on the suitability of the system. (6)

8 A company is about to replace its old *batch* processing system for the preparation of customer accounts, by an *on-line* system.

   a) Give THREE distinct advantages which could be expected as a result of this change over.

   b) Suggest TWO problems which are likely to be encountered during the changeover period. *London Board Computing Specimen Paper 1*

9 Explain these terms: (8)

   a) 32 MB RAM

   b) CD-ROM

   c) 1.44 MB Floppy disc

   d) 3 Gigabyte Hard disc

   e) back-up tape

   f) bit

   g) hardware

   h) software

10 Give two advantages of an international company using e-mail and two disadvantages. Compare with fax. (4)

11 To check data processed by an IT system, a library uses check digits.

   a) Describe briefly how check digits validate data.

   b) State one example where check digits would not be a good method of validation.

   c) State what method could be used and why. (8)

12 Using a practical example, give one advantage and one disadvantage of OCR in comparison with other forms of computer input. (3)

13 A school is considering 'hooking up to the net.' Write a brief report recommending what they should do giving your reasons. (4)

14 A firm with two offices in London and New York regularly needs to send designs from one office to the other. Describe two possible ways of sending the designs, including their relative merits. (20)

15 A school uses a LAN. What is meant by LAN and why is it better for the school to use this and not use stand alone machines? (6)

16 An electronic mail system exists for messages to confirm deals between money and share traders. Briefly describe some benefits of using this system rather than sending papers by courier. Comment on one possible disadvantage. (6)

17 Discuss the advantages of using a public phone network or a private line to link sites for data transfer. (6)

18 E-mail is replacing face-to-face contact in an office building. Give three reasons why this might not be a good idea. (3)

19 The following are applications which use either a wide area or a local area network or a combination of both. For each justify which network type is most suitable

   a) Cash dispensing and account inquiry facility for a national building society. (2)

   b) Accounting and stock control system for a department store using point of sale terminals (2) *AEB Computing Specimen Paper 2*

20 Many people now spend much of their working week at home using computers (telecommuters). Describe possible advantages and disadvantages of telecommuting for these people and for their employers. (10)

21 Show with the aid of a sketch, the structure of a simple bus network. (4)

22 State five ways a dairy farmer could make use of the Internet. (5)

23 Teletext and viewdata systems are widespread, aiming at home users and business users. Give examples of each service may help each type of user. (4)

24 A hospital has a network of many work stations in wards, consulting rooms and the administration offices. Discuss the advantages and disadvantages of such a system compared to a manual system. (8)

25 Describe a hardware configuration which could be used in the homes of disabled people who are unable to use a keyboard. The system should enable them to control actuators to switch on lights, television etc. (4)

26 Electronic money is slowly taking the place of notes and coins, if not replacing it altogether. Describe three types of electronic money, stating why the electronic money is better than cash and the role of IT. (9)

27 What are smart cards? Give two advantages and two disadvantages of using smart cards to store medical records. (5)

28 A large multi-access computer system provides on-line storage for programs and data. What precautions should staff take to protect stored files against being corrupted? How could they rebuild the files if such a problem did happen? (4)

29 Explain the difference between security and integrity of data. (2)

30 Give two arguments for and two against the belief that data stored electronically is more secure than that stored on paper. (4)

31 A file of account holders is to be established by a departmental store on its computer system. Describe briefly, in each case, two steps that would help to ensure

   a) the integrity

   b) the security of this data (2) *AEB Computing Specimen Paper 2*

32 The Data Protection Act talks of data users and data subjects. Explain what these terms mean. (4)

33 A small manufacturing firm needs software to help budgetary control and financial modelling. They want to

study the effects on overall profitability of changes in wage rates, prices of products and numbers of units sold. What type of package is likely to be most suitable for their needs? Explain the advantages of buying this package rather than having software specially written for them. (5)

34 Explain with examples two problems you may get using a spell checker apart from slowness. (4)

35 In a computer system, the operating system co-ordinates the processor and peripherals. State what is meant by peripherals and give two functions of the operating system. (3)

36 Distinguish between multi-access and multi-programming systems. (3) *NEAB Computing Specimen Paper 2*

37 The operating system controls the input and output devices of the computer.

a) Describe three input devices have you seen in school/college.

b) Describe two input devices not used in your school/college.

c) Describe three output devices have you seen in school/college.

d) Describe two output devices not used in your school/college. (10)

38 When purchasing software, it is often possible to buy either an 'integrated package' or separate applications packages that run under a common operating system environment.

a) What is meant by the term 'integrated package'? (1)

b) What applications would you normally expect an 'integrated package' to offer? (1)

c) What are the relative advantages and disadvantages of an 'integrated package' over a collection of separate applications packages running under a common operating system environment? (4)

39 A mail-order book club holds its customer data on a computer file. The club wishes to contact its customers who have not ordered a book for one year or more.

Briefly describe the stages involved in the necessary mail-merge. (3) *NEAB Computing Specimen Paper 2*

40 Why might the latest software not be appropriate for someone to use on their computer? Using a program you have used as an example, describe three possible problems that could occur. (3)

41 You have been asked to give some advice to a friend who wants to buy a word-processing package. Describe briefly the facilities you should advise them to look for in such a package, other than the ability to enter, delete, store, retrieve and print text. (4) *NEAB Computing Specimen Paper 1*

42 Human speech as input to computer systems is increasingly common. Outline two advantages and two disadvantages to society in general. (4)

43 What are the legal implications of storing personal data on computer files? (4) *NEAB Computing Specimen Paper 2*

44 Discuss some of the developments of the last decade that have led to widespread use of computers in homes, offices and factories. (10)

45 State **two** uses of bar codes in a factory data collection system. (2)

Give **two** reasons why bar codes are preferred to magnetic stripe in this system. (2) *NEAB Computing Specimen Paper 2*

46 Early computer software was not very user friendly. Describe three ways in which software has become much easier to use. (6)

47 A football ground has 25 000 numbered seats. During the season, supporters can buy tickets for matches at one of five ticket booths. The club wants to computerise ticket sales, printing tickets as they sell them.

a) Describe the hardware required.

b) What stops two booths selling a ticket for the same seat at the same time?

c) Describe the benefits the club might get. (7)

48 Give **two** facilities provided by electronic mail systems which are not provided by the ordinary postal system. (2) *NEAB Computing Specimen Paper 2*

49 A government office where much of the information is classified, plans to replace its security officer on the front door with an on-line computer system to monitor arrivals and departures of office staff. The door will in future be operated by typing in a secret code.

a) State two possible advantages of such a system. (2)

b) Suggest how such a system might lead to security problems. (2)

50 A college is to introduce a computer based public examination entry system. Examination entries are to be transmitted by an electronic mail system from the college to the examination's board. During the transmission it is important to maintain the "integrity" and "privacy" of the entries. Explain these terms and describe how each may be achieved. (6) *NEAB Computing Specimen Paper 4*

51 Give two reasons why the storing of information on a police computer constitutes a potential danger for the individual in society. (2) *AEB Computing Specimen Paper 2*

52 The use of computer controlled machinery in manufacturing and product assembly has become widespread. Give **two** benefits for the manufacturer, other than financial, and **two** implications for the employees of such automation. (4) *NEAB Computing Specimen Paper 2*

53 A video rental shop has several hundred video tapes available for members to hire. Each week new titles are added to stock. There may be several hundred copies of popular videos. The shop has a computer system with a file containing details of members and a file containing details of videos. The shop does not allow more than 2 videos to be on loan to a member at any one time and the videos must be returned within 2 days. The three most useful queries to be answered using the system are:

- list those members who have a particular title currently on loan;

- list the videos a particular member has on loan

- list those members who have not hired any videos in the previous six months.

a) Describe the file organisation and record structure necessary to enable these queries to be answered. (5)

b) Describe the processing needed to answer each query. (4)

c) State two checks that the system should make before allowing a member's record to be deleted. (2)

d) Describe the amendments that are made to the files when a video is

i) hired;

ii) returned (4) *London Board Computing Specimen Paper 1*

A2

Organisational structure

All information systems occur within the context of an organisation. So, to be able to understand how such systems work it is necessary to first look at how organisations are structured. An organisation consists of a group of people; it has a specific purpose that determines both its objectives and the activities it carries out. A public limited company, a private limited company, a civil service department and a Parent–Teacher Association are all examples of organisations.

Individuals within an organisation have defined roles. This allows for the **division of labour**, so that specialisation of tasks and skills can take place. The activities and tasks needed to meet the aims of the organisation are divided into logical groups. By specialising, individuals can develop knowledge and

expertise in a particular group of tasks. The larger the organisation, the more likely specialisation is to occur.

For example, a small business is likely to have one person in charge of all aspects of human resources, which form just a part of their total responsibilities. A very large company would have separate staff, or even separate departments, dealing with recruitment, training, contracts, and so on.

Responsibility for decision making is also defined. A structured framework shows the roles of individuals and the formal pattern of relationships between staff. These structures set out responsibilities and divide work within the organisation, enabling administrators to co-ordinate, control and monitor

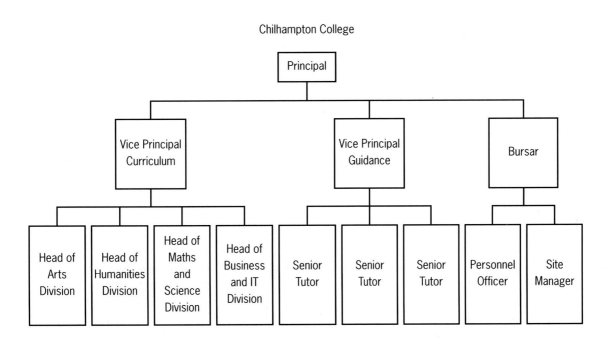

Figure 12.1 Organisational chart

their activities. An organisation chart is often used to illustrate where responsibilities lie.

Departments, where related tasks are carried out, may be created in a number of ways depending upon the nature of the organisation.

human resources, sales and finance. This is the traditional organisational form and is usually acceptable to employees. Any professional expertise, for example in public relations, is enhanced as qualified staff will be working with other professionals in the same field.

Functional specialisation

With functional specialisation, tasks are grouped together on the basis of common functions such as

Figure 12.2 Functional specialisation (a manufacturing company)

Product or service specialisation

Tasks and activities that relate to a particular product or service are grouped together. This form of specialisation is common within the public sector. A

local authority would group its activities together in departments such as Housing, Education, Local Taxation, and so on. Such specialisation allows for the development of expertise in the product or service.

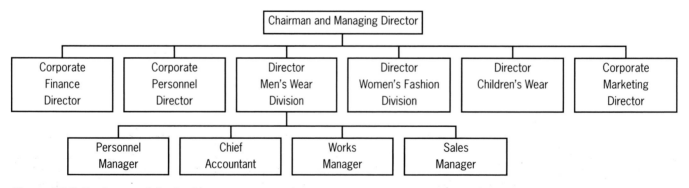

Figure 12.3 Product specialisation (a garment company)

Geographical specialisation

See Figure 12.4. Many large organisations, such as building societies or large retail chains, operate nationally or globally with operations in many locations. It is sometimes appropriate for such organisations to be structured on a geographical basis, giving local management the responsibilities for its own local activities. Such specialisation allows for decision making based on local knowledge that can lead to a better service for customers.

In practice, the structure in many organisations is a

mixture of different forms of specialisation. Every organisation attempts to develop a structure that best fits its circumstances. For example, most organisations operating geographical specialisation retain a number of functions, such as accounting and research and development, at Head Office.

Most organisation structures are hierarchical in design. The structure consists of a number of layers; moving down from the top to the bottom of the structure, the number of personnel in each layer increases and the amount of responsibility decreases.

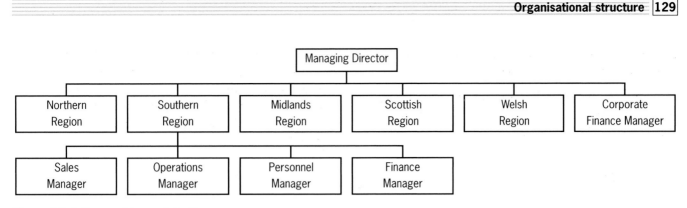

Figure 12.4 Geographic specialisation

SPAN OF CONTROL

The **span of control** (see Figure 12.5) is the number of employees who are directly supervised by one person. This should be clear in the organisational structure. Too wide a span of control leads to a lack of control and is inefficient. Too narrow a span wastes staff. The nature of the roles of the staff being supervised will help to determine the appropriate span of control in any particular circumstance. A supervisor of supermarket check-out clerks would be able to sustain a larger span of control than a personnel manager. The check-out clerks are all carrying out the same, fairly straightforward, tasks whilst a personnel manager's subordinates would have a range of spheres of work, such as recruitment, industrial relations and remuneration.

CHAIN OF COMMAND

The chain of command (see Figure 12.6) is the path through the levels of management from the managing director downwards. Instructions go down the line of authority. Problems are referred up the lines to a higher level. Long lines of communication mean messages can be distorted.

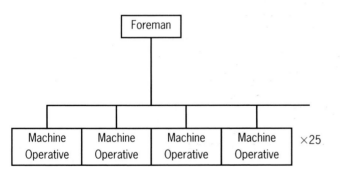

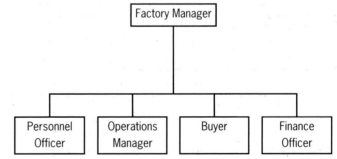

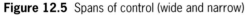

Figure 12.5 Spans of control (wide and narrow)

The pyramid or tall hierarchical structure

This is the traditional structure typified by large public limited companies, the military and the civil service. Roles are clearly defined within a large number of layers, each responsible to the layer above.

At the top of the pyramid is the managing director or chief executive who is responsible for the success or failure of the organisation. Each manager has a relatively small span of control. The chain of command down from the managing director is long.

This tall hierarchical structure is suitable for large organisations with centralised decision making by the strategic staff. However organisations with a tall

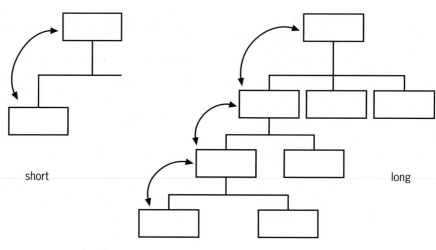

Figure 12.6 Chain of command

structure are likely to be slow to change as important decisions have to be referred all the way up the line. Policy decisions are made at board level and instructions go down the line to ensure that the decisions are implemented. Decisions take a long time to be made and take even longer to implement. Senior staff can be very remote from the lower levels of the structure.

The horizontal or flat structure

An alternative structure is the horizontal structure. In a flat structure there are fewer layers, but the spans of control are much greater. As a result, problems being referred up the line can be resolved more quickly.

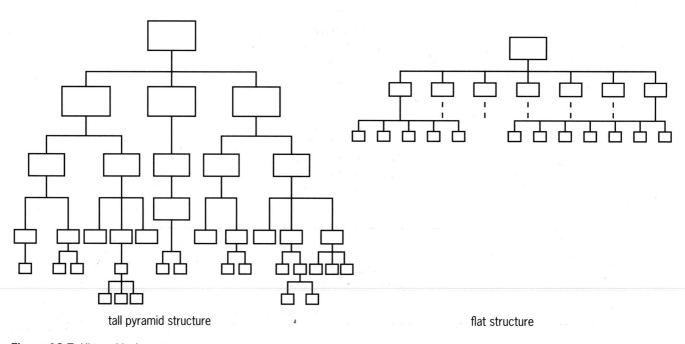

Figure 12.7 Hierarchical structures

As more people are directly answerable to the managing director, the power to make decisions for themselves will need to be delegated to middle managers. Parts of the organisation may tend to operate independently of the other parts but are still under the umbrella control of senior management. Employees have more responsibility which often leads to better motivation. It is more likely that employees can contribute more to decision making.

However, as departments are specialised, different departments may have little to do with each other which can lead to poor communication across the organisation. Control of top management could be weakened as they have a greater span of control and need to delegate more frequently. Fewer levels usually means that there are fewer prospects of promotion.

The flat structure is becoming more popular. It allows considerable independence to different units which means that these units can make decisions and change more rapidly. Tall hierarchical organisations are *static*. Flat organisations tend to be more *dynamic*.

The levels of an organisation's structure

There are generally three levels of personnel in a business organisation, although there may be considerably more layers (see Figure 12.8).

The **strategic** level that consists of senior management, responsible for long-term planning and policy making decisions.

The **tactical** level consists of middle management in charge of one particular department or area of the business. Examples would be a regional sales manager or a training manager. Planning at this level is medium term.

The **operational** level consisting of the workforce who are making the product, taking sales orders, keeping the accounts, and so on. Operational managers include foremen, supervisors and charge hands. Planning is on a short-term, often daily or even minute-by-minute basis.

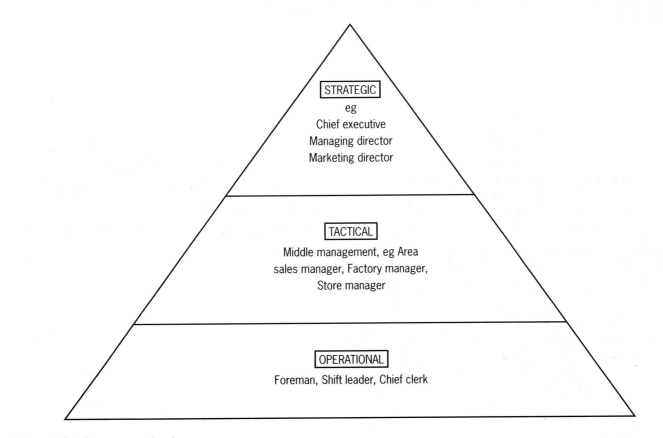

Figure 12.8 Management levels

How has the development of IT affected the choice of organisational structure?

Information Technology tends to lead to a flatter organisational structure. This is because the introduction of IT and the pace of hardware and software development means that frequent change is inevitable and businesses must be dynamic. IT systems provide better information on staff performance, thus enabling managers to monitor more people and cope more easily with a wider span of control.

Some jobs at lower levels, such as typists in the typing pool, may disappear altogether as a result of the growth of word-processing. New, direct methods of data entry reduce the number of clerical staff needed. Robots have replaced many jobs on the production line. All these changes have resulted in the reduction of the number of levels in an organisation.

Over the last decade, the jobs of middle managers have been eroded. Contributory causes include the developments in IT and communications. These have enabled information to be produced in a form suitable for the strategic managers directly from the operational level, without the need for manipulation and interpretation by middle managers. Many decisions that used to be taken by middle managers are now taken by computer based systems. For example, decisions regarding granting of loans to bank customers and stock replenishment in supermarkets can all be made by computer based systems. Increasingly, operational staff can work without needing direct middle management involvement.

IT will be involved with providing information and facilities to many different parts of the organisational structure.

Summary

Organisations can be structured in a number of ways, for example functionally, geographically or by product. They can have:

☐ tall hierarchical structure

or

☐ flat structure.

Tall hierarchical structures can be rigid and decision-making is slow. Flat structures can be flexible and autonomous.

There are three main levels of hierarchy in an organisation:

☐ strategic

☐ tactical

☐ operational.

New technologies have effected the organisation structures of many businesses and altered the way in which decisions are made.

IT provides information at all levels of the organisation's structure on which decisions can be based.

Organisational structure questions

1 Give an example of one organisation you know which uses a hierarchical structure. (2)

2 Explain why is it necessary to have an organisation structure in a business. (2)

3 In 1996 the then government started privatising British Rail, with a turnover of £3 billion and 100 000 staff. British Rail had five divisions: InterCity, Provincial, Freight, Network South East and Parcels. What sort of structure do you think British Rail used to have? Justify your answer. (3)

4 For each of the following types of organisation, suggest decisions which would be made at the (a) strategic; (b) tactical and (c) operational level

a) a school or college

b) a retail chain of shoe shops

c) a multinational bank

d) a car manufacturer (12)

5 A small company with a rigid hierarchical structure is planning to introduce a computer system. Describe four concerns a director may have on the effect on the structure of the company. (4)

6 Identify and describe the three levels of management usually found in an organisation. What types of information are needed at each level? (9)

7 List the relative advantages and disadvantages of functional, product and geographic specialisation. (12)

Information systems

Information systems within organisations

Computers enable information to be produced and manipulated very effectively. For example, wide-area networks allow information to be distributed over great distances at high speed. All organisations depend on the creation and use of information. Data must be collected and sorted, calculations and processing carried out and resulting information must be disseminated to a range of people.

The arrival of IT-based information techniques has speeded up the data collection process. Large amounts of data can be processed, analysed and the information generated can be communicated very quickly. Electronic links improve communications and reduce the need for paper. Decision making can be speeded up and an increased confidence in decisions created.

Data processing systems

A **data processing system** deals with the day-to-day transactions of an organisation, for example recording the loan of books from a library, producing bills in a utilities company or making seat bookings for a cinema.

The first commercial computing systems, developed in the 1950s and 1960s, were data processing systems that replaced the manual clerical procedures that were currently in use. Computers were able to carry out processes such as calculating and sorting more quickly, accurately and consistently than humans.

The first examples of such systems were developed for activities such as the production of the payroll for employees and billing systems. The earliest systems input and stored data on punched cards. Over the years, magnetic media (tape followed by disk) took

over the storage role and a range of input media such as OCR, MICR and key-to-disk were used to input the data. Huge, high speed line printers were used to produced paper outputs: payslips, bills and long lists of transaction details for internal use on continuous, perforated paper. These data processing functions are still carried out in the most up-to-date systems.

A data processing system is designed to process the data generated by the day-to-day business operations of a company. Examples include systems for accounting, invoicing, stock control and order entry. For example, a clerk processing a customer order needs to know whether the item is in stock, what the price of the item is, what discount, if any, the customer should be given, as well as customer details such as address and credit status.

Data processing systems are, in the main, tools for the operational level of an organisation. Their development over the years has had a major effect on the number and type of jobs at this level. Such a system by itself does not provide the kind of information necessary to help tactical or strategic decisions to be taken by middle or senior management. However, the information needed for such decisions can be produced from the data gathered for these operational systems. Information systems were developed to enable managers to make use of the information.

An example of a data processing system is a payroll system for a large company. Data concerning overtime from time sheets is entered via a keyboard. A program is run which reads the time sheet data and accesses a staff file for necessary data (such as basic salary, tax code) so that pay details can be calculated. Pay slips are printed for distribution to employees and the necessary data is transferred electronically to the appropriate banks to initiate the transfer of funds.

Information systems

An information system is a system that processes data to produce information in such a way that it can be used to help in decision making. In many systems, data processing and information systems operate together at the same time. An example of such would be an airline booking system. Flight bookings are made using an on-line system through which seats can be reserved for passengers. The system produces tickets and boarding cards as well as various lists. For example, the catering company would need to know the exact dietary requirements of the passengers for a particular flight.

The flight booking system also provides an information system for the customer, travel agent and airline. The airline can find out the percentage of seats sold. Using the on-line system, the customer can gain information on the availability of seats for a range of flights. This can help her to make a decision.

A payroll data processing system could be transformed into an information system by adding reporting facilities to the day-to-day functions of inputting timesheet data and outputting cheques and payslips. The reports could include a summary of the total wage bill, broken down into departments.

In a school or college, a timetabling system which produced summaries of class sizes, staff work load and room utilisation as well as allocating students to classes and printing individual timetables, would be an information system.

Hill Fort Community School – Summary of Class Sizes for Year 12

	Male	Female	Total		Male	Female	Total
Art	3	12	15	Further Mathematics	4	2	6
Biology	5	13	18	Geography	6	12	18
Business Studies	9	3	12	German	2	3	5
Chemistry	22	5	27	History	5	12	17
Computing	14	1	15	Information Technology	15	6	21
Design	11	3	14	Mathematics	34	18	52
Economics	8	2	10	Music	3	2	5
English Language	5	14	19	Physics	18	4	22
English Literature	3	25	28	Psychology	2	6	8
French	5	13	18	Religious Studies	1	3	4
				Total	**175**	**159**	**334**

Figure 13.1 Class size summary

Management Information Systems

> An MIS is a system that converts data from internal and external sources into information, communicated in an appropriate form to managers at different levels of an organisation. The information enables effective decisions or appropriate planning to be carried out.

A management information system (MIS) is an information system that aims to provide a manager with all the information needed to make decisions associated with her job as effectively as possible. Management information systems are a subset of information systems. Like any information system, it converts data into information. The data can come from both internal and external sources. The rapid growth in the use of database systems within organisations has allowed MIS to develop. An MIS is usually based on data from one or more databases.

Management is not an activity that is only performed by people who are called 'managers'. It is carried out at all levels of an organisation: strategic, tactical or operational. Management involves a range of functions: decision making, planning, coordinating, controlling, organising and forecasting (see Figure 13.2).

An MIS converts data into information for a manager. The data is usually already held in the computer as a result of data processing activities. It

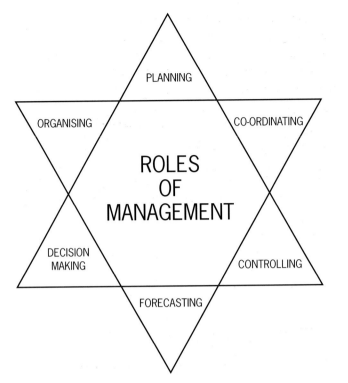

Figure 13.2 Roles of a manager

is important that the information is produced to the correct level of detail.

For example, an operations manager of a chocolate factory needs to decide the number and types of bars to be made in a particular week. The following information would help him make this decision:

- the number of each type currently in stock
- outstanding orders still to be delivered
- the sales of each type last week
- the sales of each type this time last year

This information could be created from the data collected as part of the day-to-day operational data processing system.

The sales manager for the same company will require information on products sold rather than products produced. He will need to be able to compare the performances of different members of the sales force. The information will need to be in summary form – perhaps represented graphically.

To make effective use of the information, a manager will need to receive it in an appropriate form and at a suitable level of detail. The amount of detail will usually relate to the level of management using it.

(based on an article in the Financial Times)

Your tickets are in the ether

Imagine that Mr Bell in Glasgow asks consultant Ms Rossi of Milan to visit him. Using Web technology, Rossi consults Bell's electronic diary, finds a free time that is also convenient to her, and books the meeting. Following this instruction, her computer automatically offers her a flight from Milan to Glasgow that ensures she arrives at Bell's office on time. Because Rossi is allowed by her company only to travel economy class, the system offers her an economy class seat.

Her computer also offers accommodation at a Glasgow hotel. It so happens that Rossi's company has a Europe-wide discount deal with Marriott. So, the first room the system offers is at the Glasgow Marriott. Rossi amends or approves the suggested itinerary and the system makes all the necessary reservations. The package also offers a hotel booking system, a database of 30 000 properties which takes full advantage of Microsoft's mapping technology. Travellers can call up a detailed map of a city and draw the area in which they want to stay. Within those geographical parameters, it is possible to add further criteria, such as a requirement for hotels charging less than, £60 per night. The system will then provide a list of hotels which meet all the specifications. The maps can also measure the distance from the chosen hotel to other points in the city.

- What advantages does the system bring to Ms Rossi and her employer?
- List the stages that are gone through from the point when Ms Rossi is asked to visit Mr Bell. Note which actions are taken by Ms Rossi, and which by the software.

Strategic information

Strategic information is used by senior managers such as directors and the chief executive in a business; head teacher and governors in a school or directors of a charity. Long-term planning is a key function at this level of management and most decisions made will reflect this. An overview of the operation of the whole organisation is required so that an assessment can be made of how well objectives are being met. Actual costs and profits need to be compared with forecasts for all sections of the business. An MIS can produce projections and predictions based on current data, both internal and external, that relates to the business.

The nature of strategic management means that the information that is required at this level can be very varied both in content and in timing. There will be a need for some regular reporting, but depending on the decision to be made, other, 'one-off', information may be needed. External sources will often play a major role at this level. Many organisations are developing Executive Information Systems (EIS) which allow management access to information from wide ranging sources in a variety of forms (see page 140).

For example, a company that produces and sells ice cream and other associated products has six factories located in different parts of the United Kingdom. The senior management may wish to close down one factory to reduce costs. This would be a strategic decision and would need a wide range of information. An example of internal information would be the increase in labour costs at each factory. External information would include the present site value of each factory.

Tactical Information

Middle managers, typically department heads, regional or functional managers, have roles that are tactical. Such a manager would be responsible for a certain section of a business and would be responsible to a senior manager. She would be likely to have a number of operational managers reporting to her. In an organisation with geographical specialisation, such a manager could be responsible for a sales region, a specific factory or group of shops. In an organisation that specialises functionally, a middle manager could be in charge of training, customer accounting or IT Services.

Much of the information needed by such managers relates directly to the operational performance of the organisation and is used for monitoring and controlling purposes. For example, sales figures for each of the company's sales representatives. Regular reports to assist making tactical decisions are common at this level in a variety of forms: tabular, graphical and pictorial. The information is usually prepared on a routine basis, perhaps weekly or monthly. A factory manager of the ice cream company might consider running an extra shift during the summer months. Such a decision would be a tactical one, based on tactical information.

Exception reports (see Figure 13.3), for example a list of all sales figures which fall below their target level, provide tactical level managers with a powerful tool in establishing areas for further investigation. Successful decision making at this level often depends

---

A LEVEL IT PERFORMANCE YEAR ONE

STUDENTS WHO ARE A CAUSE FOR CONCERN

(Total points less than 150)

Surname	Forename	TG	ITO1	ITO2	ITO3	Total
Lister	Peter	F56	56	43	45	144
Smith	Dan	F33	34	24	49	107
Stanley	Anthony	F23	49	54	34	137
Dawson	Cathy	F11	29	31	54	114
Green	Amy	F22	45	34	59	138
Manie	Les	F41	36	45	66	147
Thompson	Pat	F34	50	47	32	129
Davey	Charlotte	F62	32	45	12	89

Figure 13.3 Exception report displaying details of students gaining low marks in first year exams

upon accurate forecasting. (For example, cash flow forecasts.) This can be enhanced by modelling the situation with different data. Decision Support Systems (DSS) are useful here (see page 140).

Operational information

Operational managers are closely involved at the productive end of the operation. A supervisor may oversee the workforce on a particular production line. He may need to work out rotas and rest breaks, monitor the rate of production, ensure that hold-ups due to machine failure or delay in the arrival of spare parts are minimised and ensure that the quality of the finished product is maintained within acceptable levels.

Consider the role of a manager of a small retail shoe shop that is part of a large national chain. She will need to organise staffing rotas, balance the till at the end of the day, re-order stock, monitor sales on a daily, weekly and monthly basis.

Operational information is needed by such managers to help them in their decision making on a day-to-day level. Nowadays, many operational decisions, such as when to re-order stock, are made automatically by the computer software. The re-ordering can itself be initiated automatically. Simple lists and charts will play a major part in operational information. Such a list could be produced by sorting the transaction data that has been processed as part of the normal data processing function.

At the operational level, information is characterised by a high level of detail. For example, in the shoe shop chain, the local shop manager might require a daily, itemised list of all shoes sold, sorted into types, styles and quantities. The regional manager (tactical level) would require a weekly or monthly summary report showing the total sales for each shop in her region. At a strategic level, the marketing manager might wish to forecast sales trends over the next few years.

Development within MIS

As the power of computers has rapidly increased alongside the developments in communications, there has been a growth in the developments of specialised management information systems for different levels of users. Although these are becoming popular, they are by no means universally used in organisations and many problems still remain with their implementation.

Figure 13.4 Graphical displays aids middle managers gaining fast access to information from bar charts

Decision support systems

Systems called Decision Support Systems (DSS) help decision making but they do not themselves make decisions. They are particularly valuable when making 'one-off' unstructured decisions that cannot by their nature be produced on a routine basis from an information system. DSS are used at the strategic or tactical management levels. They enable a manager to explore a range of alternatives under a variety of conditions. For example, a manager may wish to know the effects on profits if sales increase by 5 to 15 per cent and costs increase by 5 to 12 per cent.

DSS are hands-on, interactive systems. They need to be user friendly and easy to use so that managers can use them with confidence whenever needed. Examples of DSS include spreadsheets with their 'what-if' capacity for modelling and testing out different scenarios, and statistical packages.

A decision support system is a system designed to help someone reach a decision by summarising all the available relevant information. Some of this information may be held in the company's database, and some may be external to the company. External information could include current interest rates, the price of oil, population trends, or details of new competitors starting up in the area.

Decision support systems often include query languages to enable managers to make spontaneous requests from databases, spreadsheet models that enable 'what if' calculations to be made, and graphics to provide a clear representation of the available information.

Executive information systems

Most management information systems will provide summary, statistical information suitable for senior management. Often, however, such summarising hides crucial detail. The need for such detail would be impossible to predict as it depends on specific circumstances. This lack of appropriate detail could result in incorrect decisions.

Executive Information Systems (EIS) (also known as Executive Support Systems (ESS)) provide aggregated information for senior managers. The manager can display the information in more detail by clicking on hot spots. Such a system would bring

Based on an article in the Financial Times

Data Warehousing

Retailers are at last starting to make full use of mountains of accumulated data – with highly profitable results. For years, many leading retailers have collected reams of item-level sales data, numerous lists of which products customers bought during a single shopping trip, information on hourly trading peaks and troughs, and enough numbers to allow comparisons by any product, any branch, or by any time of day.

Unfortunately, the number-crunching needed to pull the facts into a truly useful report generally meant that management information was limited to exceptions and trouble-shooting.

With modern client-servers and powerful data warehouses, retailers are at last starting to make full use of all that accumulated data – with dramatic and highly profitable results. In the UK, Woolworth have installed a Tandem processor, SQL database and Microstrategy's DSS Agent management tools. The system allows users to pull any data – gleaned from transactions, supplier or store records – from the central store and assemble it for further analysis.

'It allows us to measure performance across any group of products and any combination of stores. For example, it has made it very easy to establish performance in our "price crash" promotions where we cut the price of kitchenware in those stores where we were competing head-on with a particular chain of hardware discounters.' Woolworth has also been able to monitor merchandising experiments more successfully: before Christmas, various Disney lines were brought together into a special section in some branches and spread through conventional departments in other stores.

Managers were able to monitor the success of the experiment on a daily basis.

- What is data warehousing? Describe the advantages brought by data warehousing to Woolworths.

together information from a range of internal and external sources.

For example, a senior manager is reviewing company expenditure over the past year, comparing with the estimated budget. This information is displayed in a graphical form. She then notices that one department is well over budget, and decides to investigate further. A click of the mouse on the appropriate figure results in the details of the budget and expenditure of the department in question is displayed. It appears that the overspend is greatest in the raw materials expenditure, so our manager clicks on this figure to reveal that prices are as estimated but the department has purchased more raw materials than planned. The manager can investigate sales and stock levels to find out whether these extra purchases were necessary.

An EIS could also provide access to newspaper articles, economic indicators and are usually linked to e-mail systems. An EIS is a simple, easy to use interface, which sits on top of huge databases. The manager is able to view a high level, overall summary of data and then 'drill down' by selecting a particular figure or 'hot spot', and see more detailed displayed. This drilling down can be continued to greater levels of detail.

A key feature of an EIS would be the quality of its presentation. Such systems are designed to be used by managers at the highest level of an organisation. Many such people are unused to using IT and are not prepared to invest much time into learning to use complex and unfriendly systems.

Intranet

Currently there is a huge growth in the development of intranets within organisations. An intranet uses the same browser software that is used to access the Internet, such as Netscape or Microsoft Explorer. The information available on an intranet is internal to the organisation. A well-managed intranet can provide a very useful tool for sharing information.

Expert systems

An expert system is a computer program that tries to emulate human reasoning. It does this by combining the knowledge of human experts on a given subject, and then following rules that it has been given to draw inferences. To use the system, the user sits at a terminal and answers questions posed by the computer. The program eventually reaches a

Case Study

TSB

TSB is a well-known UK retail bank. Early in 1994, TSB was faced with increasing volumes of data and information on which to base the day-to-day operational decisions. The development of a management information strategy was a top priority to improve the reliability and speed of access to timely and consistent reports. It was recognised that any system would have to be both simple and intuitive to use if it was to gain acceptance with senior management.

TSB needed a way of looking at information alongside business dimensions which could also show how the business was performing and identify opportunities.

They chose an EIS system based on Gentia developed by Planning Sciences. The system allows easy development of new applications, real time access to information and offers a high standard of tabular and graphical presentation. It was first implemented as a senior management tool. □

(From promotional literature of Planning Sciences)

- Why was an EIS needed by TSB?
- Why should the system be simple and intuitive to use in order to be acceptable to senior management?
- Suggest ways in which a package can be made simple and intuitive to use.

Case Study

BT

BT has installed touchscreen kiosks using intranet technology to provide tourists around London with a range of services. These include maps, and information about sports events and entertainment. The service is extremely cost effective as BT only need to install one copy of the software and run it over the intranet. □

- What advantages are there for BT in using an intranet rather than standalone machines with individual software for this system?
- Why was a touch screen chosen as the input/output device?

diagnosis or decision and will also tell the user how it has reached a particular conclusion.

Expert systems are also called knowledge based systems and are able to store and manipulate knowledge so that they can help a user to solve a problem or make a decision.

Figure 13.5 BT information kiosk with touch screen

Cyber Bus Stops

(based on an article appearing in the Guardian August 1999)

Free standing computer terminals that can be used to gain information by bus passengers are appearing at bus stops in parts of London. The information available includes a channel for job-seekers.

The terminals are web-based, but don't yet offer full internet access. The information that is offered is free. The Jobfinder channel promises more than 500 vacancies that are updated weekly. All can be searched by inputting details such as job type and required salary. Detailed information can be printed out and taken away when the bus arrives.

Other information includes entertainment listings, details of local council services and transport news. There is a slot for credit cards to allow users to book theatre tickets directly – the receipt can be printed out.

- The kiosks are also able to deliver sound and video. What services, other than those described above, could the kiosks provide?
- Describe a suitable user interface for such a system.

Peter Symonds' College

Peter Symonds' College is a large sixth form college in Hampshire. Alistair McNaught set up the first departmental intranet in the college for the use of geography students. The intranet uses Netscape browser software and provides students with access to a wealth of material developed in house to support their independent study. This includes sections of questions and model answers and details of the course structure as well as topic information. Hypertext links are provided to suitable internet sites which have been carefully selected by the geography staff for their relevance and readability. Care has been taken to make the interface attractive and easy to use. ☐

- Define the term **browser software**
- What advantages would the use of the intranet described above bring to students
- Describe the work involved in keeping such a resource up to date.

Figure 13.6 Geography Intranet

An expert system is limited to a specific area of expertise such as the causes of a car's failure to start. It is typically rule based although it is able to work with uncertain or incomplete data. An expert system does not replace a human but delivers advice for a course of action based on the data given. This advice mimics the advice that the actual expert or experts who provided the knowledge for the system would have made. An expert system is able to explain its reasoning to the user.

An expert system is made up of a knowledge base that contains the facts and rules provided by a human expert; a means of applying that knowledge (usually called an 'inference engine') and a means of communicating with the user (the 'human-computer interface').

The knowledge base will store knowledge in different forms, namely FACTS and RULES. For example:

THE STANDARD CAR RENTAL IS £30 PER DAY	FACT
IF THE CAR IS A LUXURY MODEL, ADD £10 PER DAY	RULE
IF THE CAR IS HIRED FOR THREE DAYS DEDUCT 20 PER CENT	RULE
IF THE CAR IS HIRED FOR SEVEN DAYS DEDUCT 30 PER CENT	RULE
IF THE DRIVER HAS A LOYALTY CARD DEDUCT ANOTHER 10 PER CENT	RULE
THE SHOWROOM IS OPEN SEVEN DAYS A WEEK	FACT

Knowledge like this can be stored in a knowledge base, and the expert system should then be able to make deductions.

If we supply the information about when Mrs Miggins wishes to hire a car, the expert system should guide us through a series of relevant questions and work out the charge.

In practice, such as in a system which calculates the level of income support for an individual, there may be hundreds or thousands of facts and rules. When the program (the inference engine) runs, it does not simply start at the first rule and then run through them all in a fixed order. Instead it makes deductions as it goes along, finding out what else it needs to know before providing an answer.

Not all rules are Yes/No in type. In fact, in many systems the majority of rules are expressed as probabilities. For example:

If..........

    And..........

    And..........

Then there is a 0.4 probability that..........

■ USES OF EXPERT SYSTEMS

Expert systems are best suited to where a well-defined set of rules exists or can be written down. Expert systems have been used in such fields as medical diagnosis, fault diagnosis in a steel rolling mill, classifying, debugging, prediction, planning, geological prospecting, processing Social Security claims and quantity surveying.

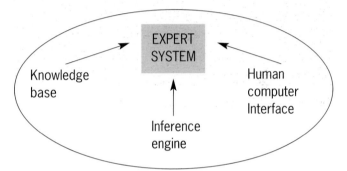

Expert systems allow expertise that is in short supply to be more widely shared. When particular experts are no longer available, their expertise is still available. An expert system that is based on the knowledge of a number of experts could provide better advice than an individual expert. The demand for human experts might decrease. For example, an expert system that dealt with all the legal aspects of house conveyancing might make it unnecessary for house buyers to employ a solicitor.

There are a number of dangers associated with the use of expert systems. If the knowledge engineer does not do an adequate job, the advice given by the expert system could be inaccurate and inappropriate. If the advice given by the system is used without any interpretation, it could be wrong.

A number of experts systems have been developed in the field of law to enable the user to pose legal problems or check the legality of a possible decision or action. The knowledge base of legal facts and rules from Acts of Parliament and case law is built up by consulting a number of law practitioners.

Developing an expert system, by prototyping

The person who carries out the tasks of defining a problem, constructing a suitable system, acquiring the knowledge to go into the system and developing the system is called a knowledge engineer. His job is similar to that of a systems analyst. Acquiring the rules and facts from an expert is not a simple task.

Many 'rules' that an expert uses to make decisions are very complex and not always consciously acknowledged. There are always many exceptions or special cases that need to be included.

Expert systems are nearly always built using prototyping, whereby a trial model is developed so that users can try it out and evaluate it.

Judges with records

Computers are being used to help judges hand out sentences. Richard Colbey of the *Guardian* argues the case.

Scottish judges are about to adopt a more scientific – and hopefully more consistent – approach to sentencing, thanks to a computerised system being tried out in the Scottish High Court.

Traditional wisdom is that judges have a 'feel' for sentencing that is more than a simple weighing up of the circumstances of the case, and that they then apply their experience to produce a result that is neither oppressive nor unduly lenient.

The trouble is, there is little consensus about what constitutes oppressiveness and leniency. Different judges hand out different sentences and even the same judge can be inconsistent. Judges are vulnerable to the same irrational misjudgments as anyone else. One who has a row with their spouse at breakfast could end up adding a year or two to a sentence they pass on a serious criminal later that day, subconsciously influenced by the bad mood that row left them in.

It is difficult for judges to resolve this without access to sufficient precedents when passing sentence. Law reports contain about only 300 cases a year and are not comprehensive enough to cover every situation.

The new system contains a database of 6000 sentences from the court's archives, which judges will be able to access from laptop computers and modems. The judge enters the relevant factors of a case, and the software list sentences passed for similar offences. These will be displayed in the order of proximity to the present matter, in the way web searchers do.

When passing a sentence for burglary, for instance, the judge will specify whether the premises were residential, whether the offence took place at night, whether the offender had any previous criminal record, and if so whether it included burglary, the amount of property taken and damage done and whether they pleaded guilty. The judge could also enter special features such as whether the burglar carried a weapon or deliberately soiled the premises.

- Discuss whether a judge should always use the sentence suggested by the database.
- What are the advantages of using such a system? (Include advantages not mentioned in this article.)

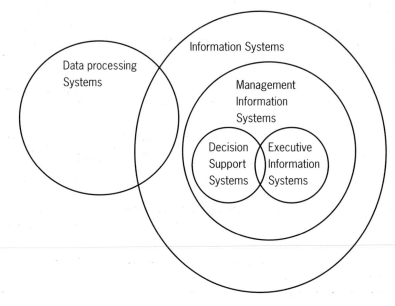

Figure 13.7 Diagram showing types of system

Success or failure of MIS

In spite of many technical advances and the investment of huge amounts of money, time and effort, many management information systems have not fulfilled their promise and have failed to provide the management with the information that they need. Just as with any information processing system, careful planning is needed if it is to be successful.

It is vital that the management is involved with the design of the system. They are to be the users, and it is therefore crucial that any system meets their real needs. However, for this involvement to be useful and realistic, the management will need to have an up to date knowledge of current information technology systems and their capabilities. They will need to be able to make informed decisions, and not be blinded by the IT experts' knowledge and use of jargon. Some form of training may be necessary. An inadequate knowledge of the capabilities of current technology may result in management making inappropriate or excessive demands from the system. When the system inevitably fails to meet their high expectations they will be disillusioned and may well not then use the system. It is important to understand that IT does not always bring benefits.

It is equally important that the IT experts spend adequate time getting to know the information needs of the managers. This can only be achieved once a thorough understanding of the organisation has been achieved and a detailed analysis made of the system. Systems development should be the result of a true partnership between management and IT professionals. Managers do not always know what information they need and IT specialists do not always know enough about the process of management and the business in question.

Many management information systems have failed to provide adequate information as too much emphasis has been placed on the lower level data processing applications. A poor system will be produced if there is too much emphasis on the computer system and inadequate attention given to the whole system and the data flow throughout the organisation. The system should be designed around the information needs of the managers rather than be based upon what the computer can easily produce.

Inadequate teamwork and a lack of professional standards can lead to unmet deadlines and a system that does not function as was intended.

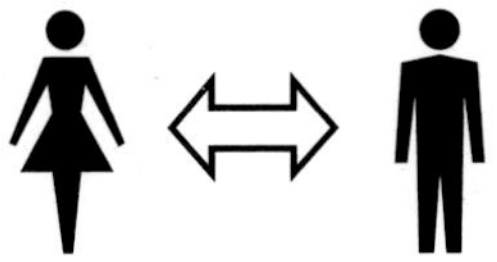

IT SPECIALISTS		MANAGEMENT
Have a real, detailed knowledge of the business and the systems involved. A system should be designed around an organisation's needs, not just what works easily on a computer!	**Work together as a team**	Acquire a basic knowledge of IT systems, especially as they relate to the business. If necessary undergo training. Do not 'leave it all to the computer experts'
Carry out careful analysis of current system and make sure that management requirements are fully understood	**Establish regular meetings**	Be involved in all stages of design. Assess prototypes and feedback comments
Consider the whole system, not just the parts that require computer use. Consider the complete information flow	**Agree and establish professional standards that should be maintained throughout the project**	Have an awareness of what can be done using IT at the moment and what cannot be done. Ensure that demands from new system are technically realistic
Look at the ways in which the new system will interact with other systems within the organisation	**Use agreed project management techniques**	Do not ask for too much and resist the temptation to keep changing your mind and demanding extras and alterations to the system
Consider information needs alongside day to day data processing needs	**Make sure to share a common goal**	Be realistic

Figure 13.8 Ensuring a successful MIS – the role of IT Specialists and management

Passport to Nowhere

In June 1999, 'teething problems' with a new computer system at Britain's Passport Agency led to a backlog of over half a million would-be holiday-makers waiting for their passports.

The problem was made worse by changes in the regulations requiring all children to have their own passport and by a twenty per cent increase in applications for passports. Offices were taking nearly 40 working days to process an application, compared to a target of ten days.

Queues formed outside passport offices. In Glasgow the first people started queuing one night at midnight. By 9.30 a.m. over 1000 people were waiting in the rain.

The new system, installed by the German company Siemens at a cost of £230 million, was needed because the old system was not Year 2000 compliant. Siemens said: 'It is misleading to suggest that the delays experienced by the public are primarily caused by failures in IT systems. It is clear that the application demand has exceeded Home Office forecasts.'

The new computer systems were installed first at offices in Liverpool and Newport. This was where the biggest backlogs occurred. Reports suggested that the need to install the whole system before the end of 1999, meant that the new system had not been fully tested.

So what went wrong?

Many reasons were given for the Passport Agency's difficulties in issuing passports.

1 The new ruling that children had to have their own passports undoubtedly made the problem worse. This obviously meant there would be an increase in the number of passports required and the number of passport applications. Was this taken into account in staffing levels and in the original hardware specification?

2 The new computer system went on-line at the start of the summer – the Passport Agency's busiest time.

3 The rush to introduce the complete new system meant that full testing and training had not been carried out.

4 The new system was piloted at two offices. It should have undergone thorough testing at one site with real data before going live.

• What lessons can be learned from this case study?

Summary

- A data processing system carries out the day to day operational activities of an organisation.

- An information system provides information for the user that can be used in decision making.

- A Management Information System (MIS) provides information in appropriate forms for managers. It converts data from internal and external sources into information. This is communicated in an appropriate form to managers at different levels to enable them to make effective decisions.

- Decisions can take place at different levels within an organisation: strategic, tactical and operational. The level of detail, form and type of information needed is different at each management level.

- Decision Support Systems (DSS) are software tools that enable managers to make 'on-off' decisions. They can provide modelling or analytical tools.

- Executive Information Systems (EIS) are information systems that provide strategic managers with access via a user friendly interface to huge amounts of data held on internal and external databases. They provide selected, summarised data, often in the form of exception reports. The use of 'hot spots' allows 'drilling down' through levels of data to gain more detailed information.

- Expert systems are software programs that are designed to behave in the same way as human experts. They have been used mostly in the field of diagnosis.

- Not all MIS systems are implemented successfully. Factors influencing success or failure include:
  - inadequate analysis
  - lack of management involvement in design
  - emphasis on computer system
  - concentration on low-level data processing
  - lack of management knowledge of IT systems and their capabilities
  - inappropriate/excessive management demands
  - lack of team work
  - lack of professional standards

Information systems questions

1 Using an example, not given in the text, explain the difference between data processing and information processing. (4)

2 Distinguish between the terms MIS, DSS, EIS and Expert systems. (4)

3 Describe the types of packages that could form decision support systems. (4)

4 Patrick Maddock is the managing director of a medium sized publishing company. He feels that it is time that an MIS were developed. However he has read horror stories about other businesses that had attempted to install such systems with disastrous results. He calls in a consultant for advice.

Write a report from the consultant, advising Patrick on what he needs to do to ensure that an MIS system developed for his business will be successful. (8)

5 In your school or college, identify three levels of management: strategic, tactical and operation.

For each level, give two decisions that would need to be made and for each decision, list the information that would be required. (9)

6 'Beautiful Buttons' is a major producer of buttons. It has four factories located in different parts of the country. The **managing director** has overall executive control of the business. Amongst those employed by Beautiful Buttons is a **sales manager** who manages a team of **sales representatives** whose job is to sell the company's products to garment manufacturers. Each factory has a **manager** who is in charge of production locally. The production workers are organised by a **shift leader** who is responsible for day-to-day matters. At head office, there is a **personnel officer** who is responsible for recruitment, staff welfare and remuneration.

a) For each of the jobs shown in bold, state whether the level is strategic, tactical or operational.

b) For each of the jobs shown in bold, give two typical decisions that would need to be made. For each decision, give two pieces of information that would be needed. (12)

7 Describe how an EIS could help the managing director of 'Beautiful Buttons' make better informed decisions. (6)

8 The manager of a cinema is facing a dilemma whether to raise admission prices by a small amount to ensure total income continues just to cover running expenses or to risk a much larger increase. Suggest a suitable *type of general purpose package* for the manager to use to study the implications of the options. Explain how the manager could use this package to assist the decision making process and the setting of admission prices. (4) *AEB Computing Specimen Paper 2*

9 Organisations often use Management Information Systems.

a) What is the purpose of a Management Information System? (1)

b) Why is such a system required by managers of an organisation? (1)

c) Give **one** example of the use of a Management Information System within an organisation, clearly stating its purpose. (2) *NEAB 1997 Paper 3*

10 With the aid of appropriate examples, explain the difference between formal and informal information flows. (6)

a) What is meant by a Management Information System? (4)

b) State **four** factors which could contribute to the success or failure of a Management Information System. (4) *NEAB 1998 Paper 4*

11 The manager of a local company complains that the company's information system continually fails to provide the correct level of information. State four possible reasons why the system is failing. (4) *NEAB IT04 Paper 4*

12 a) What is meant by the term 'expert system'? (2)

b) Describe one example of the use of an expert system. (2) *NEAB IT04 1999*

13 Information systems are capable of producing strategic and operational level information. With the aid of examples, explain the difference between these two levels of information, clearly stating the level of personnel involved in using each one. (6) *NEAB IT04 1999*

Activity

The first category of systems to be computerised are now called back office systems. These are systems such as payroll and accounting that support the business without impinging directly on the customer.

Next to be developed were the front office systems which deal with such operations as customer orders and other transactions.

Now a third type of system is being developed where companies use the Internet to allow the customers themselves to initiate the transaction. Dell Computers has developed an electronic ordering system selling directly to end-users over the Internet. For each of the types of organisation listed below

1 Suggest likely applications for
   i) back office systems
   ii) front office systems

2 Explore the possibilities for Internet use by:

   • a restaurant

   • a high street bank

   • a supermarket

   • an electricity provider

The life cycle of an information system

The process of introducing a new information system is called the 'system life cycle'. The old system may be manual or computer-based. As the term cycle implies, producing a new information system is not a one-off exercise involving a few months of activity. A system, once developed, will need maintenance and eventually will be seen as inadequate to meet the users needs so a new system will then need to be developed, and so on.

An information system might need replacing for a number of reasons. The technology used might have become outdated. This might not matter: it would be impossible and inappropriate for every organisation to change systems whenever new technology is developed. In many cases the 'old' technology and system are perfectly adequate for the user needs. However, as advances in technology are made,

organisations could become non-competitive if they continued with their perfectly functioning old system, or new technology may offer new facilities. For example, British Telecom billing systems now provide itemised details of calls, increasing the amount of information available to customers. As new telecommunication companies start operating, British Telecom needs to provide more information to compete.

The organisation of a business may change bringing a need for a new information system to service the information requirements of the new structure. This might happen for a number of reasons: the organisation might be expanding, restructuring, merging with, or taking over, another company or diversifying into new areas of activity.

Figure 14.1 An interview

What was acceptable to the user when a system was installed might no longer be so a few years later. Users become increasingly sophisticated in their understanding of IT and consequently more demanding in how they expect a system to perform.

Stages of the life cycle

The life cycle can be broken down into a number of separate stages as follows:

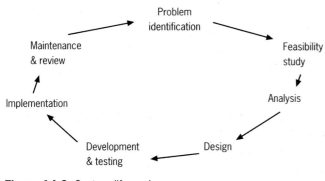

Figure 14.2 System life cycle

Problem identification

A new project is put into action when an improved information system is seen to be desirable. The reasons could be any one or a combination of those described above. An individual or team would be brought together to plan and carry out the next stages.

Feasibility study

The purpose of the feasibility study is to investigate the project in sufficient depth to be able to provide information that either shows that the development of the new system is cost-effective or shows why the project should not continue. There are three types of feasibility to consider: technical, economic and social.

If the project is not technically feasible it is not possible. It may be that a particular system would require response times that are not achievable with the current equipment available. Systems dependent upon voice input using natural speech are not yet

fully feasible. A system that is not feasible today might well be feasible in a few years' time.

A system is only economically feasible if the benefits outweigh the costs. It is not a straightforward matter to estimate costs and benefits accurately. Certain costs are obvious. These include hardware and software costs, personnel costs to develop the new system, the cost of transferring data from current files, the cost of training users in the new system and installation costs. Other costs are more hidden such as the change or even loss of some jobs, environmental effects, deskilling and loss of job satisfaction for some employees.

Some benefits are obvious: reduced running costs, faster turnaround times, reduction in staffing costs. Others that are harder to quantify include increased level of service and customer satisfaction, keeping ahead or abreast of competition, company prestige. Social costs, such as the loss of jobs or reduction in job satisfaction, should not be ignored.

All these costs and benefits have to be explored in a cost/benefit analysis.

The findings of the feasibility study are presented to management in the form of a report, including costs, benefits, alternatives and appropriate recommendations. The decision to proceed will be made at a high level, usually the board of directors. Time scales for the project will be established and agreed. It is vital that these are clear and achievable as many such projects have foundered and run months or even years over schedule as unrealistic time scales were set at the start.

When planning a project, all tasks and use of resources (human and other) have to be carefully scheduled. Certain tasks will need to be completed before others can be started, and the use of certain resources have to be carefully dovetailed. Project planning and monitoring techniques such as the use of Programme Evaluation Review Technique (PERT) and Gantt charts would be used. (See also Project Management software in Chapter 21.)

Gantt chart

A Gantt chart is a form of bar chart depicting the timings of different tasks. It can be used to monitor progress throughout the project by showing which tasks have been completed at a certain date.

Tasks	Time in weeks	1	2	3	4	5	6	7	8	9	10	11	12
Define the problem		■											
Feasibility study			■										
Gather data				■	■								
Analyse data and develop logical design						■	■	■	■	■	■	■	
Write system analysis report													■

Figure 14.3 A Gantt chart

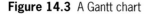

PERT

PERT can be used to show the dependence of tasks on others. The time taken to carry out each task is included as well as the relationship between one task and another. This can all be represented diagrammatically as shown.

There are a number of routes through the tasks from start to finish of the project. One route is called the critical path. If there is a delay in this route, then the completion of the whole project will be delayed. Other routes through have the capacity to absorb some delays.

Analysis

Once a project has been formally agreed, a full analysis needs to be undertaken. At this stage, it is important that the systems analysts find out in detail how the current system performs. They need to find out the detailed requirements of the users of the new system. This is achieved using a number of formal methods.

Structured interviews with the present staff will always play an important part of this investigation process. These provide a flexible means of gathering information, but are time consuming.

The use of **questionnaires** is occasionally appropriate when many users need to be asked the

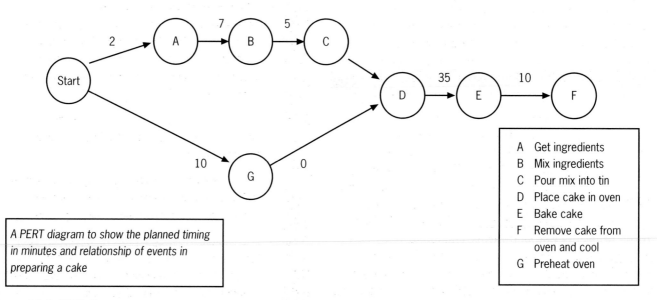

A PERT diagram to show the planned timing in minutes and relationship of events in preparing a cake

A Get ingredients
B Mix ingredients
C Pour mix into tin
D Place cake in oven
E Bake cake
F Remove cake from oven and cool
G Preheat oven

Figure 14.4 PERT diagram

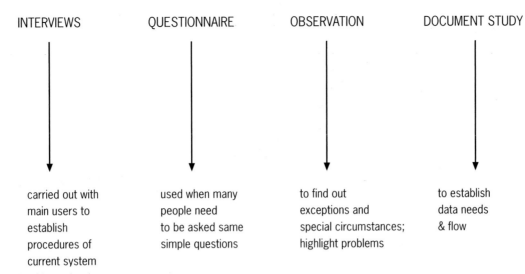

Figure 14.5 Methods of investigation

same questions. It is important that questionnaires are prepared with care and that questions produce mainly quantitative rather than qualitative data.

The analysis must include the **study of all forms and documents** to find out what information is required by the present system.

Direct observation of a system in action can provide insights that cannot be found any other way. Often exceptions to normal methods may occur which are not mentioned at an interview. For example, a small business uses a manual system for ordering stationery from central stores. Employees requiring stationery fill in a form which they take to the clerk in charge of the stores. Every few days, she goes to the stores and removes the items that have been ordered and takes them to the employees who requested them. She then enters the details of stationery allocated to different departments in a spreadsheet so that costs can be allocated appropriately. All this information is gathered by the analyst through interviewing the clerk and examining the Stationery Request Form and the spreadsheet.

However, when observing the clerk during a morning's work, the analyst discovers two exceptions to the normal processing. The first comes to light when the clerk receives an urgent phone call from a senior manager who requires five special folders immediately for a client. The clerk notes down the request on a memo pad and then collects and delivers the folders immediately. The second exception is observed when the clerk is collecting the stationery to fulfill normal orders. She finds that she has insufficient rolls of sticky tape to meet the requests. She highlights one order and writes a memo to the employee who made the request.

These exceptions had not come to light until the analyst actually observed the system in action.

The systems analyst will use formal graphical and tabular methods to represent the current system. Data flow diagrams may be drawn up as well as a number of charts showing which tables or files data items are used in.

The systems analysts use the information that they have found to produce a requirements specification for the new system.

Systems analysis Case Study

A new system at Nissan

Paperwork at Nissan's Sunderland plant had reached such proportions that labour costs were excessive and mailing costs high.

- A **feasibility study** suggested that some of this paperwork could be carried out by computer using EDI (Electronic Data Interchange). This means sending documents like orders and invoices to suppliers electronically via the telephone network, rather than by post.

- The **analysis** involved looking at the old manual system to see how it was carried out and what documents were involved.

- A new system was **design**ed cutting out the printed mailings. This saved time in communicating and removed the new for re-keying, reducing errors.

- After thorough testing, a **phased implementation** took place, originally with only a few suppliers. The system was then extended to more suppliers, cutting mail to suppliers by over 90 per cent.

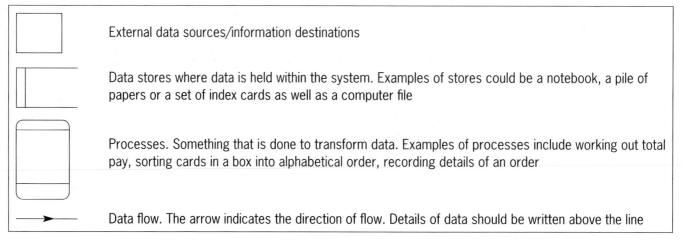

	External data sources/information destinations
	Data stores where data is held within the system. Examples of stores could be a notebook, a pile of papers or a set of index cards as well as a computer file
	Processes. Something that is done to transform data. Examples of processes include working out total pay, sorting cards in a box into alphabetical order, recording details of an order
⟶	Data flow. The arrow indicates the direction of flow. Details of data should be written above the line

Figure 14.6 Data flow diagam symbols

Data flow diagrams

A data flow diagram (DFD) is a useful tool for a systems analyst that helps him to clarify in his mind the flow of data around a system.

The four elements that make up a data flow diagram are shown in Figure 14.6

Consider the following application:

Mr Jollifant owns a children's party entertainment business and employs a number of clowns, conjurers etc on a part-time basis. All bookings are kept in a large ledger type book. Mrs Jollifant writes details of bookings made over the phone into the book. Every month, Mrs Jollifant produces a word processed list of bookings for the next month for each entertainer. At the end of the year she goes through the books, and produces a table similar to the one shown below.

JOLLIFANT JOLLY PARTIES SUMMARY OF EVENTS 1998

Entertainer	No. of Events	Total Hours	Total Pay
Cleo the Clown	5	13	319.50
Magic Marcus	2	6	117.83
Jumping Jimmy	24	65	1926.21

Drawing up the data flow diagram

First, identify the sources of data and the destinations of the information (input and output). It is necessary to decide at this stage the limits of the system. In this case, the system under consideration is the work done by Mrs Jollifant as described in the text.

◀ Booking details are input from clients;

◀ Monthly schedules are produced (output) for each entertainer;

◀ An annual statistical summary is produced for Mr. Jollifant.

Now consider the data stores in the system.

◀ The data concerning bookings is stored in the bookings ledger by Mrs. Jollifant.

Finally, the processes carried out by Mrs. Jollifant must be identified. She

◀ writes all the bookings down in the ledger;

◀ produces monthly word processed schedules;

◀ works out statistics at the end of the year.

Now put these all together in a diagram. An example is shown in Figure 14.7.

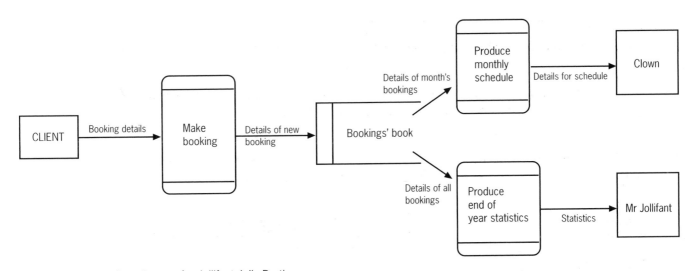

Figure 14.7 Data flow diagram for Jollifant Jolly Parties

Design

In this phase the details of the new system are drawn up.

The design stage determines how the requirements specification will be implemented. It involves designing the database, input formats, output formats, validation checks and the test plan. A system specification is drawn up in sufficient detail for the programmers to implement the system.

Prototyping is a technique in common use in systems design today. A prototype system is a working model of part of the system. This prototype is shown to the user who can then suggest additions and changes. It allows the analyst to be sure that she has fully understood the requirements and the user to have full involvement in the design. For example, a prototype data entry screen can be designed, with appropriate field and validation checks. The user can enter data, and make suggestions about layout and method of use. The use of prototyping develops a greater partnership between users and specialist staff.

The new system must be designed to cope with the demands of peak-periods. It must perform with adequate reliability and speed. Appropriate methods of processing must be chosen. The design of a new system will involve detailed consideration of data capture methods and input format, outputs and their formats, files or database design.

Consideration of input includes data capture methods, validation, input media, volumes of input and design of input documents. In designing forms, it is important to keep the design as simple and clear as possible to minimise the possibility of errors. Important considerations in form design are size, type of paper, space for information to be entered, column width, clarity and use of multi-part sets.

The output requirements of the system such as screen layouts and print-outs must be decided, including when to use hard copy and when to use screen displays. The analyst will need to consider the form, use, type, volume and frequency of output.

The design of files and database are linked to the input and output. Considerations involved in designing files are the layout of records, including primary and secondary keys, the storage media to be used, the method of file organisation and access, and the methods of ensuring file security.

Automated tools for systems analysis and design have been developed which speed up many of the repetitive tasks of systems analysis. Such software is referred to under the general name of computer-aided systems engineering (CASE) tools. These can include tools that allow for the rapid development of prototypes of user input screens, forms and output reports as well as charting tools which enable analysts to develop and modify diagrams such as data flow and flow charts quickly and easily.

Development and testing

This stage involves the development of programs and/or customisation of software packages. This could be be achieved either by writing the programs

STORAGE	HCI	INPUT
• Decide on tables needed (e.g. Client, Artist, Booking) – ensure linking fields are there. Draw E-R diagram • Select attributes (fields) • Identify primary key for each table • Choose appropriate data type for field	The 'look and feel' of the system • Menu system or icon buttons? • Colours/layout • Titles, position of date etc. • Size of print	• Design data entry screen for bookings, new clients and new artists – consider layout very carefully • Determine default values • Validation checks for each field • Is there any need for a paper form?

SECURITY		OUTPUT
• Work out back up procedures • Is a password system necessary? • Is it appropriate to have 2 modes, one for set-up and one for normal running?	**DESIGN POINTS**	• Design layouts for printed reports: end of year statistics; list of forthcoming events for each performer • Headings, date etc. page numbering • Totals required

DETAILED SPECIFICATION	TESTING
• Work out all processes that are required and how they are to be carried out • Design queries • Design macros • Design end of year archiving procedures • Design deletion file procedures to ensure that details of deleted or amended records are not lost	• Produce table of tests with data to be entered and expected outputs • Obtain real data for last 3 months to enter into system • Use extreme data to test limits of range checks • Use invalid data to check all validation checks

Figure 14.8 Design considerations for Mr Jollifant's party system

in-house, commissioning a software house to write the program or purchasing a suitable program. It is also necessary to acquire any hardware that may be required for the system. Programs are coded, tested and documented. In all but the smallest projects, a team of programmers will be involved. It is vital that the work is monitored very carefully, and that time limits are adhered to.

Testing is a crucial part of program development. Test data should test that all branches of the program perform to specification. Data should be used to test extreme cases. For example, if a temperature value can be any number in the range 0 to 25, the values 0 and 25 would be the extreme values. Testing should also include invalid data to ensure that it is rejected. (Temperature values of 26, —4 and 56 would be invalid.) When testing, it is crucial that the results produced by the program are compared with expected results. Any discrepancies should be investigated.

Implementation

The implementation takes place once all the programming is complete and follows a complete systems test. It is possible to test a system before full implementation by using historical data. The output from the new system is compared with that produced by the old system.

Hardware must also be installed and thoroughly tested. Installation may require extensive cabling and alteration of buildings. Extra security methods may need to be put in place. Files will need to be converted into a form suitable for the new system.

It is crucial that all users of the new system should be trained. They will need to believe in the new system and have confidence in their ability to use it. Such training needs to be hands-on and may take considerable time. If the users have been involved in the design of the new system, the possibility of resentment or fear of the new system is likely to be reduced.

Implementation can take considerable time and requires careful planning.

There are four commonly used methods of installation: parallel running, direct conversion, phased changeover and pilot running. The choice of the appropriate method of implementation depends on a number of factors that will include the nature of the system and the type and size of the organisation.

Parallel running

With parallel running, both old and new systems are run together for a certain time. When the new system is established and running smoothly, the old system can be dropped. This is a safe method of conversion as a back-up to the new system is available in case problems arise. If any bugs arise in the new system,

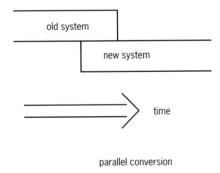

parallel conversion

Figure 14.9 Parallel conversion

they can be corrected. Users can learn the new system at their own pace, without the fear that making mistakes would be disastrous. However, parallel running can be very costly as extra time will be required to run both systems concurrently. It can also lead to confusion.

Direct changeover

With a direct changeover, an existing system is replaced by the new system at a certain time and date. From then on, the old system is no longer used.

A direct changeover costs less than any other method of conversion, as at any time only one system is operational, so no time is wasted. However, there is no back-up if things go wrong so there is a much higher risk than with parallel running. Staff will

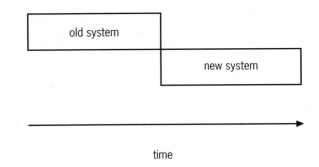

time

Figure 14.10 Direct conversion

need to have received a high level of training prior to changeover so that they feel confident in using the new system. For some systems, direct changeover is the only possible method.

Phased conversion

Phased conversion allows the changeover to occur in stages. This is only appropriate for systems that can be broken down into separate sections that can be developed one after the other. In most systems this is not possible, as different sections of a system all interrelate.

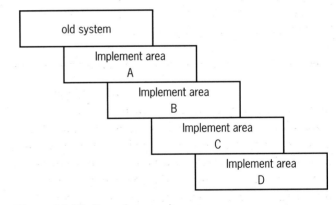

Figure 14.11 Phased conversion

Pilot conversion

The final method of conversion uses a pilot. This is where the system is implemented in one department or location in advance of the whole organisation. Bugs and running problems can be cleared up and users' reactions to the system taken into account before the whole system is implemented. Training can also be modified in the light of experience. Such a method is only possible in organisations that have discrete sections or branches. It may prove necessary for the pilot sites to run the old system in parallel.

Maintenance and review

Once a system is in full operation it is monitored to check that it has met the objectives set out in the original specification. Inevitably, changes will need to be made to the systems. These changes are known as **systems maintenance**. Systems maintenance can take three forms: corrective; perfective and adaptive.

- Corrective maintenance corrects errors and bugs found in the system.

- Perfective maintenance involves adding extra features and functions to the system. These arise when the users use the system and become aware of changes that would enhance the performance of the system.

- Adaptive maintenance makes changes to the system which result from outside modifications to the original requirements. These could be made necessary by changes in tax legislation or business procedures.

(See Chapter 21.)

Most programming hours are spent on maintaining existing systems rather than in producing new systems.

Eventually, the system will need to be examined to see if it can cope with the current requirements. At some stage further adaptive maintenance will be impractical and the system life cycle will be repeated again.

An electronic registration system

A sixth form college in the south of England decided to install an electronic registration system for class attendance. Each teacher has an A4 wallet which holds a specialised computer device. Each device is battery run and linked to the central computer by radio, a receiver/transmitter being mounted for each cluster of classrooms.

The new method of registration was chosen to replace the current, paper based system. Transmitters were installed and wallets configured during the Easter holidays. For the first half of the summer term, the system was piloted by the Biology and English departments. Some bugs and operational problems were sorted out during this pilot run. The rest of the teachers were introduced to the system and trained in its use.

After the half term break, the system went live throughout the college. However, to ensure that backup would be available in the case of failure, the paper based system was continued in parallel. Gradually the staff became more confident users and fewer mistakes were made and less support needed. The following September the old paper based system was abandoned.

1 Explore the effects that computer failure of the EARS system could have on the college attendance system.
2 Describe three other electronic means of collecting attendance data. Explain the advantages and disadvantages of each method.
3 Examine the aspects that would have been considered by the college when undertaking a feasibility study. □

Figure 14.12 EARS (Electronic Attendance Registration System) wallet in use in class

Summary

The system life cycle is the series of stages involved in replacing an old system with a new one.

The stages are:

☐ Problem identification

☐ Feasibility study

- A preliminary investigation to find out whether the required new system is technically and
- economically possible.

☐ Analysis

An investigation into the current system to find out how it works and what is required from a new system.

Techniques used include:
- interview
- questionnaire
- observation
- detailed study of documents

☐ Design

All the elements of the new system are planned:
- inputs
- outputs
- data storage
- HCI
- procedures
- testing

Prototyping is used – a model of part of the system is shared with the user for comments before the full system is built.

☐ Developing and testing

The application is built using an appropriate programming language or software development tool. It is thoroughly tested, using a realistic volume of real data as well as extreme and invalid data.

☐ Implementation

The current system is replaced with the new system. New files have to be created, hardware set up, users trained. Implementation can be:
- parallel
- direct
- phases
- piloted

☐ Maintenance and Review

Maintenance involves modifications being made to a system. These changes fall into three categories:
- perfective
- adaptive
- corrective

Life cycle of an information system questions

1 Explain the purpose of a feasibility study. Include in your answer the areas that should be considered. (5)

2 Draw a diagram to illustrate the main phases of the traditional system life-cycle. (5)

3 Describe the advantages of using prototyping in systems design. (6)

4 The management of a company wishes to introduce a computerised diary and scheduling package that is known to be compatible with the existing software base. With the aid of examples, give **three** factors that could influence the success or failure of this exercise. (6) *NEAB 1997 Paper 3*

5 Give, with justification, a suitable method of changeover for the following:

   a) a new electronic data entry system for a chain of retail clothing shops

   b) a computer controlled traffic light management system

   c) a computerised loans system for a college library to replace the current card based manual system

   d) an on line seat booking system for a cinema, allowing seats to be booked from a number of outlets.

   e) an electronic time keeping system in a manufacturing company with ten factories throughout the UK.

   f) a networked shared diary system for a large business. (6)

6 Describe three types of maintenance that might be madee to a system. Illustrate your answer with examples relating to:

   a) a payroll system

   b) a stock control system (9)

7 A systems analyst has been employed to produce a computer based system to replace the current manual one in a lending library. Describe three methods of investigating the current system; for each method explain what information the analyst would expect to gather. (9)

Activity

POTENTIAL CAUSES OF FAILURE WHEN IMPLEMENTING AN INFORMATION SYSTEM

Several stages of the system life cycle are shown in the diagram below. At each stage, a possible, unsuccessful outcome has been described. Taking each stage separately, describe a) what could have gone wrong and b) what measures should have been taken to avoid the mistakes made. Try to explore a wide range of issues at each stage.

Activity

Christine works in a large sixth form college, and part of her job is to administer the distribution of stationary to teaching staff. She takes in orders from teachers, provides them with the requested goods, (if in stock), and at the end of each month bills each department. She currently uses a manual system to administer the job, but is certain that there is an IT based system that would provide a better solution.

You have been asked to produce a suitable system. Your first job is to find out details of how the current system works and Christine's requirements from a new system.

List out the questions, together with the purpose of each, that you would ask Christine. *Remember that you questions should aim to find out how the current system works, as well as exploring what Christine would require from a new system.*

ANALYSIS
A newly trained systems analyst spends two days in an organisation where he investigates the internal supplies system. He then produces a report that outlines the objectives for a new, computerised system. The management feel that the objectives are totally inappropriate.

DESIGN
IT specialists come up with an exciting design for an ordering and billing system for a chain of restaurants. They think that it is a great improvement on the current system and meets all the objectives. The presentation of their design to management and staff is met with a frosty silence.

IMPLEMENTATION
A new computerised system has replaced the manual system for managing newspaper deliveries in a chain of newsagents. Unfortunately, the staff in one shop do not seem to know what is going on or what they are supposed to. Papers are delivered late to the wrong houses.

DEVELOPMENT
A programmer leaves a team that is producing a new personnel system for a building society half way through completing a particular program. A new programmer taking over the work cannot understand what has been written, so starts again. Programming overruns by three months.

TESTING
Once a new system for a packaging company goes live, it continually crashes and several serious errors are found every day. Testing was clearly inadequate.

Activity

HAMPTON TOY LIBRARY MANAGEMENT SYSTEM

The toy library has recently been established in the little town of Hampton. It is run by Val Gordon and has proved to be very popular with parents of young families. A variety of toys are held, from expensive items such as Fischer-Price airports to smaller items such as jigsaws, games and books. Helpers attend the library to discuss with parents how the play items can be used.

Families can borrow items for one week. Currently approximately 25 families use the library and the toy stock stands at 153 items. Both membership and toy stock are growing rapidly and it is envisaged that there will eventually be 100 members and a pool of 1000 toys.

Details of the toys are kept in a notebook. The date of acquisition, description and cost of each is written down. Two card indexes are maintained for members. Every member has a card that contains details of their name, address, phone number together with their children's names and dates of birth.

When a toy is borrowed the member's card is retrieved from the box. Details of the toy and date are then added to the card that is then stored in the 'loans' box. Both boxes are maintained in alphabetic order. When a toy is returned the member's card is retrieved from the 'loans' box and placed in the other box.

When a card becomes full up a new one is created. At the end of the year, the organiser of the library goes through all the cards and removes the cards of any members who have not borrowed any toys and writes a letter to all these members.

1. a) list all the inputs to the system
    b) list all the outputs from the system
    c) identify the data stores involved in the system
    d) identify all the processes
    e) Draw a data flow diagram to represent the current system

2. What problems can you identify with the current system that limits its use?

3. You have decided to organise an eighteenth birthday party in 3 months' time. List all the tasks that need to be undertaken. Draw out a Gantt chart for your plan.

    Produce a Gantt chart for your major Project.

Data, information and information flow

Information

Managers need information that is relevant to the task in hand. It must help them to plan, to control and to make decisions. This relevant information should increase their knowledge of the situation and reduce uncertainty.

All managers need information. However, the information that they need, and the form that it should be in, will be different in different situations. The type of information needed depends on a number of factors, including the level of the manager, the actual task being carried out and its urgency.

Information may be classified in many ways including:

By Source:	for example internal, external, primary, secondary
By Nature:	for example quantitative, qualitative, formal, informal
By Level:	for example strategic, tactical, operational
By Time:	for example historical, present, future
By Frequency:	for example continuous (real time), hourly, daily, monthly, annually
By Use:	for example planning, control, decision making
By Form:	for example written, aural, visual, sensory
By Type:	for example detailed, sampled, aggregated

Figure 15.1 Classifying information

Sources of information

A **source document** is the original document bringing information into an organisation. When an electricity company calculates customers' electricity bills, the source document is the meter reading card storing details of the customer's meter reading. This is an **internal source of information**.

When a mail order company receives a written order from a customer, this is an **external source of information**.

Both these examples are filled in directly, they are **primary sources of information**. Sometimes the information has to be transferred on to another piece of paper or on to a computer file. This is a **secondary source of information**.

■ EXTERNAL SOURCES OF INFORMATION

External sources of information are those which are outside an organisation. Business people may use external surveys and annual reports from other organisations, as well as statistics and research reports. There are many information sources available: newspapers, magazines, radio, television, teletext, local authority departments and government agencies.

Many local authorities have advisory services which can provide help to people who want to set up small businesses. Any organisation which wishes to extend its buildings must consult the local Planning Office before building. Information will be needed from the Department of Environmental Health about issues such as noise levels and waste disposal.

The public library contains a wide selection of business and commercial directories which can be a useful source of information. Information can also be purchased from outside sources, such as research houses and public opinion organisations like Gallup.

Nature of information

■ QUANTITATIVE AND QUALITATIVE INFORMATION

Information can be of two general types; **quantitative** or **qualitative**. Quantitative

Figure 15.2 Forms of information

information is that which can be measured numerically. For example, 3476 cars were sold last week; Mike German's net salary was £1243.44 last month. Quantitative information can be presented in numerical or graphical form. A balance sheet or a graph showing sales trends both contain quantitative information.

Qualitative information cannot be measured in numerical terms. For example, the different colours that a model of car can be produced in is qualitative information. Qualitative information may have come from data based on value judgements.

INFORMAL AND FORMAL INFORMATION

Formal information is created and disseminated as part of the predetermined information systems of the organisation. It can consist of reports of all types provided by a MIS. Committee meeting minutes and agendas, and internal publications such as a company newsletter are also examples of formal information.

Informal information is very often communicated by face-to-face contact or through telephone conversations. It can also be conveyed in memos or even scribbled notes written on a piece of paper.

Level of information

Strategic information is that which is required at the highest levels to make decisions. A director or chief executive would use strategic information to decide on future policy, for example the decision of a supermarket to offer a banking service.

Tactical information is used by middle managers. For example, a sales manager may need information related to products and performance.

An example of **operational information** is that used by a foreman in a factory to decide on the day's production and staff rotas.

Timing of information

Management information systems attempt to ensure that managers can obtain accurate, relevant information at the right times, to improve their decision making. It should provide information on the past (**historic** information about production levels, and so on), the present (that is, **current** production figures, markets, and so on) and the **future** (that is, projected or forecasted profits, and so on).

An example of **historic** information could be last December's sales figures, needed for a toy shop just before Christmas to be used to plan for this Christmas's stock levels. Speed in production of the information is not important as it is not needed for nearly twelve months. Accuracy is important, as the shop needs to order the right goods in the right quantities.

Current information could be information from time sheets for this week's wages which have to be paid on Thursday. This needs to be accurate (or the workers will not be paid the right amount) and quick (or they may be paid late). Speed and accuracy are thus both important.

Future information could be sales figures and population details for an area, helping the store decide whether to open a new store or expand an existing store. These figures are for future planning so speed is not important. Population figures are constantly changing so total accuracy is impossible.

Frequency of information

The use to which information is put determines the frequency of its production.

Real time information is produced immediately transaction data is processed. When a shop assistant swipes a customer's credit card to check the credit ratings, he requires feedback at once before carrying on with the sale. When booking a flight at a travel agency, information on availability and cost are needed so that a decision can be made whether or not to book.

Other information is needed at regular intervals: perhaps hourly, daily or monthly. Percentage figures for student attendance in class can be obtained on a weekly basis from an electronic attendance registration system.

Uses of Information

Information helps managers in several ways.

Information is used by a manager to monitor and control the work done under his span of control. For example, a sales manager may use the monthly sales figures of different regions as the basis for making changes where individual performances are below target. In a factory producing plastic mouldings for car interiors, the foreman can make technical adjustments to the machinery on the basis of quality control information on faulty goods.

Information can be used for **planning**. For example, a marketing manager, planning the launch of a new product, would use historic sales information on other company products together with their projected sales. External information gathered from consumer surveys would also be valuable in deciding how to market the product, which market sectors it should be aimed at and what price to charge.

Decision making takes up a large part of the work of a manager. The appropriate information helps the manager make good decisions.

Form of information

Information can be presented in a variety of forms. Much business information is **written** – in the form of reports, memos or tables. However, for many

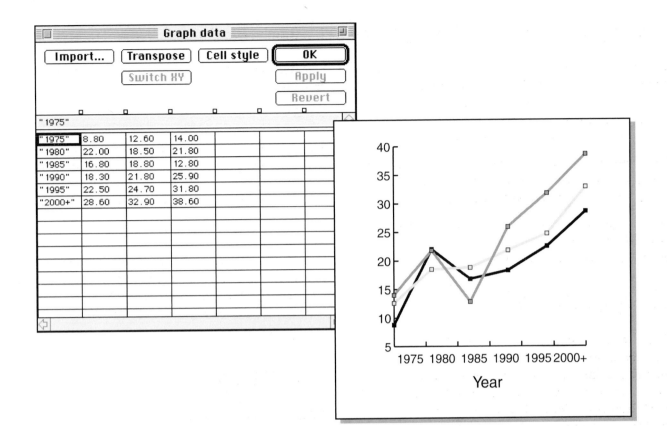

Figure 15.3 Different ways of showing information

purposes, a more **visual** representation can convey the same information with more impact; charts, graphs and pictures are often used to convey information more clearly. Some information is received aurally, in casual conversations, at meetings or over the telephone. Information can also be obtained through other senses.

Type of information

Detailed information, such as an inventory list showing the stock level of every item held is most often used at the operational level.

Aggregated information consists of totals created when detailed information is summed together. For example, the details of all purchases from all the customers added together would be aggregated data.

Sampled information refers only to selected records, for example, details from just a few of the customers.

Internal and external information requirements – an organisation has to produce information for a number of reasons. Most information is used internally within the organsiation, but there are certain needs external to the organisation. Examples of such would be company reports to shareholders and product information for customers.

SURNAME	FORENAME	AREA	JOB	AGE
Morrison	James	Eastleigh	Driver	33
Cohen	Harry	Southampton	Gardener	65
Wilson	John	Winchester	Guide	34
Sutton	Tony	Eastleigh	Driver	39

SURNAME	FORENAME	AREA	JOB	AGE
Sutton	Linda	Basingstoke	Canteen	54
Smith	Louise	Basingstoke	Kitchen	44
Goldfield	Chris	Salisbury	Guide	26

Figure 15.4 Detailed information

HATHERLEY HATS UK SALES – 1995

	Trilby	Boater	Bonnet	Topper	Bowler	Cap	Total
1st quarter	255	1276	329	711	3226	2421	8218
2nd quarter	265	1045	389	654	5657	2545	10555
3rd quarter	278	1097	456	644	5534	2621	10630
4th quarter	269	995	468	438	5434	2599	10203
Quarterly average	266.75	1103.25	410.5	611.75	4962.75	2546.5	9901.5
Quarterly Total	1067	4413	1642	2447	19851	10186	39606
% of Total Sales	2.69%	11.14%	4.15%	6.18%	50.12%	25.72%	100.00%

Figure 15.5 Aggregated information

Orders over £1000 in value

Order No.	Order Date	Co. Ref.	Co. Name	Value
14003	11-Mar	1289	Gilchrist Products	4456.00
14009	12-Mar	1289	Gilchrist Products	1652.54
14001	08-Mar	2413	Smith Industries	1466.00
14005	11-Mar	955	Thompson Transport	1678.00
14010	11-Mar	955	Thompson Transport	1346.78
14007	09-Mar	1453	Bradshaw Garages	2654.00
14000	10-Mar	1453	Bradshaw Garages	3200.00

Figure 15.6 Sampled information

Characteristics of good information

Good information is that which is used and which creates value. Experience and research show that good information has numerous qualities.

Good information is:

- **relevant** for its purpose
- sufficiently **accurate** for its purpose
- **complete** enough for the problem
- **reliable** – from a source in which the user has confidence
- communicated to the **right person**
- communicated in **time** for its purpose
- that which contains the **right level of detail**
- communicated by an appropriate **channel of communication**
- that which is **understandable** by the user

■ RELEVANCE

What is relevant for one manager may not be relevant for another. The user will become frustrated if information in any way contains parts that are irrelevant to the task in hand.

■ COMPLETENESS

Ideally all the information needed for a particular decision should be available. However, this rarely happens. In reality, good information is incomplete, but complete enough to meet the needs of the situation.

■ CONFIDENCE IN THE SOURCE

There is no point in producing information if it is not going to be used. Managers will not use information if they do not have confidence that it is likely to be good information. If the source has always been reliable in the past the user is likely to use it.

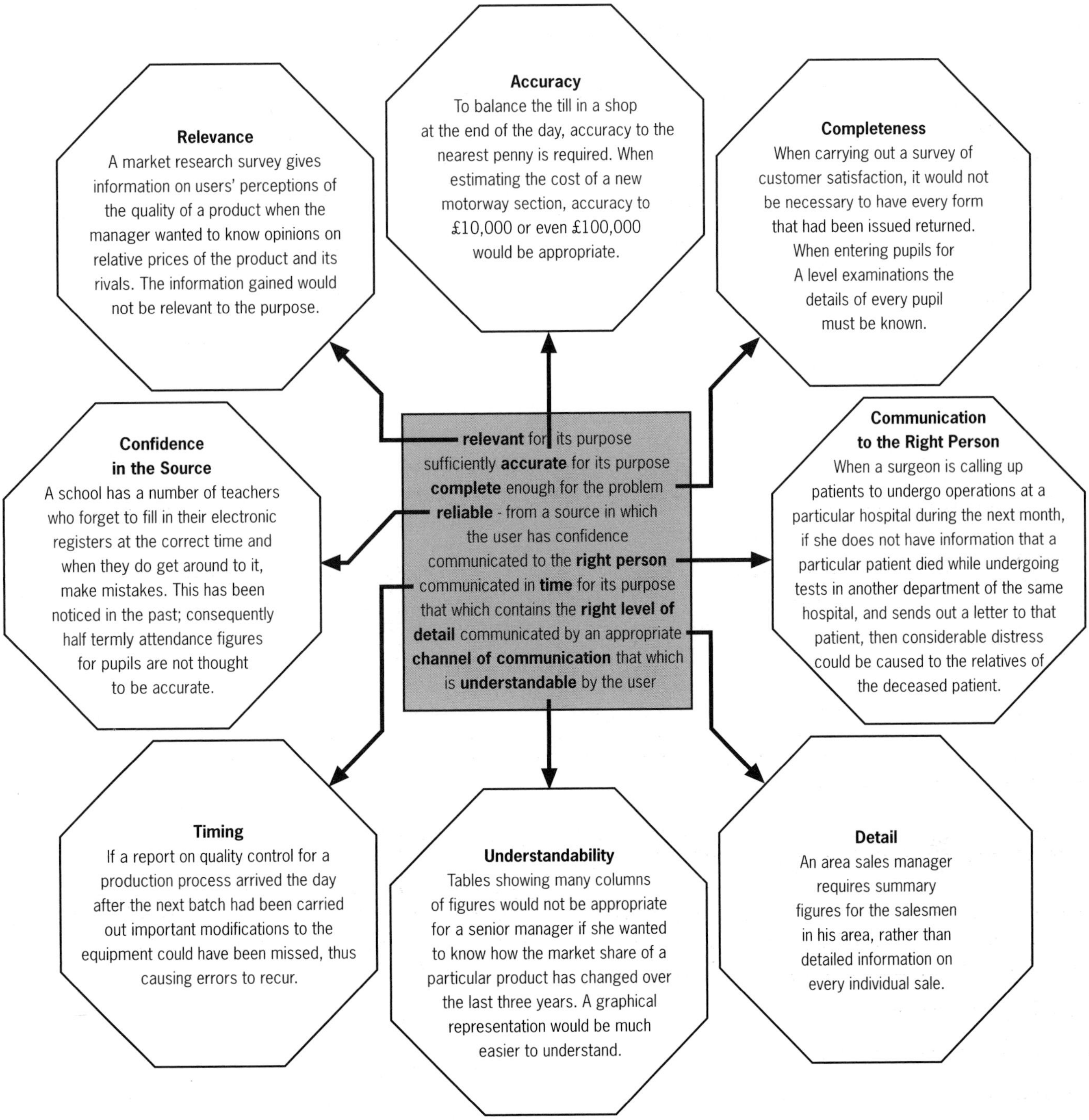

Relevance
A market research survey gives information on users' perceptions of the quality of a product when the manager wanted to know opinions on relative prices of the product and its rivals. The information gained would not be relevant to the purpose.

Accuracy
To balance the till in a shop at the end of the day, accuracy to the nearest penny is required. When estimating the cost of a new motorway section, accuracy to £10,000 or even £100,000 would be appropriate.

Completeness
When carrying out a survey of customer satisfaction, it would not be necessary to have every form that had been issued returned. When entering pupils for A level examinations the details of every pupil must be known.

Confidence in the Source
A school has a number of teachers who forget to fill in their electronic registers at the correct time and when they do get around to it, make mistakes. This has been noticed in the past; consequently half termly attendance figures for pupils are not thought to be accurate.

relevant for its purpose sufficiently **accurate** for its purpose **complete** enough for the problem **reliable** - from a source in which the user has confidence communicated to the **right person** communicated in **time** for its purpose that which contains the **right level of detail** communicated by an appropriate **channel of communication** that which is **understandable** by the user

Communication to the Right Person
When a surgeon is calling up patients to undergo operations at a particular hospital during the next month, if she does not have information that a particular patient died while undergoing tests in another department of the same hospital, and sends out a letter to that patient, then considerable distress could be caused to the relatives of the deceased patient.

Timing
If a report on quality control for a production process arrived the day after the next batch had been carried out important modifications to the equipment could have been missed, thus causing errors to recur.

Understandability
Tables showing many columns of figures would not be appropriate for a senior manager if she wanted to know how the market share of a particular product has changed over the last three years. A graphical representation would be much easier to understand.

Detail
An area sales manager requires summary figures for the salesmen in his area, rather than detailed information on every individual sale.

Figure 15.7 Characteristics of good information

■ ACCURACY

Information needs to be accurate enough for the use it is going to be put. To obtain information that is 100 per cent accurate is usually unrealistic as it is likely to be too expensive to produce on time. The degree of accuracy required depends upon the circumstances. At operational levels information may need to be accurate to the nearest penny, £, kilogram or minute. A supermarket till receipt, for example, will need to be accurate to the penny. A Regional Manager comparing the performance of difference stores at the end of a month would find information rounded to the nearest £100 most appropriate.

■ COMMUNICATION TO THE RIGHT PERSON

Each manager has a particular area of work within the organisation and needs to be provided with the necessary information to help him do his job.

■ TIMING

Good information is communicated in time to be used. If information arrives too late or is out-of-date by the time it does arrive, then it cannot be used by the manager when decisions are made. The frequency of which information is produced is important and needs to be driven by the needs of the manager who is using it.

■ DETAIL

The amount of detail needed in information should be determined by the purpose to which it will be put. More detail than is necessary will confuse the recipient. Too little detail will provide an incomplete picture of the situation.

■ UNDERSTANDABILITY

Understandability, or putting into context, is what changes data into information. If the information is not understood then it has no meaning and cannot be used.

Information flow

Good communications

Communicating information is a vital part of the work of all organisations. Without information transfer, business could not function; without the information, managers might make the wrong decision, delay a decision or not even know that a decision needs to be made. Communication has taken place successfully if the recipient has gained and understood the necessary knowledge.

■ METHODS OF COMMUNICATING

Information transfer may take many forms. It can be verbal or even conveyed by body language with a nod of the head or a shrug. Communication also takes place in writing and increasingly, electronically, using IT. The speed and ease of use of information technology means that it is playing an increasing role in communicating business information. Invoices and letters can be produced using a word-processor. The mail-merge makes the production of standard letters automatic. Global communications are much easier using electronic methods such as fax and e-mail.

■ CHANNELS OF COMMUNICATION

To be usable by the manager, information must be transmitted by means of a communication process. Communication involves the interchange of facts, thoughts, value judgements and opinions. The communication process may take many forms including: face-to-face conversations, telephone calls, informal and formal meetings, conferences, memoranda, letters, reports, tabulations or VDU transmissions.

Whatever the process, good communication results where the sender and receiver are in accord over the meaning of a particular message. Good information must be delivered via the correct channel of communication.

Internal information transfer

Internal information transfer is the communications within the organisation. These communications are affected by the organisational structure. In a two-person business, there is little need for formal communication.
In a larger organisation, the communication needs are greater. No one person can be expected to deal with and know everything. Decisions are delegated, so that the managing director only deals with major items, at a strategic level. Middle managers deal with tactical level decisions – how best to achieve a particular objective. Operational level matters are dealt with by operational staff. However it is still important that staff at all levels have the right information to perform their functions effectively.

Speed versus accuracy

So, good information must be accurate, relevant, complete, go to the right person, at the right time, in the right detail, via the right communication channel; the user must be able to understand it and have confidence in it; speed and accuracy are both important. There is a danger that the need for speed will lead to a loss of accuracy and the need for accuracy will delay the information.

Which is more important, speed or accuracy? This

depends on the type of information and what it is to be used for.

It is likely that quick, inaccurate information will not be as good as slow, accurate information but the need for speed and accuracy depend on the use and the cost of providing the information. Measures to increase accuracy such as validation and verification are likely to decrease speed.

Why delays occur

Information has to be entered into a computer, processed and then output. Delays can occur at all three stages due to large volumes of data to be entered by hand, complicated processing taking a long time or vast volumes of output and slow printers.

There are many other factors that influence how well data flows through an organisation, including:

- the structure of the organisation
- the size of the organisation
- the amount of information
- the number of levels the information has to pass through to reach its destination
- whether the information has to go from one department to another
- the methods used to communicate, for example verbal, formal meeting, IT, e-mail, memo
- the importance of information; strategic, tactical, operational
- the urgency of the information
- staff training – do staff know how to access the information?
- the quality of the design of the information system
- the competence of staff concerned

It is likely that delays are due to human weaknesses such as sickness, lateness, laziness, bad planning, incompetence, wrong equipment or software choices.

Figure 15.8 Bar code scanner in use in a supermarket

Instantly-generated bar codes are used by lottery operators Camelot to identify individual tickets. When the customer's coupon is inserted into the ticket machine, the details are transmitted to the central computer which then allocates an individual transaction code, which in turn is relayed back to the ticket machine which prints a unique bar code. Winning tickets are scanned and verified, and the machine can display how much cash the customer has won.

When shopping from home finding the product you want to order from a very long list can be frustrating. Instead, it is planned that consumers with portable scanners can scan the bar code on the tins and packets and so on, when they are nearly empty. The information would then be transferred to a PC and e-mailed off to the supermarket. Small, pen-sized scanners to be used with home shopping catalogues that display bar codes alongside pictures of goods are also being developed.

Considerable use of bar codes is made in the manufacturing industry. Here bar codes are used to identify components, to track goods in the warehouse, and to co-ordinate the delivery of the right parts at the right time along an assembly line.

It is 20 years since bar codes were first used in British Supermarkets. They are now being used in many different ways.

BAR CODES

Couriers use bar coded stickers to track the progress of documents from despatch to signed-for receipt. Portable scanners are used on a package as it is delivered. The data is then transmitted to the central system. This also allows customers to track their package's progress, via the Internet.

A bar code generally does nothing more than identify the product. The up-to-date price information is usually stored in a central computer. When an item is scanned the product is checked against the database which generates a price at the till. One month a special issue of The Banker magazine was priced at £9.95, rather than the normal price of £7.50. Unfortunately WH Smith's database was not updated so when a magazine was scanned at the checkout, the price came up at the lower price. This was a costly mistake for WH Smith.

Hospitals use bar codes for tagging new born babies so that no unfortunate mix-ups can occur. Bar codes are also used to identify both books and borrowers in a library. Delegates at trade fairs are often given an ID badge with a bar code, so that exhibitors can scan delegates' badges, thus collecting vital contact details.

Organisations that have to deal with very large amounts of paper, such as the Inland Revenue and the electricity boards, use bar coded tax forms and bills to allow office staff to call up the correct customer details when necessary. Such a system can also be used to improve security: access to an account can be denied unless the appropriate bar code is read.

Figure 15.9 Uses of a bar code

Avoiding delays

To avoid delays, an organisation must firstly review its information flows to see why delays have occurred. This can be done by examining existing procedures, interviewing staff about how they implement these procedures, observing staff at work, auditing how particular pieces of information have flowed through the organisation or reviewing the effectiveness of present hardware and software.

As a result of this review, it may be necessary to appoint extra staff, train or dismiss incompetent staff, introduce new procedures or acquire better hardware and software.

How information flows

Sharing information within an organisation can often prove difficult. Huge amounts of information can be generated as a result of day-to-day processes. An organisation must establish appropriate methods and routines for communicating information internally.

Within the same department, information usually flows up and down the chain of command. It is unusual for information to go up or down more than one level at a time. The production line in a factory breaks down. The production line workers tell the foreman; the foreman tells the production supervisor; the production supervisor tells the production manager, the production manager tells the production director who may make the decision. This may mean that messages get distorted and decisions take a long time.

Transferring information from the originating department to others can prove difficult. For example, the details of the development of a new product in the Research and Development department would be of interest to members of the Sales and Marketing. Many large organisations build up their own library and information service to collect information, mainly from external sources, which will be useful to managers throughout the organisation. The recent growth in intranets within organisations is providing a coherent structure for sharing this kind of information.

As we have seen, it is crucial that information arrives at its destination in time to be used. For this to

happen, it is vital that the data and information flows within an organisation are carefully planned. These flows will differ between organisations and are dependent upon a number of factors.

The size, type and structure of an organisation will all play a large part. Within a small business the close proximity of employees means that formal systems for sharing information are not necessary. Within a large organisation, the flow of certain information will be formally defined as part of a planned system. The stages of flow will be fully documented and there will be a list of agreed procedures for control and distribution. Rules for dealing with exceptions will also be laid down in the documentation. Without such planning a report destined for a particular manager could end up sitting on the desk of another for several days by mistake.

As we have seen in Chapter 12, the more levels there are in an organisation structure, the longer the chain of command. The more levels that information has to pass through, the greater the likelihood of delay. An organisation can be based in one site or spread over several sites, sometimes located around the world. If the organisation is spread out geographically, then methods of information flow need particularly careful planning and the use of electronic methods becomes crucial.

The nature of the data will have a major affect on information flow. How and where the data originates is a factor. It could be electronically generated, for example through POS terminals and processed by powerful tools to produce information that is disseminated over a network, available to managers on their desk top computer.

Alternatively, it could have originated from hand written notes and telephone conversations. The use of e-mail can dramatically speed up such information flow, ensuring that the information is copied to all necessary people in one operation. However, users in many organisations complain of e-mail overload where the indiscriminate copying and forwarding of messages has caused an unmanageable volume.

Care must be taken when designing a new IT system to include all stages in the design. The dissemination and distribution of reports to the appropriate people at the appropriate time will be a crucial factor in a system's success. Reports must arrive at the right manager's desk at the time when a decision needs to be made, in a suitable format to be useable.

Many decisions are made on a regular basis, and a system can be designed, by ensuring that all the above factors are taken into account, to allow a manager to deal with these. However, many decisions are ad-hoc in nature and require one-off information. The way in which a manager can make an informal request to the information providers and receive a timely response will be a measure of the success of information flow within the organisation.

When making decisions, a manager draws on information generated from formal flow such as reports and printouts. He may have access to a range of graphical displays on his desk top computer via an EIS system. He will also have gathered information informally. He will himself have observed events, had face to face and telephone contacts with his colleagues and business contacts, both in one-to-one situations and in meetings. Thus formal and informal information flow will feed into his decision making.

Data

Sometimes it is possible to get information as an automatic by-product of a routine data processing operation. In a supermarket POS terminals are used with bar code scanners to capture data on goods purchased. An itemised bill is produced with a calculated total for the customer. The data collected can also be used in the stock control and re-ordering system, for marketing decisions as trends of product sales can be analysed and for personnel tracking as the rate of working of POS operators can be calculated.

Data capture

Data is very often typed in on a keyboard but it can be collected by other means such as by using computer readable documents (using OCR, OMR or MICR), voice recognition or sensors (for example under the road at traffic lights, PIR sensors in burglar alarms).

If data is not captured automatically, it may require translation or transcription before to entry into the system. This can affect the accuracy of the data. A series of controls may need to be put in place in order to ensure that the quality of the data is maintained.

Errors in data entry and processing

GIGO (Garbage In, Garbage Out) is a commonly used acronym in IT. The information that is output is only as good as the data input. Errors in data can occur at all stages of processing and they need to be planned for and controlled.

Data can get lost before it is even input, or it might contain errors. The likelihood of errors will depend on the method of processing input data. Any form of human involvement will open up the possibility of error.

Case Study

Example of Information flow

A mail order company takes a telephone order for a product. (**external source**) The details must be recorded, either on paper or on computer. The information must be sent to the dispatch department in the warehouse. This could be done on paper or by e-mail. (**internal operational information transfer**)

The dispatch department find the goods, write the dispatch note (**external operational information transfer**) to send with the goods. This could also be done on computer. A copy is sent to the accounts department (**internal operational information transfer**). The goods are posted.

The accounts department prepare an invoice (**external operational information transfer**) to send out and record that the payment is due (on computer). The payment is awaited (**external operational information transfer**).

This is all common day-to-day operational information. If the dispatch department noticed that this particular item was out of stock, they may inform their manager. (**internal operational information transfer**) He or she may need to make a tactical decision based on this and other information, to order more stock or to order from a different supplier or to find out if there are problems with the supplier. (**tactical information transfer**)

If the supplier is unreliable, the manager may need to tell the managing director who may need to take the strategic decision not to buy from this supplier any more. (**strategic information transfer**)

■ ERRORS IN RECORDING DATA

When data is being written down or keyed in, a number of categories of potential error can be identified. It is common to get the order of characters (especially numbers) wrong. For example, keying in an account number of 58762 instead of 57862. Here the operator has swapped around the 7 and the 8. This is called a **transpositional** error as the position of the characters has been misread. **Spelling mistakes** are common, especially with data such as a customer's name. Other, less common, mistakes can occur when measurements are being made (for example reading the current value from an electricity meter dial) or incorrectly coding information from a source document.

In fact, any form of human copying of data, called **transcribing**, is liable to cause mistakes. If the source document is handwritten, it may not be easy to read the handwriting. Whenever data needs to be transcribed or translated into another format before entry into the system there is a risk that the accuracy of the data will be reduced.

Controls over data capture

Many data entry errors can be picked up by the process of **verification**: entering the data twice and comparing the two versions to highlight errors. If input is done using a keyboard, the input data will be shown on the VDU screen and a visual check on the data can be made.

The greater the human involvement in transcribing data, the greater the chance of making mistakes. One way of reducing transcription errors is to use turnaround documents which include as much pre-printed information as possible that can be read by a method of optical character recognition. Identifying goods with bar codes removes the likelihood of an incorrect code being entered. Where human recording is necessary, the chance of error can be minimised by paying careful attention to form design and by giving clear instructions about how the documents should be filled in.

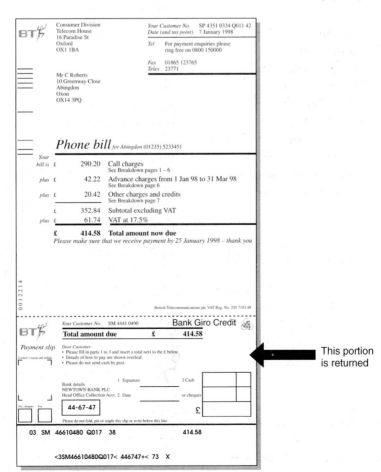

Figure 15.10 Example of a turnaround document

The input record will often have a key field identification code. For example a sales ledger file will consist of customer records, with each customer having a code number for identification. When a transaction record is keyed in, the customer code would be part of the input data and the program might search for the customer record on the sales ledger file, and display it on the VDU screen. The input operator could then check visually that the correct customer record is being processed.

Some checks on the validity of input data can be written into the system's programs. These **data validation** checks might be performed by a separate data validation program in a batch processing system. Alternatively, any program can incorporate validation checks on input data, for example on data keyed in from a terminal into an on-line system.

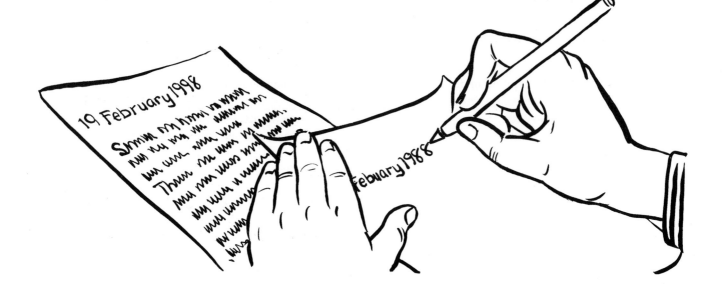

Figure 15.11 Transcription error

Summary

Information is necessary to make informed decisions. Information transfer is important to any organisation. Information flows must be clear so that information arrives at the right person, at the right time.

Information can be:

☐ historic/current/future

☐ strategic/tactical/operational

☐ detailed/sampled/aggregated

☐ from a primary/secondary source

☐ qualitative/quantitative

Communication can be:

☐ internal/external

☐ verbal/non-verbal

☐ formal/informal

Information Flow

Information flow is affected by a number of factors. Delays can occur at all stages of flow. Factors include:

☐ structure, size and geographical distribution of the organisation;

☐ how and where the data originates within the organisation;

☐ the validity and volume of data;

☐ the formal and informal communication structure;

☐ the nature of the processing cycle.

Data

Data errors can occur at all stages of processing.

Control over data capture can reduce data entry errors.

Data, information and information flow questions

1 A supermarket cashier scans a bar code on a product. The code is sent to the store's computer. The price and name of the product is sent to the till and printed on the receipt. Details of the sale are immediately stored in a transaction file. Every hour the transaction file is sent to the store's warehouse to up-date stock levels.

 a) What is the source document?

 b) What level of information is the code going into the computer?

 c) What level of information is contained in the transaction file going to the warehouse? (3)

2 For each of the following examples say if the communication is internal or external, and if the information is strategic, tactical or operational; historic, current or future.

 a) the managing director's diary for next week

 b) details of a car's former owners from the Police National Computer

 c) a delivery note when a washing machine is delivered to a customer

 d) news that George will be off sick today

 e) six thousand widgets need to be delivered immediately (15)

3 Give an example of strategic information, tactical information and operational information in a school or college. (6)

4 An IT department uses a computer program written in-house which cannot draw graphs but can export to a spreadsheet. The IT manager then uses the spreadsheet to draw the necessary graphs. This takes three days. On presenting his report, the IT manager is told that the figures appear to be out-of-date. How did problem arise and how could it be overcome? (5)

5 A company has decided to open a computerised distribution warehouse in the Midlands, handling deliveries for the whole country. Give an example of strategic information, tactical information and operational information needed in starting the warehouse. (6)

6 A customer of a bank wants to withdraw some money from a cash machine. Describe the information flow. (6)

7 The manager of a company complains that the Management Information System (MIS) continually fails to produce the appropriate information at the right time. The person responsible for the MIS responds by blaming the 'inadequate data and information flow' within the company and requests a review of 'data and information flows'.

 a) State **six** factors which influence the flow of information and data within an organisation. (6)

 b) With the aid of examples, describe **three** techniques which could be used to review the current information flows. (6) *NEAB 1997 Paper 3*

8 A school is planning the introduction of a computer based attendance system for classes and registration groups. The purpose of the system is to produce information for the following end-users:

 i) class teachers

 ii) tutors/Head of Year

 iii) senior managers (for example Deputy Head)

 a) Describe **three** alternative methods of collecting the attendance data for the system. (6)

 b) For each of the different end-users describe, with the aid of an example, information that the system might produce in relation to their requirements. (6) *NEAB 1997 Paper 3*

9 A company with two office buildings 200 metres apart is having problems with internal communications. They are considering internal e-mail to solve the problem. Explain three reasons why this may help and three reasons why it may not. (6)

10 'The quality of management information is directly related to its timing.'

 Discuss this statement paying particular reference to: the different purposes for which the information may be required;

 the relative merits of speed versus accuracy. (6)

In planning the information flow within a system, where are the delays likely to occur and why? (6) *NEAB Specimen Paper 3*

11 'The nature, quantity and quality of data often dictates the method by which it must be captured for use within an IT system.

a) Name **four** different methods of data capture.(4)

b) For each method:

i) give an example of the type of data which may be captured; (4)

ii) state the reason why this method is particularly appropriate. (4) *NEAB 1998 Paper 4*

12 A sales manager claims that he is always provided with 'quality' management information from his Management Information System.

With the aid of examples where appropriate, describe **five** characteristics of good information. (10) *NEAB 1998 Paper 4*

13 A particular organisation is upgrading its computer-based stock control system. The previous data collection system was OMR based.

One function of the system is to allow stock levels to be monitored on a regular basis.

a) State three other alternative methods of collecting stock control data. (3)

b) What factors, other than cost, will determine the method of data collection? (4)

c) The software used to control the system must support an audit trail. Explain what is meant by the term 'audit trail', and state why this functionality is necessary. (6) *NEAB IT04 1999*

Activity: Categories of Information

1 For each of the tables (a, b, c and d) state, together with a brief explanation of your decision, their:

- type (detailed, summarised, aggregated, exception);

- level (strategic, tactical operational);

- time (historical, present, future);

- use (planning, control, decision making);

- occurrence (at planned intervals, occasional; on demand).

You may think that a table fits into more than one category.

a

HATHERLEY HATS
UK SALES – 1999

	Trilby	Boater	Bonnet	Topper	Bowler	Cap	Total
1st quarter	255	1276	329	711	3226	2421	8218
2nd quarter	265	1045	389	654	5657	2545	10555
3rd quarter	278	1097	456	644	5534	2621	10630
4th quarter	269	995	468	438	5434	2599	10203
Quarterly	266.8	1103.3	420.5	611.75	4962.8	2547	9902
Quarterly Total	1067	4413	1642	2447	19851	10186	39606

b

A LEVEL IT. Students with ALIS < 5.9

Surname	Forename	TG	Alis	Core 1	Core 2	Core 3	Total
Cloke	Roger	F49	5.7	45	56	81	182
Davey	Charlotte	F62	5.3	32	45	12	89
Davidson	Ian	F12	5.7	54	48	78	180
Dawson	Cathy	F11	4.9	29	31	54	114
Ferguson	Ian	F32	4.9	60	34	76	170
Green	Amy	F22	5.4	45	34	59	138
Hall	William	F12	5.7	56	44	66	166
Kinder	Jackie	F34	5.8	48	54	75	177
Laidlow	Hunter	F27	4.5	55	61	78	194
Lister	Peter	F56	4.5	56	43	45	144
Manie	Les	F41	5.3	36	45	65	147
Mitchell	Mary	F53	5.7	47	45	80	172
Mountrie	Harry	F45	5.8	62	54	106	222
Murdock	Colin	F38	4.5	66	68	75	209
Nice	Andrew	F27	5.2	56	46	65	167
Renn	Anne	F61	5.7	59	61	89	209
Smith	Dan	F33	4.8	34	24	49	107
Stanley	Anthony	F23	4.3	49	54	34	137
Steele	Jeremy	F34	4.7	68	68	78	214
Thompson	Pat	F34	5.1	50	47	32	129

c

LEEMING SUPPLIES PLC
Orders up to 13-Mar 1999

Order No.	Order Date	Co. Ref	Co. Name	Value	Vat	Total
14008	12-Mar	2245	McElroy Stables	123.85	21.67	145.52
14003	11-Mar	1289	Gilchrist products	4456.00	779.80	5235.80
14009	12-Mar	1289	Gilchrist products	1652.54	289.19	1941.73
14001	08-Mar	2413	Smith industries	1466.00	256.55	1722.55
14004	10-Mar	2413	Smith industries	567.00	99.23	666.23
14006	10-Mar	2375	Smith kitchens	55.54	9.72	65.26
14005	11-Mar	955	Thompson Transport	1678.00	293.65	1971.65
14010	11-Mar	955	Thompson Transport	1346.78	235.69	1582.47
14007	09-Mar	1453	Bradshaw Garages	2654.00	464.45	3118.45
14000	10-Mar	1453	Bradshaw Garages	3200.00	560.00	3760.00
14002	11-Mar	1453	Bradshaw Garages	98.76	17.28	116.04

d

PERSONAL FINANCES – TERM 1   (PLAN)

INCOME	Week 1	Week 2	Week 3	Week 4	Week 5	Week 6	Week 7	Week 8	Week 9
Opening Balance	£0.00	£1,015.00	£885.00	£755.00	£625.00	£495.00	£365.00	£235.00	£105.00
Grant	£500.00								
Loan	£400.00								
Parents	£300.00								
Total Income	£1,200.00	£1,015.00	£885.00	£755.00	£625.00	£495.00	£365.00	£235.00	£105.00
EXPENDITURE									
Accommodation	£60.00	£60.00	£60.00	£60.00	£60.00	£60.00	£60.00	£60.00	£60.00
Food	£30.00	£35.00	£35.00	£35.00	£35.00	£35.00	£35.00	£35.00	£35.00
Books	£75.00	£15.00	£15.00	£15.00	£15.00	£15.00	£15.00	£15.00	£15.00
Other	£20.00	£20.00	£20.00	£20.00	£20.00	£20.00	£20.00	£20.00	£20.00
Total Expenditure	£185.00	£130.00	£130.00	£130.00	£130.00	£130.00	£130.00	£130.00	£130.00
CLOSING BALS.	**£1,015.00**	**£885.00**	**£755.00**	**£625.00**	**£495.00**	**£365.00**	**£235.00**	**£105.00**	**−£25.00**

2 'BettaShu' is a small, retail chain which specialises in the sale of high quality shoes. The company are currently investigating the feasibility of replacing their current information system. It is important that all information needs are considered. Can you sugest examples of:

- information that is needed in real time;
- information that is needed daily;
- information that is needed monthly;
- information that is needed annually.

For each, identify who would use the information (include **level**) and why they would need it (include **use**).

3 Plumper Pets plc has traditionally produced a range of cheap and medium priced cat and dog tinned food brands. It has recently launched a new, luxury brand for dogs, and is now considering launching a similar brand for cats which they intend to call 'Kat-i-dins'.

Describe the range of information that Plumper Pets would need to enable them to make their decision. Explain where they would obtain the different information (distinguish between internal and external).

4

**A level Information Technology
Course Evaluation**

Did you attend the induction course? Y/N ☐

Did you study IT at GCSE? Y/N ☐

Please select the type of Computer you have at home.

486 ☐ Pentium ☐ Pentium II ☐
Pentium III ☐ Mac ☐ Other ☐

I find the turnaround time for homework marking satisfactory

Agree | 1 | 2 | 3 | 4 | 5 | Disagree

Please write below any comments or suggestions concerning homework.

- What information could be produced if the questionnaire shown above was given to all students studying A level IT in year 12.

- Itemise which information is **quantative** and which is **qualititive**.

- Which data items require value judgements?

- Explain who would use the different information produced, and why it would be useful for them (**use**).

Activity – Data capture 1

A new anti-hooligan football smart card scheme has been introduced in Belgium. At first, some clubs refused to take part, but, after pressure was put on them by the Belgian Interior Ministry, which included clubs having to pay for extra police for certain games and being fined if there was trouble at matches. All clubs are now committed to the scheme.

A smart card is a small, wallet sized piece of plastic which contains a microchip. Data can be both read from, and stored in, the card. The Belgian smart card contains 1 Kbyte of memory, and can be used as an electronic purse and a loyalty card. An agreement has been made with the oil firm, Fina Belgium, to accept it at their petrol stations.

The smart card also acts as an ID card. Once it is used to purchase a ticket the fan must use the smart card to gain entry to the match. Trouble-making fans can have theirs de-activated so that further tickets cannot be bought. A controversial issue which worries civil liberty groups is that police have access to the networked smart card computer scheme which is connected to servers at all clubs. This gives police advance access to the identities of all those who will be attending a game.

1 Name and describe two other data capture methods that could be used for an identity card to allow entry into a sports ground instead of a smart card. (6)

2 Explain why civil liberty groups might be concerned about the introduction of these cards. (8)

3 Describe two other situations when the use of a smart card would be appropriate. (4)

Activity – Data capture 2

Copy the grid below. Research and fill in the gaps. Several boxes have been filled in for you to indicate the level of detail required.

Method of capture with description	Typical use	Reasons for choosing this method
Optical mark reader (OMR) *reads marks made in predetermined positions on special forms by light sensing methods*		
Bar code		*Carries a uniquely defined code. Many reading devices on the market from scanners that are hand held to those that are built in to equipment such as a supermarket check out.*
Smart card		
Optical mark Recognition (OCR)		
Magnetic Ink Character Recognition (MICR)	*Used on bank or building society cheques: the bank sort code, account number and cheque number are held.*	
Magnetic stripe		
Sensors (for example voice, electronic counter)		
Digital camera		
Manual: touch screen		
Manual: direct data entry (DDE)		

Activity – Information flow 1

1 Investigate the flow of information relating to student progress within your school or college. Work in a group, each member exploring a different aspect. Example activities to investigate could include: student enrolment on to courses and any subsequent change of course; entry to examinations; class attendance. Find out what information is needed by teachers, tutors and various senior members of staff.

2 Draw a diagram, or diagrams, to represent the flow.

Activity – Information flow 2

A company produces and sells animal feed to farmers. It is organised into 3 divisions: one each for cattle, pigs and poultry. Each section employs a number of salesmen who visit farms. They are paid expenses for car mileage, meals and overnight accommodation while they are away from home. Every week they submit an expenses claim that has to be authorised by their Sales Manager before being forwarded to the finance department. All receipts must be included and the salesmen are chased up for any missing ones.

The salesmen receive a monthly cheque to cover the expenses. Each month the Sales Managers receive a summary of expenses for each salesman in their Division. Every 6 months the Managing Director is given an overall summary. It shows the last 3 periods of 6 months. If there are any major changes in these costs, he will discuss the issue with his managers.

1 Draw a diagram to demonstrate the flow of information in this system. At each stage, suggest possible methods for information transfer.

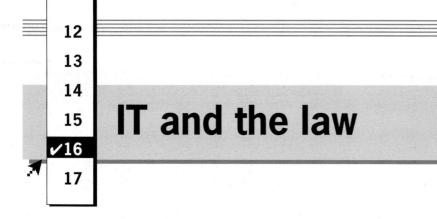

IT and the law

Computer Misuse Act 1990 (part)

Section 1

A person is guilty of an offence if –

a. he causes a computer to perform any function with intent to secure access to any program or data held in any computer;
b. the access he intends to secure is unauthorised; and
c. he knows at the time that he causes the function that that is the case.

Figure 16.1 Extract from the Computer Misuse Act

Legislation

Laws are a major influence constraining the operations of businesses. Laws come from three sources – common law (developed over hundreds of years' custom and practice), statute law (passed by Parliament) and the European Union. The procedures within a company must reflect the requirements of the legislation to ensure that all laws are being adhered to. When new legislation is passed, an organisation will need to look at current practice to check that new requirements are being met and, if necessary, modify procedures accordingly.

An organisation has a responsibility to ensure that all its employees are aware of laws relating to IT and their responsibilities under these laws, which in particular are the Data Protection Act 1984, the Copyright Designs and Patent Act 1988, Health and Safety at Work Act 1974 and EU Health and Safety Directive 87/391.

New employees usually undergo an induction course which provides them with a background to the organisation as well as giving specific training about the job. This course should include a discussion of the legal

requirements of the post. All employees should be given a handbook which lays out their legal responsibilities. It is useful to provide a list of 'dos and don'ts'.

It is vital that managers take an active role to ensure that legislation is enforced within an organisation. Employees should be reminded of the law through individual memos, public notices posted on walls, and the organisation's intranet. Software purchasing and control should be centralised. This enables a register to be kept that holds details of all software that is installed. Such centralisation also allows reliable, known suppliers to be used and makes the checking that no unauthorised software has been installed a relatively straightforward matter.

Every organisation must ensure that they register all data stores that fall under the Data Protection Act with the Data Protection Registrar. Enquiries should be made to the Registrar in any cases of doubt about the need to register personal data. The organisation should draw up a written data protection policy which should make clear what data can be kept and for how long. In large organisations, a Data Protection Officer should be appointed who is responsible for monitoring practices and making sure they are following the requirements of the act.

Every organisation needs to take positive steps to ensure that the Copyright Designs and Patent Act is not being broken. No software should be used without the appropriate licences being in place. Particular care needs to be taken when LANs are used as sufficient licences for the number of users must have been obtained. The use of software should be monitored. Software should only be installed with permission: spot checks can be made to check that employees have not installed programs illegally. Security measures should be in place and enforced.

Organisations must maintain a healthy and safe environment for work. The role of safety officer, who checks that the appropriate laws are complied with, must be established. He should review health and safety issues on a regular basis. A safety committee, with representatives from all parts of the organisation, should discuss safety matters on a regular basis. Management should encourage and give recognition to a trade union health and safety representative who could act on and report the concerns of colleagues. Such representatives should be given a very thorough training.

All staff need to be reminded of the importance of health and safety issues on a regular basis. Posters can be displayed which show potential hazards and precautions to be taken. A safety policy should be produced and a copy given to all staff. In workplaces where mistakes can be life threatening, such as oil refineries, safety incentive schemes are often introduced. In one such scheme, the team or location with the best safety record each year is rewarded with a bonus payment.

Keeping within the law

An organisation needs to take measures to keep within the law. Appropriate actions are listed below.

■ THE DATA PROTECTION ACT

- Register with the Data Protection Registrar
- Appoint a Data Protection Officer to monitor systems
- Ensure all new systems take data protection into account at the design stage
- Use a variety of methods to remind all staff of their responsibilities in keeping data private
- Include matters of Data Protection in the organisation's Security Policy (see chapter 20)
- Put measures in place to make it easy for customers or clients to obtain their rights

■ SOFTWARE COPYRIGHT

- Centralise the purchase of software
- Carry out spot checks as well as regular audits to ensure that no unauthorised software is stored on individual computers
- Remove all illegal software as soon as it is found
- Maintain records of all licences held
- Discipline employees who break the rules

■ HEALTH AND SAFETY

- Establish a Safety Committee and ensure that computer-related matters regularly occur on the agenda
- Follow space guidelines when designing office layouts
- Carry out Risk Assessments on a regular basis
- Make sure that the correct equipment and furniture is used
- Inform all staff of likely hazards using a variety of methods
- Establish a Health and Safety policy and ensure that all staff are familiar with it
- Appoint a Safety Officer

Software theft

Software theft can be divided into two categories: piracy and counterfeiting. Piracy occurs when more copies of software are made than the number of licences purchased. Users who do not realise that it is illegal can sometimes do this. Counterfeiting is when software is illegally copied for sale to other users. Counterfeit software comes without manuals, user guides or tutorials. The software cannot be registered, so there is no technical support or upgrade service available. An added problem for users is that such software carries a high risk of bringing a virus.

The illegal use of computer programs in companies costs software providers world-wide billions of dollars. In some Eastern European countries the rate of pirate copies reaches 95 %.

BSA

The Business Software Alliance (BSA) is an organisation set up to combat software piracy by trying to promote the use of licensed software. The BSA is an international organisation which represents leading software companies. It works in conjunction with software sellers, governments, law enforcement bodies, and users. The BSA attacks the costly problem of software piracy in three ways.

Firstly, it aims to educate users. Its educational programme concentrates on making end users and IT managers aware of the legal implications of software theft and highlighting the importance of correct software licensing.

Secondly, the BSA aims to enforce current legislation through software audits, raids on premises and legal action against software thieves. So far over 600 legal actions have taken place as a result of the BSA's work. Sentences imposed have ranged from heavy fines to imprisonment.

Finally the BSA lobbies governments at both national and European level to introduce legislation to protect software copyright.

Audit requirements

Company accounts have to be audited by law. An auditor's task is to check to make sure that no financial mistakes have been made and that no fraud has taken place. The auditor needs to be able to access all relevant records within the organisation: for example records of customer transactions, orders, payroll details as well as the overall end of year statement of accounts. Most companies use information technology systems to record financial information. Only very small companies record customer orders in a paper-based system these days. The computer-based applications are subject to audit in the same way as paper-based systems. However, the auditor will need to have a knowledge of IT and the particular risks of fraud and error associated with computer-based systems. It is likely that an IT-based financial information system will store much more data than a paper-based system. Auditing such systems is therefore a considerably more complex operation.

The auditor needs to have a sound knowledge of the organisation and responsibilities of the data processing staff and the methods of control of systems development, programming and operations. An understanding of the data collection and validation techniques, file organisation and processing techniques and systems controls used in the systems are all needed.

To carry out her job, an auditor will need to examine the data files. However files on magnetic media such as disk or tape cannot be read as easily as traditional

Figure 16.2 Auditor examining records

paper ledgers. Old files are not normally kept for more than a few days before a tape is re-used. The sorting of data from source documents is performed by the computer, therefore these documents, if they are kept, are not filed in order and so are difficult to retrieve. The sheer volume of work means that it is not possible for the auditor to check every transaction. The auditor will therefore probably only check a sample of transactions.

The types of error occurring in clerical ledgers were mainly mistakes in arithmetic and copying. Such errors are almost unknown in computer systems. The hardware and software errors which may occur in computer systems are very difficult to find. The auditor must be satisfied that the hardware and software has been fully tested and is bug-free. This is why an auditor should be involved with systems development, including the design stage of a new system. The system must be designed with the work of the auditor in mind, so that auditing is an integral part of the system. The auditor should supply a set of test data which includes deliberate errors. These are then processed by the computer and the auditor examines the output and compares it with the expected results. This will check the processing method.

It is then possible for the auditor to concentrate on the inputs and outputs of the computer and ignore the processing involved. Doing this should give sufficient information for the audit and the auditor would not need a detailed knowledge of the processing methods used. This is known as 'auditing around the computer'.

Test packs are used to test a system before it goes live, to test a system after it has been amended and to test a system to make sure no unauthorised amendments have been made.

Audit package

A range of auditing software packages can now be bought which save the auditor a lot of routine work. They enable the auditor to check computer files.

Main facilities of an audit package:

- verification of file control totals (these are validation totals stored in the file)
- verification of individual balances in records
- verification that all data is present in records

- selection of records for checking, for example random records, overdue accounts, non-active accounts, payments over a certain value (these values are selected by the auditor)
- analysis of file contents, for example debts by age, payments by size, stock by value
- comparisons of two files to show up any differences.

Audit trails

An **audit trail** is an automatic record made of any transactions carried out by a computer system (for example all updates of files). This may be needed for legal reasons so that auditors can check that the company accounts are accurate.

An audit trail is a means of tracing all activities relating to a piece of information from the time it enters a system to the time that it leaves. An audit trail should provide sufficient information to establish or verify the sequence of events. It enables the effects of any errors in the accounting information to be traced and the causes determined.

Traditional audit trails follow transactions through a written ledger. This will include handwritten details of ordering, payments, sales, and so on initialled by the clerk. This is not so easy on electronic media. It is possible to print out data during processing but this is not much help in trying to find the history of some transaction.

The systems designer may be asked to design a system to provide special audit trails on demand. This is a mechanism built into the software to allow

Joanna Dewey took her car to the local garage for a service. When she picked her car up, she was told that the brakes were worn and that two pads had been replaced. Joanna was convinced that there was something seriously wrong with her car's braking system as she could remember paying out to have them repaired on several occasions. She talked to the manager who was able to view the trail of past transactions relating to Joanna's car and the specific repairs that had been carried out. With this information he was able to assess whether or not there had been an underlying problem with the braking system.

the auditor to trace a transaction from input through to output. By typing in the record details, the auditor can find out all about the transaction and check it. An auditing log would also record who has been using the computer, when, how long for and what they did with the data.

An example of an audit trail in use can be seen when the Police National Computer is used to trace the history of owners of a motor vehicle. Before computerisation, every car was issued with a paper log book which had to kept by the car owner. The log book had details of all the previous owners of the car recorded in it. The current computer system has been designed to hold the same information so that the names of all past owners can be found on request.

On-line systems provide problems for the auditor for a number of reasons. Transaction details can be entered at many points on a WAN. Source documents may not exist; this could occur if an order were made by phone. Controls such as validation and verification may not be used as they may waste time in a time critical system. Very often, immediate processing may make an audit trail impossible.

The auditor must check the software thoroughly. In particular he or she must pay attention to the validation checks made on input data. Careful checks should be made that protection against unauthorised

inspection of files is maintained, that passwords are used properly and that any suspicious transactions are reported. Operating system software can keep a record of all activities at network stations. The auditor can check this and make unannounced visits to terminal locations.

FE Colleges are funded by a government body, the Further Education Funding Council (FEFC). Colleges receive funds on the basis of units of activity achieved. Units of activity are awarded for such events as a student attending an induction course, a student having enrolled on a course at a certain date, a student completing a course and a student gaining the qualification which was the aim of the course. Careful and accurate records have to be kept of all such events within a college.

Every student must have signed a formal document called a 'Learner's Agreement'. This lays out exactly which course or courses the student is studying. Any changes to this agreement must be carefully recorded and dated.

The auditors will scrutinise many records including those relating to enrolment, exam entry and the results obtained. It is likely that they will pick several students at random and check through every record relating to them. Class attendance records will be examined to cross check that claims for course completion are correct.

Summary

An organisation has a responsibility to ensure that all its employees are aware of laws relating to IT and their responsibilities under these laws, which in particular are:

☐ Data Protection Act 1984

☐ Computer Misuse Act 1990

☐ Health and Safety at Work Act 1974

☐ EU Health and Safety Directive 87/391.

Many IT applications are subject to audit. The auditor needs to be familiar with data-processing techniques.

☐ An auditor should be involved in systems development.

☐ Audit packages are computer programs to help the auditor.

An **audit trail** is an automatic record made of any transactions carried out by a computer system (for example all updates of files). This may be needed for legal reasons so that auditors can check that the company accounts are accurate.

IT and the law questions

1 An insurance brokers is introducing computers so that all members of staff have access to a new computer database. What advice would you give to the owner of the business about:

   a) legal requirements relating to the keeping of customer information on computer file. (2)

   b) how to ensure that members of staff are aware of these requirements. (2)

   c) regulations governing the working conditions of staff. (2)

2 Explain why have auditors had to change their practices in recent years. (6)

3 Using an example, explain what is meant by an audit trail. Describe what would be found in it. (5)

4 Some software packages can be set up to monitor and record their use. This is often stored in an access log. Name **four** items you would expect to be stored in such a log. (4) *NEAB Specimen Paper 3*

5 A particular college uses a computer network for storing details of its staff and students and for managing its finances. Network stations are provided for the Principal, Vice-Principal, Finance Officer, clerical staff and teaching staff. Only certain designated staff have authority to change data or to authorise payments.

   a) What are the legal implications of storing personal data on the computer system? (4)

   b) What measures should be taken to ensure that the staff understand the legal implications? (3) *NEAB Specimen Paper 3*

6 Many retail organisations have developed large databases of customer information by buying data from each other.

   a) Describe two possible uses these organisations could make of the data they purchase. (4)

   b) Some customers may object to data held on them by one organisation being sold to another organisation. Describe some of the arguments which either of these retail organisations may use to justify this practice. (4) *NEAB Specimen Paper 3*

7 Many accounts packages have an audit trail facility. Explain why such a facility is necessary, what data is logged and how this information can be used. (6) *NEAB Specimen Paper 3*

8 Some IT applications use software which maintains an audit trail. Name **one** such application and state why this facility is necessary. (3) *NEAB 1998 Paper 4*

9 'Legislation will have an impact on the procedures used within any organisation.'

   Discuss this statement. Particular attention should be given to:

   • the different aspects of IT related legislation which affect organisations;
   • the types of formal procedures which are Used to enforce legislation;
   • the potential differences between legislation and company policy.

   Illustrate your answer with specific examples. (20) *NEAB IT04 1999*

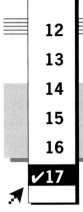

Disaster recovery

It is important that commercial IT users recognise the potential threats to their information systems, plan to avoid disasters leading to loss of data and have contingency plans to enable recovery of any lost data.

If disaster recovery plans and management are inadequate then commercial enterprises face serious losses. They will be unable to process transactions which are at the heart of their business. Much day to day functioning would cease. This would lead to loss of trade as customers are forced to go elsewhere. Serious, extensive or repeated failure would lead to loss of confidence in the business. A high proportion of businesses never recover from serious failure to their information systems.

Many supermarkets have to close their doors and cease trading if their point of sales terminals fail to function.

Financial Times December 1996

Lloyds Cashpoints fail for five hours: Krishna Guha

A power failure paralysed Lloyds Bank's entire national cash machine system yesterday, leaving thousands of Christmas shoppers without cash. For more than five hours, Lloyds' customers were unable to obtain money from any automated teller machine.

Almost all the Lloyds' machines flashed up the message 'sorry, cash point service closed'. Bank staff in shopping centres reported big queues as shoppers struggled to get money from cashiers inside branches. The timing could not have been worse: nine days before Christmas demand for cash is at its yearly peak.

The system was brought down by a power failure at the computer in Peterborough which authorises cashpoint withdrawals nation-wide. Damage to a single power cable may have been to blame.

- What contingency plan could have avoided the 5-hour down time?

Figure 17.1 Lloyds Bank cashpoint

Figure 17.2 Buildings after terrorist attack

Software failure

Software can contain bugs which only occur when a particular combination of unusual events occur. Such bugs may not be detected in testing and lead to systems breakdowns at any time. They can be hard to locate and put right and cause considerable damage to data as well as delay in processing. Viruses can alter the way that programs function and lead to breakdown. Software can fail because it is unsuitable for the task. The volume of data may grow too big.

Potential threats to information systems

The threats to an information system are far ranging. Some of the major threats are described below.

External failure due to fire, floods, earthquakes or terrorism

Physical disasters may be relatively rare, but when they do occur they can be devastating. As well as equipment, files containing vital data could be destroyed. Without far-sighted disaster planning, many businesses would be unable to recover from the data loss. Companies and many organisations employ specialist disaster recovery companies to manage their plans.

Hardware failure

Hardware failure is a major cause of system breakdown, as the Lloyds' Bank problem described earlier illustrates. Failure can arise from processor failure or disk head crash. Computers are dependent upon a constant supply of electricity. The failure of one hardware component can cause the whole system to crash. The growth in networks and distributed systems have in some ways made disaster recovery easier as it is possible for alternative sites to take over the functions of a site which has a hardware failure. On the other hand, as sites become more dependent upon each other, a failure at one location could cause universal shut down.

The terrorist attack: Royal & Sun Alliance

On Saturday 15 June 1996 one of the biggest bomb blasts in peacetime Britain ripped through the Arndale shopping centre in Manchester, injuring 206 people. Longridge House, home of the Royal & Sun Alliance insurance company bore the full brunt of the blast.

The building was almost destroyed, and as it contained the company's mainframe, it was initially feared that core business operations could be seriously jeopardised. In spite of extensive damage to all seven floors, Royal & Sun Alliance staff in Liverpool detected signs of life in the IT system. But when the fire brigade cut off power to the site later in the evening, all systems effectively died.

Fortunately for Royal & Sun Alliance they had a contract with data recovery partner Comdisco. Key Royal & Sun Alliance staff were rushed to Comdisco's Warrington centre. With the help of recovered back-up tapes, Comdisco's mainframe was able to mimic the characteristics and requirements of the insurance company.

Meanwhile, 200 Royal & Sun Alliance staff were switched to the company's Liverpool centre. When the sun came up on Monday morning the company had a makeshift, temporary switchboard which was up and running for 9am. Luck and effective disaster recovery planning meant that not a single day of trade was lost.

- List the features mentioned that were taken into account in the contingency plan
- The final sentence attributes part of the successful outcome being due to luck. Where did luck come in?

Problems associated with communication

The growth of data communications as wide area networks become a wide spread poses data security problems as well as expanding the possibilities of breakdown. Data is very vulnerable to illegal access. The company's IT security policy should state exactly how to prevent problems occurring and what to do if they do occur.

Risk analysis

Risk analysis involves determining what the risks are and designing appropriate counter measures. The risks will be different for different systems and will change over time.

The risk review may be made either on a quantitative basis. For example: expected annual loss = Probability of fire over 10 years (0.02)* cost of fire (£1,000,000); or on a subjective basis by consulting with staff and using knowledge of the business.

Fire: William Jackson, food manufacturer and retail group

July 6 1995 was pay day at food manufacturer and retail group William Jackson, based in Hull. Anticipating her busiest time of the month, payroll administrator Diane Rush was at work by 7am preparing wages data to be transferred from the company's AS/400 computer to BACS, the system for paying wages directly into banks.

Diane noticed smoke billowing from the food factory next door and phoned Safetynet, the company's disaster recovery partner at 7.45am. All staff had to be evacuated from the offices as fire swept through neighbouring buildings.

Although tapes and a modem had been recovered, the continuing blaze meant that the payroll could not be processed on site. Under protection from the fire brigade, Safetynet successfully rescued the AS/400 from the ashes and installed the charred machine alongside their own mobile unit. Using a parallel recovery process, the payroll was successfully relayed to BACS and all 2500 staff were paid on time.

(*Business Computer World*, March 1997)

- List features from the William Jackson contingency plan

Software packages are available that provide a checklist of all recognised dangers for a particular type of installation or activity that can be filled in. The package attaches weights to risks and provides an index rating of risks.

A security incident can lead to loss of data confidentiality, integrity or availability which in turn may give rise to impacts of direct or consequential harm. Security management means reducing risks to acceptable levels by implementing procedure to lessen the likelihood or impact of a threat.

Computer security involves reducing the risk to electronic data. A risk is made up of three factors: the potential threats to the data; its vulnerability and value to the organisation. Risk will increase if any of these factors increases. Thus risk analysis is the process of assessing vulnerability to threats, the potential losses, the current security controls and identifying possible counter measures to reduce risk. It compares the cost of the potential loss with the cost of countering it.

Determining the threats against and vulnerabilities of a particular computer system is no easy task. The value of the data to the organisation will vary. Each situation is unique, and the overall risk will differ. To undertake a risk analysis, a manager would need to know a considerable amount about the organisation: its aims and objectives; its history and future plans. This knowledge is as important as information concerning risks to the hardware and software. Managers may have statistical data on power failures, crime levels, and so on when making decisions. However, it is necessary to consider whether they reflect the risk to this particular organisation.

Risks are assessed as:

- **Now** – happening and tolerated

- **Tomorrow** – not happening but likely

- **Never (we hope)** – unlikely but potentially catastrophic

The criteria likely to be used in selecting appropriate measures are:

- the cost of measures to protect against and recover from failure

- the potential cost of the loss of data

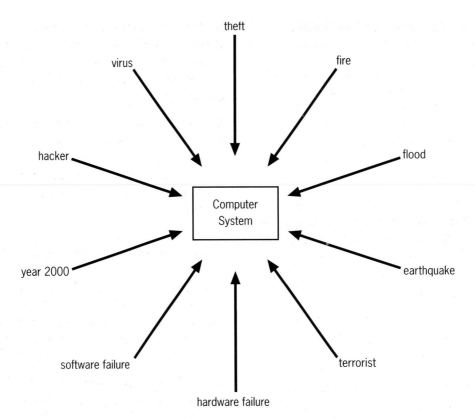

Figure 17.3 Threats to a system

- the inconvenience to staff – Security measures are useless if everyone by-passes them

- the statistical likelihood of the problem occurring

Disaster Recovery Plan

Any organisation that is in any way dependent on a computerised information system, needs a plan which details how operations can be resumed after a disaster such as fire, flood, power disruption, or sabotage has disrupted its normal processing. It is usual for a disaster recovery plan to provide for immediate access to alternative computer hardware. In setting up a successful disaster recovery plan, it is essential to identify the most critical business functions and work out how each is vulnerable. It is then possible to establish the hardware, software, files, and human resources required to resume processing of these critical applications if disaster occurs. Personnel must be trained to follow the recovery plan correctly and a step-by-step course of action for implementing the plan must be drawn up.

For successful recovery from disaster, data for critical applications must be backed-up and storage media taken off site. A business will have many back-up tapes or disks, including the most recent global back-up and several incremental back-ups. Backup tapes must be date stamped as the order in which data is restored after any data loss is vital. The disaster recovery plan must include the order for restoring.

Distributed processing facilities can be used for disaster recovery. If a business has distributed processing to local branches then the branches can operate for several days if the main computer centre is lost. Another option available to large organisations is the use of multiple computer centres. Disaster recovery firms are used by businesses that do not wish to build up their own back-up facilities.

Contingency plans and avoiding trouble

A **contingency plan** is a plan for recovery from a failure. It is a planned set of actions that can be carried out if things go wrong so that disruption is minimised. It is necessary first to identify what could go wrong and then what should be done if it did. For example, if you are organising an event to take place out of doors, it would be sensible to have a contingency plan in case of rain. Your plan could be to make a provisional booking in a local hall or provide guests with umbrellas!

There is much that can go wrong when using an IT based system, and it is important that any potential problems are identified before they occur. Some failures in a system can be avoided. Those that cannot be avoided, can, with an appropriate contingency plan in place, be prevented from having disastrous consequences.

Disastrous effects from hardware failure might be addressed by having a maintenance contract with a specified call out clause to a specialist company. Alternatively, a business could make an arrangement with another firm or bureau to run their software in the event of major problems.

The use of **fault tolerant computer systems** provides protection against hardware failure. A fault tolerant computer has extra hardware such as memory chips, processors and disk storage in parallel. Special software routines or built-in self checking logic detects any hardware failures and automatically switches to the back-up device. Some systems automatically call in the maintenance engineers. Faulty parts can be removed and repaired without disruption to the running of the system.

The chances of damage from fire, tempest and flood can be minimised by having detectors in the computer room with CO_2 extinguishers available; by placing the computer room on the upper floor of the building and using fireproof safes for disks and back-up tapes. Breakdown of power supplies can be avoided by having an uninterrupted power supply and a standby generator.

Malicious damage to the system can be addressed in part by having computer equipment in rooms protected by swipe card or other security methods. Strict codes of conduct need to be enforced to avoid illegal access to systems. For example, rules banning the use of personal floppy disks on the work stations can be established.

Hacking and associated problems could be countered by such measures as checking all accesses to the system, and only allowing three attempts before shutting down a terminal. The encryption of all data sent along communication channels should be considered.

Recovering lost data

Whatever precautions are taken, a disaster might still occur. It is only when such a disaster does occur that the adequacy of contingency plans can be seen. Loss of data should be avoided by keeping back-up copies, on tape or on disk in case of problems.

General loss of data from a variety of causes can be addressed by backing-up the files on a regular basis and storing copies off site. Backing-up daily, that is overnight when the computer is less busy, is normally sufficient. More frequent back-up may be needed in the case of constantly changing data and this can cause a problem. The rise in use of distributed systems can be of use here, as data can regularly be backed-up by sending copies to an alternative site.

Large organisations such as supermarkets and banks, have more than one computer site in case of hardware problems. Contingency plans are set up so that the critical tasks of one site can be carried out at another. An example of such a contingency put into action is illustrated in the description of the terrorist attack which caused problems for Royal and Sun

---

A system is designed to avoid failure occurring. However, if failure does occur, contingency plans should be in place to allow for recovery that will prevent loss or disaster.

Examples of disaster AVOIDANCE measures

- Use of virus scanning software
- Use of fault tolerant components
- Use of smoke detectors in buildings
- Uninterrupted power supply or standby generator
- Strict password management policies
- Extra network links
- Regular maintenance

Examples of elements of CONTINGENCY PLANS to prevent disaster in case of failure

- Back up strategies
- Contract with a company specialising in disaster recovery (see Royal and Sun Alliance case study page 192)
- Arrangement to use the hardware of another firm or bureau in case of failure
- Distributed systems maintaining duplicated data on different sites

Alliance. Alternative network routes are also maintained so that, if one cable is severed, for example, an alternative route can automatically be used.

Nearby companies can come to an arrangement to provide emergency facilities for each other in case of disaster, so that vital systems can be run. Specialist disaster recovery firms will help in the drawing up of a disaster recovery plan. They will have the expertise and the equipment that will them to act swiftly when necessary.

What to remember
Implementing a backup and recovery strategy takes a lot of planning.
Here are some tips on getting started.

Backing up
Backing up data is the process of making a copy of data stored on hard disk drives, be it on tape or other portable media. The purpose of backing up data is to ensure that the most recent copy of the data can be recovered and restored in the event of data loss.

Archiving
Archiving data is the process of copying data from hard disk drives to tape or other media for long-term storage and can be used to free hard disk space by off-loading seldom used data to tape.

Disaster recovery
Archived data can be used to recover from disasters such as fires or floods, where data stored on primary devices is destroyed. In a well-planned scheme, a copy of the business data is delivered to a secure offsite storage facility. A tape rotation system can provide data version history, depending on the needs of the business and the criticalness of the data.

Storage methods
Although most business data is stored on hard disk, some data such as customer records can be stored on tape. Using tape has many benefits including cost savings. However a high data-transfer rate is essential for maintaining productivity.

Strategy
A well -planned back-up, restore and archival strategy needs to embrace the following
- A plan and schedule for regular backups of critical data.
- Short and long term archival plans.
- A disaster recovery plan that includes offsite storage of archived data.

This in itself is not enough and a backup strategy should include the following:
- Verification that backup data can be recovered and read at any time.
- A management plan to ensure that backups are performed at the appropriate intervals
- The design and use of a rotation scheme that provides as much version history for data as an organisation needs.

The first step in a backup strategy is to verify that data backed up on tape can be restored to hard disk when needed. It is important that all recording and playback equipment is compatible.

Next, a good management plan will ensure the back-ups are performed at the appropriate intervals. Most organisations backup daily but back-ups may need to be performed more or less frequently.

Lastly, a rotation scheme that provides an appropriate data version history must be selected. The Grandfather, Father, Son scheme uses 12 tapes to allow recovery of three months of data on a daily, weekly and monthly basis.

The longer an organisation keeps its data, the more portable media are needed. When determining how long your data must be retained, consider any legal obligations for your type of business as well as all other business needs. If your data needs to be stored for a long time, you should periodically inspect the media for signs of damage. Finally, choose a facility that is climate-controlled, secure and easily accessible.

Grandfather, Father, Son Rotation Scheme

Son 1 Monday	Every Monday	Weekly
Son 2 Tuesday	Every Tuesday	Weekly
Son 3 Wednesday	Every Wednesday	Weekly
Son 4 Thursday	Every Thursday	Weekly
Father Week 1	First Friday	Monthly
Father Week 2	Second Friday	Monthly
Father Week 3	Third Friday	Monthly
Father Week 4	Fourth Friday	Monthly
Grandfather Month 1	Last Business Day Month 1	Quarterly
Grandfather Month 2	Last Business Day Month 2	Quarterly
Grandfather Month 3	Last Business Day Month 3	Quarterly

from Storage, VNU Business Publications November 1999

Crest

Case Study

Shares have been bought and sold on the London Stock Exchange since 1773. In 1986 the 'Big Bang' introduced a computerised share-trading system, called Talisman. The system did not solve the problems caused by increasing volumes of trade. Talisman was out-of-date before it went on-line and could not guarantee settlement of accounts within ten days of a sale.

A new system, called Taurus, was needed. It was planned to be introduced in 1991 but costs rose from £50 million to £80 million and the implementation date kept being put back. Taurus was eventually abandoned in 1993 – a costly and embarrassing mistake.

In 1997 another system, Crest, went on-line, involving 69 market makers, dealers, fund managers and stockbrokers, responsible for 94 per cent of the trading. It aimed to reduce the paperwork in share-buying and to cut the time taken to pay for shares to around three days.

The Crest system was created in three years at a cost of around £35 million. It only uses known, fully-tested, systems so that there is less danger of break-down. There are two mainframe computers at different sites – one at London's South Bank and the other near Heathrow Airport. If a computer crashes, one site backs up the other and can take over the other's role within half an hour.

The brokers use very high specification PCs to buy and sell shares, connected by a wide area network to the Crest computers. Each PC contains a small tamper-proof circuit board that gives the PC a unique authentication code, called the key. This key means that Crest can be sure that messages are from legitimate brokers and not from hackers. The key is not known to the user. If you try to hack into the hardware that generates the key the contents of it will be destroyed.

The Stock Exchange believe the Crest system, with a very secure network and extensive encryption, is as secure as possible. Forging the old paper share certificates would be much easier.

Some people in the computer security business are not so confident. They believe that Crest's encryption keys could still be broken by hackers working in collaboration on the Internet. It is interesting to note that Crest took out a £20 million insurance policy to cover just about all eventualities, including security failures.

- What could have caused costs to rise from £50m to £80m in Taurus?
- Why was Crest successful when Taurus failed?
- List the threats identified in the article and the security measures taken to avoid them. □

Figure 17.4 The Stock Exchange

Summary

IT users must

☐ recognise the potential threats to their information systems

☐ plan to avoid disasters leading to loss of data

☐ have contingency plans to be able to recover any data lost

Dangers include hardware failure, software failure, incompetence, malicious damage and accidental failure, for example fire. Measures taken to prevent problems will depend on the perceived risk and the cost of potential loss of data.

Disaster recovery questions

1 A fire service relies on computers to direct the driver of the fire engine to the fire by the quickest route. Experience suggests that a second back-up computer will be needed at a cost of £30 000, although it may never be used. Councillors making the decision on what to do are reluctant to spend the money. Advise them what to do. (8).

2 'Only 60 per cent of companies in the UK have adequate disaster recovery plans'. Discuss this statement, including in your answer:

   why such plans are necessary

   the potential threats to information systems

   the contingency plans needed to combat these threats (12)

3 'Information stored in a typical computer system is more secure than information stored in a typical manual filing system'. Discuss this statement, including in your answer the threats to security to a manual filing system and those to a computer system and the ways of minimising the threats to a computer system. (12)

4 Define the term risk analysis. Explain the circumstances under which such analysis is important. (8)

5 Suggest some essential elements of a disaster recovery plan for

   a) a major high street bank (10)

   b) a small corner shop (8)

   c) your School or college administrative system (10)

6 Define the term fault tolerance computer system (2)

7 'Information Systems are mission critical, the consequences of failure could prove disastrous.'

   Discuss this statement, including in your discussion:

   • the potential threats to the system

   • the concept of risk analysis

   • the corporate consequences of system failure

   • the factors which should be considered when designing the 'contingency plan' to enable a recovery from disaster

   Quality of language will be assessed in this question. (20) *NEAB 1998 Paper 4*

8 List five distinctly different potential threats to an information system. Give one way of countering each potential threat. (10) *NEAB IT04 1999*

Activity: Disaster Avoidance

Dawson and Mason Ltd is a medium sized manufacturing company which, over the last few years, has become more and more dependent on IT. The company has decided to adopt the following disaster avoidance plan, which consists mainly of common sense practices – reasonable and inexpensive measures to avoid a disaster that could cost the company thousands of pounds in loss of revenue.

Hardware and Software Inventory

Each department of the company must keep a detailed inventory of all computer equipment and make sure it is up-to-date. The inventory should cover all hardware, software, communications equipment, peripherals and back-up media, including model and serial numbers.

IT Facilities

Administrative procedures are a vital part of security and disaster avoidance:

- All perimeter doors must be kept locked if the room is unattended.
- Windows and other access points should be kept locked if unattended.
- Access should be restricted to authorised personnel.
- Strangers seen entering office areas should be challenged and asked for identification.

Local Area Networks

- Back up of server files is automated on a nightly basis.
- A rotation schedule for back up tapes should be used and several generations of back-ups kept.
- Two copies of the server back-up tapes are generated. One back-up copy is available on-site in case recovery is necessary. The other copy is stored off-site.
- Disk mirroring is used to duplicate data from one hard disk to another hard disk. Mirrored drives operate in tandem, constantly storing and updating the same files on each hard disk in case one disk fails.

- In the case of vital and sensitive data hot back-up is used. Two file servers operate in tandem and data is duplicated on the hard disks of both servers. If one server fails, the other server automatically takes over.
- A UPS (uninterruptable Power Supply) has been installed for every LAN server. The batteries should be checked regularly to ensure that they are not drained and that they are charging properly.

Storage Media

- Magnetic media should be kept away from sources of heat, radiation, and magnetism.
- Back-up media should be stored in data safes.
- Vaults used for storing critical documents and back-up media should meet appropriate security and fire standards.

Preventing Theft

- If a computer is used to store sensitive data, the data should be encrypted so that the data cannot be accessed even if the equipment is stolen.
- Anchoring pads and security cables are used to prevent equipment from being stolen.

Employee Awareness

Security and safety awareness is critical to any disaster avoidance program. A lot of problems will be avoided if employees have been trained to look out for conditions that can result in a disaster.

1 Disasters can be caused deliberately or accidentally. Give one example of a disaster caused deliberately and one caused by accident. (2)
2 Explain the reasons for keeping an inventory of hardware and software. (4)
3 Dawson and Mason's plan does not mention computer viruses. Describe the actions that should be taken to prevent infection by a virus. (4)
4 Describe the process of encryption. (4)
5 List further measures that could be taken to prevent theft. (3)

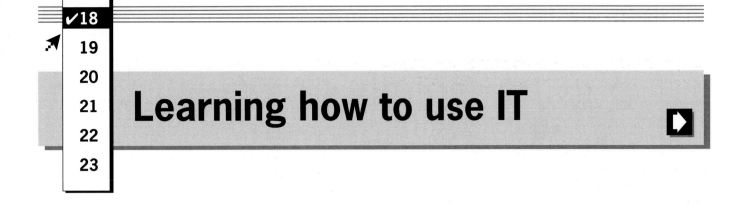

Learning how to use IT

Training

Training is *'the acquisition of a body of knowledge and skills which can be applied to a particular job.'*

Today the job market is very flexible. People do not stay in one single job for the whole of their working life, but are likely to make one or more major career changes. As well as this, the nature of a particular job changes as new technological advances are made. This is particularly true for IT users. New hardware is appearing every few months. New versions of software appear every few years. Employers need to give IT training to their work force on a regular basis: training should consider both the needs of the company and the needs of the individual.

There is a range of training methods available to employers, involving employees being taught new skills by some one else.

- **On the job training:** an employee learns new skills while doing the work. This would be appropriate when a new employee is recruited to work on a till at a supermarket. He might watch an experienced operator at work, and then work under the guidance of the experienced colleague.

- **In house courses:** short training courses are run, usually for small groups, on-site. Such a course would be appropriate when a software upgrade has been installed, or a new system implemented. These are only appropriate if the appropriate hardware and software is available as well as a training area away from the stress of the workplace.

- **External courses:** organised by specialist training companies or FE colleges. Such a course would be appropriate when one or more employee needed to learn to use a specific software package for the first time. The course would probably be held at the college but the trainers could provide the course at the workplace.

Different users require different levels of training depending on their job and its requirements. Some jobs involve IT tasks that are repetitive and specific; others call for a more open-ended use. It is crucial that the level and pace of the training fits the user and the task. Someone who has not used a computer before will need initial training. A more experienced user may need training in higher level skills. Users of special equipment (for example a bar code scanner), or special facilities (for example e-mail) will need specific training.

A database package such as Microsoft Access can be used at a number of different levels. Thus training needs to be available which meets these differing needs. For example, an operator of a database whose job is to enter data, may only need to be taught how to access an existing database, add and modify records. The tactical manager who uses the information from the database as a tool in decision making may need to be taught how to produce standard reports and carry out a range of queries. A database programmer will need to learn much more about the package. She must know how to set up a new database and amend an existing one; how to write reports and macros and much more besides. It would be inappropriate for all the above users to attend the same course.

Methods of acquiring skills

SKILL-BASED/TASK-BASED TRAINING

Some training is based on learning a skill, such typing on a keyboard, using Windows or using a program like Word or Access. This skill could be used in many circumstances. This type of training is often offered in standard courses which teach participants how to use a range of facilities according to their current skills level. Such training can be fairly open

Figure 18.1 Methods of gaining expertise in software use

ended, leaving the trainee to decide exactly how to incorporate the skills learnt into her current job.

Other training is based on learning how to do a particular task. Examples of this could be: how to use a hand set for recording electricity meter readings; how to process a sale with a Visa card and how to load transaction and master file tapes in a batch processing system. In such circumstances, the training is designed specifically for the occasion. It is more likely to take place in-house. Skills acquired will be very specific and will often not be transferable into other situations.

■ SKILL UPDATING

Employees will need to update their skills on a regular basis, particularly as new software or hardware is installed. Old skills may be superseded when new systems are installed. They may also need to refresh old skills if they have not used some piece of hardware or software for some time.

■ FORMAL TRAINING COURSES

Training courses, run by IT training companies or FE colleges, tend to be expensive. If the courses are held off-site, then employees will incur travel and subsistence expenses as well. Usually class sizes are small and the participants will receive individual help. They will be able to ask questions and benefit from human interaction. This is particularly useful for beginners. The hardware used should be similar to that used by the trainee at work.

There are other ways of acquiring certain IT expertise other than attending courses, particularly for popular commercial packages. Some alternatives are given below.

■ READ THE USER MANUAL

These comes free with the software. They claim to teach you all you need to know, but they vary in quality and are not always very easy to follow. A manual can prove to be a good reference book in case of a problem. Manuals can be used when and where the user wants and progress can be made at the individual's own pace.

Training Courses offered by Burtwhere Technical College
PC training for local industry and commerce

Microsoft Excel Introduction
Microsoft Excel Intermediate
Microsoft Excel Advanced
Microsoft Word Introduction
Microsoft Word Intermediate
Microsoft Word Advanced
Microsoft PowerPoint
Microsoft Project
Introduction to PCs/Windows
Microsoft Works Introduction
Microsoft Works Intermediate
Lotus 1-2-3 Introduction
Lotus 1-2-3 Intermediate
Lotus 1-2-3 Advanced

Oracle courses:

Introduction to SQL
SQL Report Writer
SQL Forms
Oracle database Administration
SQL programming for Finance
Using SQL RPT
Oracle DBA
Using Oracle via Microsoft Access
Introduction to UNIX
Intermediate UNIX

Places strictly limited to 6 per course

Course are tailored to the needs of your organisation

Group bookings available (£495 + VAT per day)

Figure 18.2 IT courses available in a local college

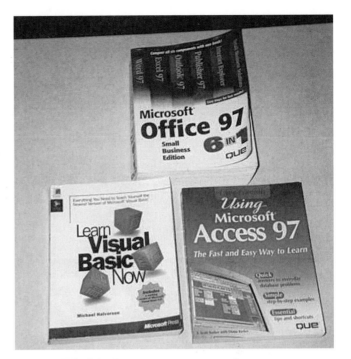

Figure 18.3 IT books

■ BUY A BOOK

These are commonly available for popular software packages from book shops. Many books are available at an introductory level. Titles include *Access for Dummies*, *Field Guide Access* and *Ten Minute Guide to Access*. These books take the reader step-by-step through the basic functions of the program. There are also books available which introduce the more advanced features of such packages. Books tend to be more user friendly than the user manual and include good screen shots. They, too, can be used when and where the user wants. The book also provides a useful guide to dip into from time to time when a particular problem is met.

■ ON-LINE TUTORIAL

This is usually provided as part of the software package. For example, the user could load up an example of a database and the tutorial would demonstrate examples of tables, forms, reports and queries using animated graphics. Such tutorials are usually interactive and require the user to make responses which will re-enforce their understanding. Most tutorial programs are easy to use and user friendly. Some have to be gone through in order which makes them inappropriate to use when a particular problem is to be investigated. On-line tutorials require access to a computer, but allow the user to go at his own pace.

■ ON-LINE HELP

This is available with most software today. Many software packages on a PC allow the user to press the F1 function key, or choose from a menu to enter the Help program. The user is able to search for a topic

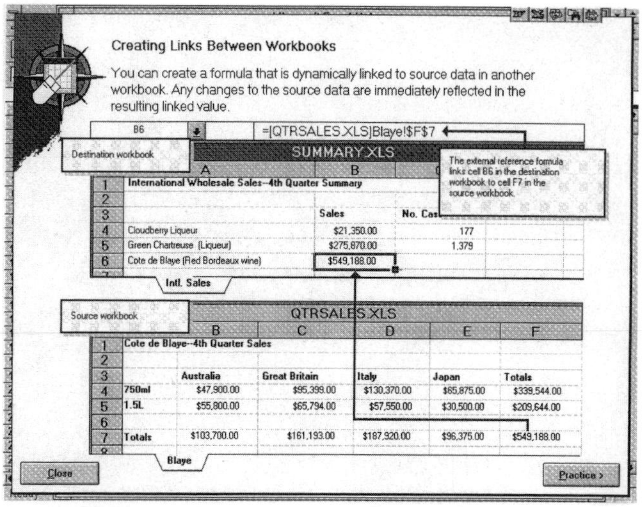

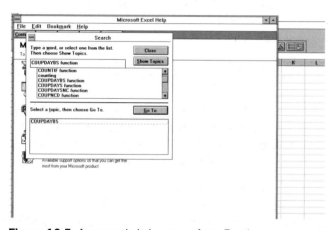

Figure 18.4 An example screen from an on-line tutorial

Figure 18.5 An example help screen from *Excel*

and be given an explanation as well as examples of use. Sometimes animated demonstrations and cue cards are also available. Cue cards are small help windows that appear over the application screen to help the user. On-line help has the advantage of being always available while the software is in use.

Context-sensitive help provides information on the specific function currently being used when the Help is invoked. Such help is easy to use, but explanations in many packages can be confusing for beginners. It is very useful for the experienced user who wishes to explore a feature that she has not used before. An extensive Help facility uses a considerable amount of disk space (4 Megabytes for Microsoft Access). On-line help has no additional cost and can be used whenever information is sought with the minimum of effort.

■ INTERACTIVE VIDEO INSTRUCTION

Such a training video could be on CD-ROM or on video cassette. Some software packages include a free training video. Videos can provide a useful means of training a number of people at the same time. They can give training away from the distractions of the job without incurring the cost of hiring a trainer. For individual use, they can prove frustrating to use if a video player is not available next to the computer.

■ COMPUTER BASED TRAINING

CBT would use a special program that teaches a specific skill. Such programs exist for teaching software. The programs are interactive and the user can go back to parts that he has not understood fully. Modern CBT uses multi-media. Every user would take a different route through the program depending upon their learning needs. CBT could be used in the workplace if compatible hardware were available. The trainee is able to progress at her own pace.

Developing IT training strategies

IT training in a company needs to be planned and strategies developed, based on that company's objectives. Training must be planned to coincide with the installation of new hardware and software. The training needs of specific jobs must be established. Employees are becoming more computer literate, so training needs are changing over time.

Strategies will include who needs to be trained, what training they need and where this training will take place. Large companies may have their own training suite and in-house trainers while small companies can probably only use outside agencies.

User Support

Most respectable software producers provide support for users. Support is sometimes provided free under the product warranty, or an entitlement to help can be bought for a fixed period of time. When software is bought, the purchaser should send off a card to register with the publisher. His details are then added to a database of users.

The extent of the support available with a product will be a major factor in determining its credibility. Corporate users are unwilling to take on unsupported

Training for use of a college's new management information system

A college management has decided to purchase and install a new management information system. Its successful implementation and use will depend upon adequate and appropriate training being provided for all the users of the system.

Users of the system can be divided into the following categories:

- the MIS team;
- data entry clerks;
- student services staff dealing with student enquiries;
- heads of subject and other senior staff.

Each category will require a specific training plan to meet the particular needs of their job.

The MIS team will need to acquire an in depth knowledge of all aspects of the software. They will need to be able to customise the software to the particular needs of the college. This is likely to involve some form of programming in a language supported by the system. Although already experienced IT users and programmers, to gain the depth of knowledge of the new software that they will need, they are likely to have to attend a training course of several days' or weeks' duration.

All other training can be provided in house. The data entry clerks will require training in limited parts of the system, and this can be provided on the job.

For the student services staff, a number of in-house training sessions should be arranged, introducing them to different aspects of the package in a planned fashion and allowing plenty of time for practice in between sessions. At this stage it will be vital to provide ready support.

The management users will require an overview of the whole system, delivered in house to raise their awareness of the facilities available to them. Brief courses should be offered on specific aspects of the system.

software, however good it appears to be. Without adequate support, software can be virtually useless if a user is unable to work out how to implement a particular feature. Well supported software tends to be more expensive than that which is unsupported; the degree of support is reflected in the cost.

User support is needed when a user gets stuck in a program, finds a bug or wants to do something but is unable to find out how to do it from the manual or anyone around him in the workplace.

Support is provided in a variety of ways. The user guides and on-line help described in the previous section are both examples of user support which are freely available to the user. Publishers of widely-used software produce newsletters which are sent to all registered users. These include articles on how to use advanced features, tips on shortcuts, suggested ways to solve tricky problems as well as ways of extending the package's use. These newsletters provide a forum for users to share ideas and problems.

For complex software, user groups are set up, where users can get together to share problems and ideas. Such groups can either meet physically, or via bulletin boards on the Internet. Software houses provide a considerable amount of information on packages via such bulletin boards, often in the form of Frequently Asked Questions (FAQs). (See Figure 18.6.)

Help desk

Perhaps the most common form of support is the help desk. With this, someone is available at the end of a telephone line to answer users' queries. Such help can often be available for 24 hours a day.

When a user phones up she will need to give some information. This should include the name of the software, the version number, the type of computer it is being run on (the type and speed of processor, and the size of memory will be needed here), the netware or operating system in use, her name and telephone number and the nature of the problem. The problem description might include the error number being displayed. The software registration number will also be required. The help-line operator will record this information together with the date and time. Ideally, the operator will be an expert in the program with a computer in front of them so that he can try to mimic the user's problem on the screen.

A number of problems can arise when using telephone help desks. The waiting time on the phone can be considerable at certain times of day. Many of these services are popular and it is not unusual to spend a lot of time listening to 'music' whilst a call is queued. Some problems are common, and are easy to answer, but other, complex ones will not be able to be answered on the spot as several experts may need to confer. In these circumstances, it will be necessary

[Word for Windows Support Options]
What's Hot!

Microsoft Word Newsgroups

Peer-to-peer discussion area for Microsoft Word. Reference Free Software Free drivers, utilities, up-dates and information to help you use Microsoft Word to its fullest.

Word for Windows Knowledge Base

Why reinvent the wheel? Chances are that someone has already asked us the same question that you have, and you'll find the answer here.

Word for DOS Knowledge Base

Why reinvent the wheel? Chances are that someone has already asked us the same question that you have, and you'll find the answer here.

Microsoft Word makes word-processing tasks easier and faster. Support When You Need It

The best way to get answers fast is to visit Microsoft Support On-line. There you will find answers to frequently asked questions, responses to technical questions via the Microsoft Knowledge Base, free drivers, service packs, code samples, and other useful files. In addition, you can get further information on specific support options, phone numbers, and services. Click the links below to visit

Microsoft Support On-line

Microsoft Word for Windows Support
Microsoft Word for the Macintosh Support

This site is best viewed with:
Microsoft Internet Explorer 2.0

[Frequently Asked Questions]
Microsoft Word for Windows, version 6.0

Click on the question to display the answer, or click here to view all the questions on a single page.

- In Microsoft Word 6.x, I imported data using ODBC. Why can't I do this anymore?

- Why do I get messages about 'Corrupt' documents from the FindFast program?

- What do I need to do to convert my Word 6.x documents into Word for Windows 95?

- What does the tab for 'Shortcut to Old Templates' mean in my File New dialog box?

© 1996 Microsoft Corporation

[Frequently Asked Questions]
In Microsoft Word 6.x, I imported data using ODBC. Why can't I do this anymore?

Up-dated Wednesday, November 08, 1995

The Word stand-alone package does not include the ODBC files and drivers. The files are included in the Microsoft Office retail package and in the Microsoft Access stand-alone retail package. If you are trying to import dBASE or FoxPro data files, you can use the dBASE file converter. If the dBASE/FoxPro converter is not currently installed, run Setup and use the Add/Remove button to add the converter.

© 1996 Microsoft Corporation

Figure 18.6 FAQ

for the help desk operator to return the call of the user at a later time.

If the problem is not time critical, then e-mail could be used as an alternative to the telephone. This has the advantage of smoothing out the demand, so that the operator can answer queries in order, throughout the day. A priority system could be used which would ensure that critical enquiries were answered first. She or he will be able to spend all their time finding solutions to problems without being interrupted by a ringing telephone. From the user's point of view, the use of e-mail avoids wasted time on the telephone. However, instant answers to simple problems will not

The following information is likely to be recorded when a user contacts a help desk:

- Unique log number (issued to user by help desk)
- Date and time of call
- User name/phone number
- Software serial number
- Version number of software
- Details of problem
- Details of hardware platform
- Details of operating system
- Description of any error messages shown

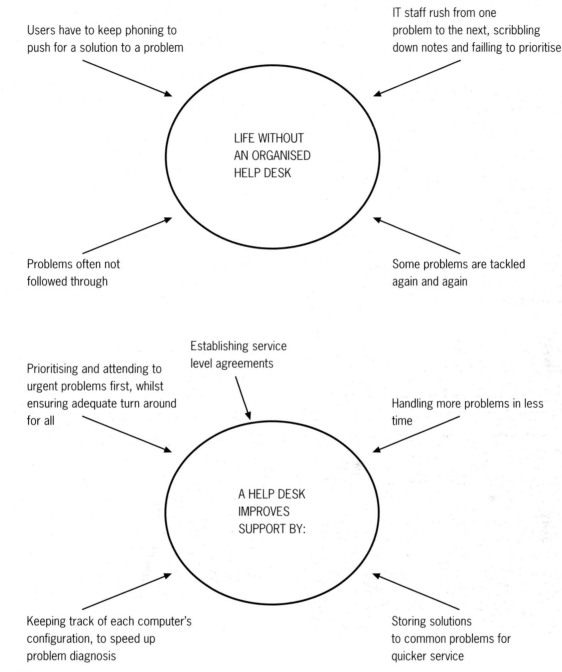

Users have to keep phoning to push for a solution to a problem

IT staff rush from one problem to the next, scribbling down notes and failling to prioritise

LIFE WITHOUT AN ORGANISED HELP DESK

Problems often not followed through

Some problems are tackled again and again

Prioritising and attending to urgent problems first, whilst ensuring adequate turn around for all

Establishing service level agreements

Handling more problems in less time

A HELP DESK IMPROVES SUPPORT BY:

Keeping track of each computer's configuration, to speed up problem diagnosis

Storing solutions to common problems for quicker service

Figure 18.7 The advantages that an organised help desk brings

be possible. E-mail lacks the opportunity for human interaction offered by a telephone conversation.

A help desk is likely to use a computerised call logging system, giving a unique call reference number for each user query. This allows performance to be monitored as well as providing a reference for follow-up calls. The help desk employees should have access to a file of registered users, to enable a check that the caller is entitled to help whenever a phone call is received. A computerised database of known

errors and their solutions, together with frequently asked question should be available. This could take the form of an expert system.

The performance of the help desk should be monitored to ensure that it provides a high level of service. Service performance indicators could include: the number of calls logged daily or per hour; the response time to the initial call and the time taken to resolve the problem for the user; the number of repeat calls on the same problem for a particular user. It may also be possible to record the level of the user's satisfaction of the problem resolution, using a qualitative code.

Help desks can categorise problems as follows:

■ 1st LEVEL

User problems that can be solved immediately, usually over the phone. These are recognised, standard problems that are well documented and have a straight forward, clear response.

■ 2nd LEVEL

These are also recognised problems, but ones that require physical work and need a visit to the user's computer.

■ 3rd LEVEL

These are new, unrecognised problems, which can take some time to solve. Once solved the solution should be documented so that it is available if a similar problem occurs elsewhere.

Documentation

A range of documentation is available for a major package. A typical package will contain a 'Getting started' booklet which contains basic information on installing and beginning to use the program. A 'User guide' will take the user step by step through the features of the package. This will be useful for a first time use. A 'Reference guide' may also be included; this will allow the more experienced user to find out how to perform specific tasks. Functions are usually described in alphabetical order. Some packages also include a 'Quick tips' guide which summarises the most common features of the package.

Industry standard packages, such as Microsoft Office, have a huge user base. Such a base means that extensive support options will develop, independent of the software house that produced the software. A range of books will be written, as well as articles in magazines. Specialist bulletin boards will be established to allow users to exchange tips and hints.

It is very likely that a user of such a package will not be the only user within her organisation. Thus help can be gained from other users and problems and solutions shared.

All types of software should be supported with appropriate documentation. In fact, the quality of the documentation will be one of the criteria considered when choosing software. Documentation is as necessary for bespoke software as for off the shelf packages.

Different types of user will have differing documentation needs. The technical support team will need documentation that provides installation instructions including disk, peripheral devices and memory requirements. They will need to have documentation of back up routines and recovery procedures. An explanation of all technical error messages, together with the necessary action to correct them, will be required.

A data entry clerk, using the same system, will need clear instructions in how to use the functions needed for his tasks. Details of appropriate error messages due to incorrect data entry should be included, together with a list of useful keyboard short cuts.

Summary

Training is vital in a highly skilled area of business like Information Technology. The rapid pace of change means that training is not something that happens when you start a new job but is continuous.

Different levels of training are required for different situations, for example beginner, intermediate, refresher course.

Training courses may be:

☐ on the job training

☐ in house

☐ external

Training can be skill based or task based.

Other methods of training include:

☐ reading user manual

☐ buying a book

☐ on-line tutorial

☐ on-line help

☐ interactive video

☐ CBT

Knowledge of software is backed-up by user support. Often this is a telephone line providing access to immediate advice for a particular problem. Software providers may also provide help on the Internet.

A range of support can be available to users. This can take the form of:

☐ on-line help

☐ a help desk

☐ documentation – different forms available for different categories of user

Learning how to use IT questions

1 For a word-processing package such as Word or WordPerfect, discuss the possible contents for three different training courses. For each course, describe the type of user for which it is aimed. (9)

2 A company has just bought a copy of a new presentation program, which is to be used by two employees. Outline three possible ways in which the employees could learn to use the software giving relative merits of each method. (6)

3 You are asked to advise an organisation on the introduction of a new software package.

a) With the aid of **three** examples, explain why different users may require different levels of training. (6)

b) Following the initial training you advise subsequent training for users. Give **two** reasons why this may be required, other than financial gain for the training agency. (4) *NEAB 1997 Paper 3*

4 Using examples, describe the difference between skill based training and task based training. (4)

5 Each day a software house logs a large number of calls from its users to its support desk.

a) Describe how the support desk might manage these requests to provide an effective service. (3)

b) Describe **three** items of information the support desk would require to assist in resolving a user's problem. (3)

c) The software house receives complaints from its users that the support desk is providing a poor service. Describe **three** reports that the software house could produce in order to examine the validity of this claim. (6) *NEAB 1997 Paper 3*

6 Many software companies offer a user support line.

a) Describe three items of information a user support line would log when taking a call from a user. (3)

b) Many user support lines need to share problems and potential solutions between a number of operators who are answering calls. Describe one method of achieving this. (3)

c) Some user support lines also offer a mailbox facility to enable users to log their problems using e-mail. What advantages does this have for

i) the software user

ii) the user support staff. (4) *NEAB Specimen Paper 3*

7 The Head of a school decides to adopt an IT package to maintain pupils' records of attainment. The package will be used throughout the school.

a) (i) Identify **three** different potential users of this package. (3)

ii) With the aid of examples, describe the different types of documentation that each user will require. (6)

b) Training in the use of this package may be provided by a variety of methods other than formal training courses. Describe **two** possible alternative methods. (4) *NEAB 1998 Paper 4*

8 A school is considering the introduction of an IT based display system to replace the existing daily newsletter read out at registration. The new system will have several display screens at various locations throughout the building. The system will be operated via a PC which is connected to the school network.

The existing daily newsletter is currently typed by a member of staff in the school office who will be responsible for the new system. The introduction of this system will cause considerable change for the member of staff involved.

a) Describe two alternative ways of collecting the data on which the messages will be based. (6)

b) The member of staff will require training in the use of the package in order to create an effective display. List three ways in which training could be provided. (3)

c) The system may also be used to display urgent messages. Give an example of one such use, and describe one safeguard which should be put in place to prevent misuse of this feature. (3) *NEAB IT04 1999*

9 A software house has a user support department that provides a range of services to customers including telephone advice and the supply of data fixes for corrupt files. The department uses a computer-based logging system to store details of incoming telephone calls from users (a call management system). The system is capable of producing a variety of reports via a report generator.

a) The software house receives complaints from its users that this department is providing a poor service. Describe three reports that the software house could produce to examine the validity of this claim. (6)

b) The department currently uses traditional mail to receive disks containing corrupt files and to return them with the data fixed. However, the department now wishes to use electronic communications based on ISDN. Describe two potential advantages and one potential disadvantage to the customer of this proposed change. (6) *NEAB IT04 1999*

Activity: IT training at Ellis Paints

Over 800 office staff at Ellis Paints, from senior management to office juniors, have been trained in *Microsoft Office* as part of the company's switch from MS-DOS to *Windows NT*.

The company used a training organisation with experience in IT training that did not exist in-house. The training organisation analysed individual needs and found a very wide range of skills. Some staff had hardly ever used a computer, while others had a very high level of IT competence. A programme was developed to cater with the different individual needs.

The programme was based on a series of seminars and one to three days of classroom-based training designed to cater for the different levels of competence. The training included one-to-one tuition for some staff and workshops looking at specific professional requirements. According to the trainers 96% of the company's staff are now trained to the initial level of *Microsoft Office* proficiency.

The training has not finished. Future plans include the provision of on-going support, lunch-time user clinics and the provision of on-line training in *Microsoft Office* via an intranet.

1 Ellis Paints chose to bring in outside trainers to train their staff in *Microsoft Office*. Describe other methods the company could have used for staff training, for each method indicate its appropriateness for Ellis Paints. (8)

2 The senior management of Ellis Paints feel that they need to draw up an IT training policy.
   a) Explain why this is necessary. (2)
   b) Describe what such a policy should contain. (6)

Activity: Sources of help

a) Write a paragraph about each of the sources of help shown in Figure 18.8.
b) Users can be broken down into three levels: expert users, frequent users and occasional users. For each type of user, describe, with justification, the most appropriate sources of help.

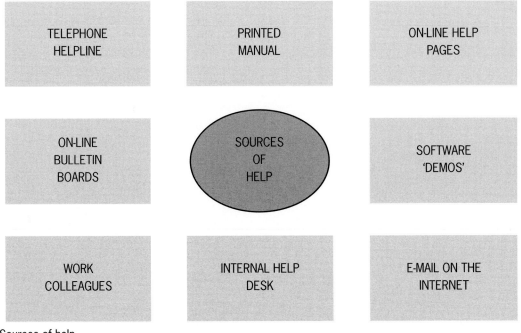

Figure 18.8 Sources of help

Activity: Training in context

Middleton College is a large FE college that offers a wide range of courses. For many students IT forms a large part of their course; there are also a number of students whose course is totally IT based.

Middleton College has over 25 administrative staff, and the turnover of staff is quite high. Recently the whole college has been networked. The new principal has made it a priority that all his senior managers make greater use of IT as a tool for decision making – this has caused some anxiety for several managers.

Before the network was installed, a variety of word processing software was used by different sections of the college. The new system will only be supporting Microsoft Office 2000.

Different groups of staff have different word processing training needs to allow them to use Microsoft Office 2000 appropriately.

1 Identify these different types of need, and for each suggest an appropriate training programme.

Managing IT

Project management

There have been many horror stories of IT projects that have been unsuccessful. Early attempts to computerise the stock exchange had to be abandoned. It is important to explore why some projects go wrong while others are successful.

IT projects are usually so large that they cannot be implemented by just one person. They are normally undertaken by a team of people working together. Examples of such projects may be the computerising of the functions of a warehouse or the installation of an internal e-mail system.

Figure 19.1 Air Traffic Control

Financial Times

A new £350m UK air traffic control centre, which was due to open last year, will not begin operating until November 1999 at the earliest, airlines were told yesterday.

The airlines were told the news at a meeting with National Air Traffic Services, the traffic control subsidiary of the Civil Aviation Authority. The centre, at Swanwick, Hampshire, which is intended to be the most advanced in the world, has been plagued software problems.

The airlines have estimated that since 1995 they have been paying a total of £10m a in depreciation charges on the building at Swanwick, even though it is not yet in operation.

The computer software system at Swanwick, which cost £120m, is now the responsibility of Lockheed Martin, the US defence and electronics group. The software contract was originally awarded to IBM of the US, which sold this part of its business to Loral, another US company, which was then acquired by Lockheed Martin.

The Swanwick centre was built entirely out of public funds, although Lockheed Martin will have to bear the cost of solving the software problems. NATS has said that while most of the software difficulties have now been sorted out, it needs to devote additional time to systems integration and the training of controllers.

The operations room at Swanwick will be half the size of a football pitch, making it the biggest in the world.

• Find out about the current progress of the system at Swanwick.

No one person could do all the work on their own, even if they had the time, as a wide range of skills and knowledge will be required. The project team should be selected with care to complement each other so that altogether they possess the drive, skills and knowledge necessary for implementation.

When implementing a project it is usually necessary to divide it up into smaller tasks and allocate these

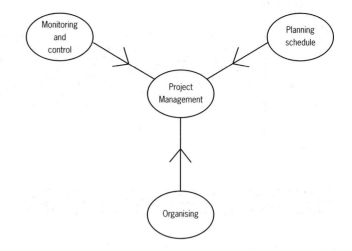

Figure 19.2 What is project management?

assess progress and the inter-personal skills to deal with clients.

Project management has three aspects: planning and scheduling; organising; monitoring and control.

Planning is crucial to the success of any project. It is very important that adequate time and thought is given to realistic planning. Scheduling involves allocating resources and facilities so that they are available when required. Resources can be human or physical. Delay in a resource being available may lead to slippage in deadlines which could incur extra expense. For example, the network cable needs to be in place before the computers can be installed. Under floor power cables need to be installed early on before the floor is laid. The project planning techniques and associated software (discussed in Chapters 13 and 21) are used with large projects.

An IT project should include:

- aims and objectives
- a project leader
- a project team
- a deadline for work to be finished
- a budget for completing the work

This budget will need to cover staffing, hardware and software costs; outlay for training, expenditure on data entry as well as other expenses such as travel and subsistence.

At the start, the client and the project manager will agree acceptance criteria for the completion of the project. These are agreed targets that must be met.

tasks to members of the team. Many of the tasks can be carried out concurrently. This makes the project more manageable as each team member has a clearly defined set of tasks. Teams can be made up of similar personnel each of whom performs a small part of the whole, for example a team of programmers, or a number of different skill holders each providing separate expertise in their own field.

The project will be managed by a project leader who will need to direct the resources to be used in the best possible way to get the job completed successfully and on time. These resources include people, time, money and hardware. The project leader will need to possess the skills of leadership and diplomacy if the project is to be successful, sufficient technical knowledge to

Examples of such criteria are that all deadlines must be met, work should be completed within budget, the system should be working fully and all agreed security measures should be in place.

The project manager will need to monitor progress as the project proceeds and report back to clients. Inevitably, unexpected problems will arise, and modifications will need to be made.

Based on an article in the Financial Times

Bugged by Failures

In theory, IT helps improve productivity, responsiveness and communication. In practice, IT projects are often dogged by management problems that result in delays, cost overruns and failure to meet the original objectives.

IT projects continue to have an extremely high failure rate, according to a recent survey by Oasig, a group supported by the Department of Trade and Industry.

It concluded that between 80 per cent and 90 per cent of IT investments do not meet their performance goals, 80 per cent of systems are delivered late and over budget and about 40 per cent of developments fail or are abandoned.

The increasing sophistication of technology can accentuate the problems. The recognition of these types of problems is focusing attention on how companies should evaluate, plan and implement an IT project. Few companies take these issues seriously. Evaluating the performance and impact of IT developments are not easy tasks, partly because it is not always clear what criteria should be used to judge the value of IT investments. For example, introducing electronic mail in an organisation may transform its internal communications, but it may be hard to justify in terms of specific financial benefits.

The Oasig study says that the main problems with IT projects stem from managers' narrow focus on technological capabilities and efficiency goals. It says that companies often fail to consider how work should be organised and jobs designed, following the introduction of new systems. Users rarely have enough influence on systems development.

■ Facilities management

'**Facilities management** is the contracting of the computer operations to an outside organisation. The facilities management company employ the staff, run the operation and often own the company hardware. The contract for this kind of service will specify what the computer system must provide for the fee charged.' (BCS)

When a business decides to invest in a computer system, it will often decide to buy or lease the equipment. They would then be responsible for the maintenance of the system. Any breakdowns would mean that costs would be incurred for replacement parts and labour.

It is difficult for such businesses to predict maintenance and repair costs. There might be a long wait for an engineer or for parts to be delivered, during which time the computer would be out of action. Such a delay could be disastrous for a business. Large businesses may employ their own hardware and software engineers in case of problems but this would not be a cost effective option for smaller organisations.

An alternative is to use facilities management. Facilities management is a complete contracting out of IT support to a third party, in other words, the company employs a second specialist company to manage computers for them. The specialist company has its own engineers (possibly on site) and guarantee that breakdowns will be repaired quickly. Any or all aspects of IT support can be contracted out: software development, hardware support and maintenance, help desk, software evaluation and maintenance. The company pays a fixed monthly amount which means they know exactly how much they have to spend. Existing IT personnel may be transferred to the facilities management supplier.

▶ Thomas Cook signs IT to Cap Gemini

Case Study

Thomas Cook has signed over 55 IT staff and a substantial slice of its IT operations to facilities management firm Cap Gemini. The five year deal covers application, desk-top and network management across Thomas Cook's travel and financial operations. Cap Gemini will manage the company's 1050 PCs.

(From Computing)

- Discuss the reasons that could have caused Thomas Cook to use facilities management. ☐

Camelot

The Camelot Group plc was awarded the licence to run the National Lottery on 25 May 1994. Within six months, over 10 000 retailers were selling tickets and the first draw was held, yet the first company building, the head office and main data centre, was not finished until August 1994.

The National Lottery sells around 100 million tickets every week and receives over 60 000 calls per week to its National Lottery Line, nearly all of them on Saturday nights and Sunday mornings. This requires a complex computer system with high security. With such a daunting task and a tight time scale, good project management was essential.

The project was split up into smaller tasks including:

- installing hardware in the head office and the two data centres

- training staff

- installing hardware in the retailers shops

- training retailers

- installing the corporate network linking the head office and the two data centres, the warehouse and the eleven regional offices for both voice and data

- testing

- implementing the call centre for the National Lottery Line for customers to claim prizes

- implementing the call centre for the Retailer Hot Line for retailers with technical problems

Geoff Pollock, Telecommunications Manager at Camelot, said, 'In normal circumstances we would have expected a project of this nature to take around eighteen months, but we did everything in only five months'.

- Draw out a Gantt chart for the project. Do not worry about times, but consider overlap and order of tasks. (See Chapter 14)

- Suggest reasons for why the project was implemented so successfully. □

Figure 19.3 National Lottery ticket

Although facilities management may appear to be more expensive, it avoids the worry about things going wrong and the cost of repairs. The specialist company can employ experts with an extensive level of knowledge of potential problems and their solutions whose experience means that a problem is dealt with efficiently and quickly. They have a flexible approach to problem solving. Costs can in fact be reduced due to economies of scale. There will be greater access to the latest hardware and software and flexibility for future development.

Outsourcing, as the use of facilities managers is known, can be used in many ways and to varying degrees. Services range from the relatively small, such as an annual system upgrade and maintenance; support for communication systems or local area network management, to full management of computer operations.

An organisation using facilities management is dependent upon the company providing that management. This could result in a lack of responsiveness to user problems and needs.

Advantages of facilities management

- A fixed cost can be agreed for services provided by a facilities management company. This means that the purchasing company has a greater control of its budget as IT costs can be difficult to control. If service costs go above the level agreed, the excess has to be borne by the facilities management company.
- The company will no longer need to recruit and train specialist IT personnel. Facilities management companies are able to recruit highly qualified professionals with expertise in specific areas of IT who can use their knowledge and skills for a number of clients.
- Service level agreements that specify such things as response time for help or breakdown calls and regularity of maintenance work, can be specified more easily with an external facilities management company than with an internal Information Services department.
- Off-site hardware can be available for back-up and disaster recovery contingency at a fraction of the cost of ownership.
- Economies of scale can often be achieved in hardware purchase as the facilities management company can acquire better discounts for large quantity purchases.

Disadvantages of Facilities Management

- If an organisation, in making a decision to move to facilities management, has to make IT staff redundant, then it could be vulnerable if the facilities management company failed to meet its obligations. It would be hard and costly to set-up an Information Services department again.
- The organisation could be considerably dependent upon the facilities provider. This could allow the provider to increase prices in subsequent years.

- The facilities provider could fail to provide the service agreed and the organisation might have to seek legal compensation.
- Friction could develop within an organisation between its own employees and those of the facilities management company.
- Sensitive data could be felt to be at risk when handled by an external facilities management company.

Effective IT Teams

A team is a small group who have come together with the aim of completing a project. Just as in a sports team, they will be allocated different roles, but need to work together rather than as individuals. They should develop to the stage where they are able to perform effectively, each member adopting the role necessary to work with others, using complementary skills. Teamwork and co-operation help to produce **consensus** and **avoid conflict**.

The essentials of an effective team

A good team needs clear and consistent leadership. A good leader will bring the best out of the individual members and encourage co-operation and the exchanging of ideas. Tasks should be allocated to the members appropriately, so that every member has work of which they are capable and which, if possible, will help develop their individual skills.

At the start of the project standards need to be set and agreed by all members. These standards should be adhered to and monitored throughout the project. An example of such standards in program writing could be the use of particular coding conventions in the choice of variable names, the layout of code and the inclusion of comments.

A good team will keep a careful watch on costs, and should complete the project within budget. All aspects of work should be carefully costed. Progress should be monitored and alternative action taken whenever necessary. Regular meetings together with the use of charts and suitable software can be crucial. (See Figure 19.4.)

The leader should have the appropriate seniority for the task, adequate understanding of the project and the ability to see the project through to successful

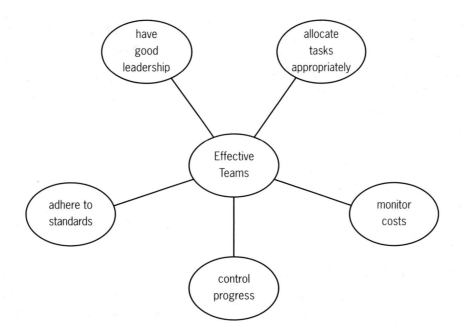

Figure 19.4 Essentials of a good team

completion. He should have the skills adequately and systematically to monitor and control progress and costs.

If possible, there should be a balance of skills between team members who could have different backgrounds – in systems, business operations or technical fields.

A team requires good communication skills, both written and verbal. Members need to be able to communicate well with each other, but also with end users. They need to be able to ask the right questions in such a way that they can establish end user requirements.

British Computer Society

The British Computer Society is the chartered body for Information Technology professionals. Formed in 1957, the society was incorporated by Royal Charter in 1984. It has nearly 34 000 members.

The BCS sets standards of professional, moral and ethical practice for the IT industry in the UK. Mr R. J. McQuaker, the Vice President of the BCS said, 'The public expect the same standards of competence and conduct from the members of the BCS as they expect from doctors, lawyers or architects'.

Professional bodies

Many professions have professional bodies who lay down codes of conduct for the profession. These include General Medical Council (Doctors), Bar Society (Barristers), Law Society (Solicitors) or Royal Institute of British Architects.

A doctor's code of conduct includes the Hippocratic oath, which states that they must not divulge information about a patient's health to anyone else. Professional bodies have the power to expel members for a breach of their rules. A doctor, for example, may then be 'struck off' for misconduct and not be able to practise.

Codes of conduct and practice

The BCS has a code of professional conduct, setting social, moral and ethical standards for its members.

- It would be unethical to sell a customer equipment that could not do the tasks required.
- It would be immoral to use information gained while working for a client for personal gain.
- It would be anti-social if the introduction of a new system was not completed on time and kept members of the public waiting for a service

The code does not have legal status, but is merely the rule of the society. Breaking the code of conduct is not illegal, but anyone doing so may face punishment from the society, for example expulsion. This may affect their professional position or their chances of getting promotion or another job.

■ SOME LEGAL REQUIREMENTS FOR IT PROFESSIONALS

- You must register with the Data Protection Registrar when storing personal data
- You must not use a computer or alter data without the owner's permission
- You must not copy software without the owner's permission
- You must provide tilting screens and screen filters in offices

---

The Ten Commandments for computer ethics

1 Thou shalt not use a computer to harm other people.
2 Thou shalt not interfere with other people's computer work.
3 Thou shalt not snoop around in other people's files.
4 Thou shalt not use a computer to steal.
5 Thou shalt not use a computer to bear false witness.
6 Thou shalt not use or copy software for which you have not paid.
7 Thou shalt not use other people's computer resources without authorisation.
8 Thou shalt not appropriate other people's intellectual output.
9 Thou shalt think about the social consequences of the program you write.
10 Thou shalt use a computer in ways that show consideration and respect.

from the Computer Ethics Institute, rinaldi@acc.fau.edu

■ EXTRACTS FROM THE BCS CODE OF PRACTICE

- Members shall endeavour to complete work undertaken on time and to budget and shall advise their employer or client as soon as practicable if any overrun is foreseen. (Clause 6)

- Members shall not disclose or authorise to be disclosed, or use for personal gain or to benefit a third party, confidential information acquired in the course of professional practice, except with prior written permission of the employer or client, or at the direction of a court of law. (Clause 8)

- Members shall not misrepresent or withhold information on the capabilities of products, systems or services with which they are concerned or take advantage of the lack of knowledge or inexperience of others. (Clause 10)

- Members shall not purport to exercise independent judgement on behalf of a client on any product or service in which they knowingly have any interest, financial or otherwise. (Clause 12)

Employee code of conduct

All employers have codes of conduct for their employees. They may be communicated verbally or formalised in a written document. For example they may specify the type of clothing an employee should wear, where smoking is allowed and forbid the drinking of alcohol on the premises.

Breaches of the code of conduct may lead to sanctions; verbal warnings, written warnings, loss of pay or dismissal; depending on the frequency and the severity of the breach.

Many jobs will involve secret or sensitive information which must not be divulged, for example because the information could be personnel or be of benefit to a competitor. For IT personnel there are other requirements, in particular due to legal requirements of copyright law and the Data Protection Act. Employees should also take no action which may subject a computer system to viruses or compromise security in any way.

A number of IT related items might appear in an employees code of conduct, for example:

No unauthorised disks may be used. (It is possible to install software that means that only the company's specially formatted disks can be read.)

No unauthorised software can be used on the company's computers.

Employees may not copy software for personal use.

How the organisation fulfills requirements of the Data Protection Act.

How often passwords should be changed.

What possible passwords are acceptable, e.g. only combinations of letters and numbers – not dictionary words or names.

Passwords should not be written down.

No action should be taken against the interests of the company.

Southampton University Code of Conduct for use of Computers

Southampton University issues regulations for use of computer systems and networks by staff and students. The following are excerpts from these regulations:

1 The use of computers is regulated by three Acts of Parliament: the Data Protection Act 1984, the Copyright, Designs and Patents Act 1988 and the Computer Misuse Act 1990. Similarly, the use of the public data telephone networks is regulated by the Telecommunications Act 1984.

2 The following regulations are framed to remind all members of the University of their legal obligations. In addition, the use of computer software may also be subject to the terms of licence enforceable in the civil courts.

3 These regulations have the status of Regulations for Discipline which apply to all members of the university. Any breach of these regulations will automatically be considered a breach of discipline.

4 **Data Protection.** Members of the university are only allowed to hold, obtain, disclose or transfer personal data (as defined by the Data Protection Act 1984) as permitted by the university's current registration with the Data Protection Registrar and in accordance with Data Protection Principles as set out in that Act. If in doubt, the university's Data Protection Officer should be consulted before any personal data is stored in a computer system.

5 **Copyright.** Members of the University will comply with the provisions of the Copyright Designs and Patents Act 1988

in relation to any computer program or data set and shall not act in any way contrary to the terms of any licence agreement applying thereto. (This is a formal way of saying 'thou shalt not use pirated software.')

6 **Computer misuse.** Members of the University are only allowed to use those computing resources, data or voice communications facilities, which have been allocated to them. Computing resources may only be used for properly authorised purposes. (Students do not have authority to grant anyone else access to the facilities that they have been given.)

7 Members of the University may not access, alter, erase or add to computer material which has not been generated by them unless they are explicitly authorised to do so by the originator of the material.

8 Authorised users of computer systems must take reasonable care to prevent unauthorised use of the computing resources allocated to them.

9 Members of the University may not use computer systems or networks in such a way as to compromise the integrity or performance of the systems or networks. (These regulations cover the activity commonly known as 'hacking'. Breach of any of these regulations is *prima facie* evidence of an offence under the Computer Misuse Act.)

10 **Networks.** Members of the University must abide by any 'conditions of use' of networks which are published by the responsible computing management for the protection of the integrity and efficiency of the network. *Continued*

11 Members of the University must not cause obscene, pornographic, discriminatory, defamatory or other offensive material to be transmitted over the University, national or public networks, or cause such to be stored in University computer systems. (It is a criminal offence to publish pornographic material, for example by including such material in a Web page.)

12 **Withdrawal of service.** The responsible computing management may withdraw access to facilities from any user found to be guilty of a breach of these regulations.

13 The University will hold the individual user personally liable for any costs or claims which may arise from any use of University computing and/or communications facilities, whether authorised or not by the responsible computing management.

14 **Passwords.** All staff and students are given an account with a unique login name and password. The account is only to be used by its owner. A password is assigned by the System Manager when the account is created: the user must change this immediately. The password must not be divulged to anyone.

15 A password may be changed by its owner but the new password must conform to security measures in force at the time. (The password must be at least six characters, it must not be a dictionary word, name, telephone number, car registration number or anything likely to be associated with its owner.)

16 **Account access.** Users must take reasonable steps to protect their own accounts from access by others: the owner will be held responsible for any improper use. In particular you must not allow any other person to login to your account. (If you need to share access to files and data with another user you should use e-mail.)

17 The Head of Department may authorise the system support staff to examine any file if there is reason to believe that these regulations are being contravened in any way.

18 **Use of the Internet and the World Wide Web.** Any reasonable use of the Internet is permitted, although users should bear in mind that the load placed on the network by excessive use (for example downloading large graphics files) will degrade its performance for other users, not only in the Department but also the University and even on a national scale.

19 Staff and students are allowed to maintain personal home pages on the Web, hosted on Departmental systems.

Management of change

Change is necessary in life, often due to biological factors such as ageing or parenthood. Organisations change too, often due to external economic circumstances. The introduction or development of an information system within an organisation must lead to change.

Organisations will cope best with change if they plan ahead. They should be able to anticipate change when a new information system is introduced, plan for it and, if possible, control of the situation.

The introduction of a new information system is likely to change the nature and content of many jobs associated with the system. Many old skills may become redundant, and employees will need to be taught new skills. In some cases these changes could be relatively minor and simply require a few hours of training. In other cases the changes to the job might be more radical and much training and reassurance will be needed when re-skilling staff.

The introduction of information systems can cause a shift from jobs which require basic, manual skills to those that contain a greater component of problem solving. The introduction of networked PCs has resulted in one person performing a much greater range of tasks. The role of many telephonists has been enhanced. Instead of having to pass callers on for help, the telephonist is empowered to answer many questions, concerning availability of stock or product lines for example, through having access to an on-line database.

The change to a new system of working can provoke fear and resistance from employees. There is likely to be a fear of the unknown; an employee may feel that he will not be able to operate the new system. Many people still have a distrust of computers. The possibility of redundancy, de-skilling or loss of job satisfaction can all lead to resistance. The tasks required by the new system might result in a loss of status or re-grading which could pose a threat to an employee's ambitions.

The introduction of new information systems can result in major changes in an organisation. Some jobs may disappear entirely. Changes in the work done may result in the need to modify the organisation structure. There has been a shift towards flatter, leaner structures as middle management jobs have been eroded. MIS make it easier for strategic management to monitor operations more directly. Some decisions, previously taken by middle management can be performed automatically by new systems. An example of such a decision could be the re-ordering of stock.

Case Study

Changes in a College Library

A traditional college library consisted of stacks of books, organised in order according to the Dewey Decimal system. Card indexes of books stocked were maintained in cabinets, one in author order and the other in Dewey Decimal order. Borrowers were issued with tickets and each book held a tag which, when the book was on loan, was stored together with the borrower's ticket, in a card file. Information on other books, not held by the library, was available on microfiche.

The modern librarian needs many new IT skills to cope with the new computerised systems that are in use. Books are likely to be bar-coded and all loans will be held on a computer based system. Library users are able to search on-line remote databases to find details of books relevant to their studies. The library facilities are likely to have been extended to include access to CD-ROMs and the Internet, both of which provide the student of alternative ways of gathering information.

- Describe how you imagine the library of the future to look like.
- Do you think that any books will remain? Justify your answer.
- What skills does the new type of librarian need?

Figure 19.5 IT in the library

Case Study

Barry

Barry left school in 1972 and was a printing apprentice at a local newspaper for seven years, learning how to set type as a compositor. On finishing his apprenticeship he became a full member of the trade union, the National Graphical Association (NGA).

In 1981 the newspaper introduced computerised typesetting. All Barry's skills were out-of-date and he had to be re-trained in the new skills of touch typing and desk-top publishing. At this stage, articles and stories were typed by journalists and then retyped by the NGA printers.

In 1983, after long negotiations between the management, the National Union of Journalists (NUJ) and the NGA, a new system was introduced where the journalists typed their stories directly on to the computer and the page layout was completed by the printers. As a result, fewer printers were required. Some printers re-trained as journalists while others were made redundant.

- Identify 5 other areas of employment that have been replaced by IT. ☐

Working hours or the location of work may need to change for some employees. Many computer systems run 24 hours a day and staff are needed to operated and maintain the computers. The development of communications and IT has led to a growth in teleworking.

Major operational changes can result from the adoption of a new system. As point of sale terminals are installed in shops, stock control, ordering and sales analysis can all be computerised.

If change is to be introduced with the minimum of distrust and upheaval, it must be managed. Careful plans need to be worked out well in advance. It is often possible to achieve necessary manpower reduction through natural wastage over a period of time. Management should set out to involve the personnel involved, and where appropriate, the trade unions, from the start and communicate fully and frankly with all those who are likely to be affected by changes.

Figure 19.6 Job changes caused by IT

Training and re-training schemes should be set up and personnel should be shown that routine, boring work can be eliminated and job satisfaction maintained or even enhanced.

There are a number of factors that influence how successfully change is managed within an organisation. The structure of the organisation and the key roles are crucial. Change may bring about a re-structuring which may require the loss of some jobs and substantial changes in others. The conditions of service under which the workforce are employed will be an important factor in determining the ease with which change can be undertaken. The attitude of both management and the workforce will be influential as well as the overall organisational culture. An organisation whose management has an open style, where there is mutual trust and support, is less likely to fear or resent change. If the skills needed far outstrip the current skills level of the workforce change will be difficult to bring about.

Industry Structure Model

The BCS has produced a comprehensive set of standards for the training and development of all those working in IT, called the Industry Structure Model (ISM). The ISM defines over 200 different jobs in IT – including programmers, software engineers and network specialists – at one of ten different levels of responsibility and technical expertise.

This means that if someone applies for a job, they know exactly what level the job is at. Therefore

- applicants know what jobs to apply for

- applicants can plan career development

- employers have an idea of the applicants' experience and abilities

- it creates a career path for IT professionals which is recognised in Britain and abroad

- a relationship exists between the job description and the individual's experience and responsibility however the employing organisation is structured

- employees can compare their salary with others at a similar level

The ISM is a nationally recognised scheme and is also used as a means of planning training and measuring its effectiveness against an independent, industry-accepted benchmark. The ISM is regularly reviewed to ensure it is reflecting the changing nature of the IT industry.

■ PROFESSIONALISM

The concept of professionalism implies taking responsibility for one's work and performing the work to the highest possible standards. In order to be able to fulfil these requirements, it is vital that the professional is trained to a high standard. Training cannot be seen as something which just takes place only in the early stages of a career because of the rapidly changing nature of IT and the speed with which new developments are introduced. Training has to be part of a life-long development process.

The British Computer Society plays a leading role in the setting of standards for the training of IT professionals.

■ OTHER ROLES OF THE BCS

The Society
- has responsibilities for education and training and conducts its own examinations

- has responsibilities to increase public awareness

- advises Parliament, the Government and its agencies on IT matters

- represents the views of its members on topical issues such as computer misuse and the impact of IT on society

The society can be contacted at:

BCS, 1 Sanford Street, Swindon, Wiltshire SN1 1HJ

Telephone: 01793 417417

Fax: 01793 480270

E-mail: bcshq@bcs.org.uk

The BCS is a registered charity: No 292786

Summary

IT projects require careful management to ensure successful completion.

☐ Introduction of IT into a company is too large a task to be performed by one person. As a result IT professionals often need to work in teams.

☐ Facilities management involves the employment of outside agencies to carry out tasks ranging from the maintenance of hardware to full management of a project.

☐ Introducing IT or any change can lead to resistance from existing staff, mainly through fear of the unknown. It is vital that any company plans ahead and is open in its introduction of new ideas.

The BCS is an organisation for IT professionals. In particular it

☐ sets standards throughout the industry with its Codes of Practice

☐ provides a model for training and career progression called the Industry Structure Model

Managing IT questions:

1 Explain why work is often sub-divided and done in teams. (3)

2 Describe the characteristics of a good team. (6)

3 Describe three possible causes of inefficiency in the working of a team. (6)

4 Hopkins, Morgan and Hopkins is a firm of surveyors. There are nine surveyors in the firm and four secretaries whose jobs include the maintenance of records of all clients and surveys carried out; typing of letters, reports and memos; keeping track of appointments. Each secretary has a stand-alone PC on the desk which has a word-processor installed. An office junior/receptionist is employed to act as a runner, carry out filing and answer the telephone. He often has to take messages for the surveyors when they are busy.

A new, networked system is to be installed which will provide every employee with a workstation. A record system will be will be installed, together with a range of drawing and financial packages, e-mail and a diary system.

a) Describe how will the new system affect the jobs of the:

- surveyors

- secretaries

- office junior. (9)

b) In what ways, and for what reasons might any of the above personnel be unhappy with the changes? (8)

c) Describe the steps that should be taken to ensure that the organisation runs successfully with the new system in place. (6)

5 State **three** different problems a company is likely to face in computerising its business. (3) *AEB Specimen Paper 2 Question 2*

6 An information system was introduced into an organisation and was considered a failure. The failure was due to the inability of the organisation to manage the change rather than for technical reasons.

With the aid of examples describe **three** factors which influence the management of change within an organisation. (6) *NEAB 1997 Paper 3*

7 Explain what is meant by 'facilities management'. (3)

Describe **two** advantages, and **one** possible disadvantage to a company of adopting this approach. (3) *NEAB 1997 Paper 3*

8 A multi-site college is considering the introduction of an IT based system to log visitors. The current system is based on a manual log at reception. The new system will capture visual images of visitors together with details of their visit. The introduction of this system will cause considerable change for staff and visitors.

In the context of this example describe **four** factors that the management should consider when introducing this change. (8) *NEAB 1998 Paper 4*

9 A firm is creating a team to plan, design and implement an IT project. Describe **four** characteristics of a good IT project team. (8) *NEAB 1998 Paper 4*

10 A company uses a computer network for storing details of its staff and for managing its finances. The network manager is concerned that some members of staff may install unauthorised software onto the network.

a) Give reasons why it is necessary for some software to be designated as unauthorised. (2)

b) What guidelines should the network manager issue to prevent the installation of unauthorised software onto the network? (2)

c) What procedures might be available to the company to enforce the guidelines? (2) *NEAB 1997 Paper 3*

11 As the IT manager for a large company, you have been asked to develop an employee code of conduct. Describe **four** issues which might be included in such a code. (8) *NEAB 1998 Paper 5*

12 A professional organisation for Information Technology practitioners has developed what is referred to as the 'Industry Structure Model'. What is the purpose of this model and how does it work? (6) *NEAB 1996 Paper 1*

13 'Codes of practice' exist for professionals within the Information Technology industry separate from any legal requirements. Explain, with the aid of an example, the distinction between a legal requirement and a code of practice. (3) *NEAB Specimen Paper 1*

14 An IT employee discloses confidential information to a third party. Have they broken the law, the code of conduct or both? (Explain your answer, depending on the type of information.) (4)

15 How does an Industry Structure Model help the development of the career of an Information Technology professional? (3) *NEAB Specimen Paper 1*

16 Bill Gregson is an IT consultant advising people on computerising their work. Bill recommends to clients that they should buy their computers from the computer manufacturers *Cheapo*, run by Bill's friend Clive Baxter. These computers are usually out-of-date 486 models. 'These are the bee's knees. They'll never let you down. I swear by them. You don't want that modern stuff – too unreliable,' says Bill. State which laws or codes of conduct have been breached (if any) and why. (4)

17 Professional progression within the IT industry requires more than just technical skills. Give **three** other necessary qualities and explain why they are important.(6) *NEAB 1997 Paper 1*

18 Study the ten commandments of computer ethics on page 220.

a) Which commandments are legal requirements and which ones are purely ethical suggestions?

b) In your answers to part (a), if you feel the commandment is a legal requirement, explain why you feel it is against the law.

19 Organisations and IT professionals are required to comply with a legal framework when introducing and using IT systems. In addition there will normally be a code of practice.

a) Define what is meant by a 'code of practice' (2)

b) Describe **two** ways in which institutions, such as the British Computer Society, promote professionalism for individuals within the IT industry. (2) *NEAB 1998 Paper 1*

20 A company has three departments to handle finance, buildings and equipment maintenance. Each department currently operates a separate IT system. The company wishes to improve the efficiency of the operations by implementing a common corporate system across all three departments. In order to achieve this improvement, the company has decided to select members of staff from each department to form a project team to plan, design and implement the new system.

a) Describe three corporate level factors the team should consider when planning the new system. (6)

b) At their first meeting the team decide to sub-divide the project into a series of tasks. Describing two advantages of this approach. (4) *NEAB IT04 1999*

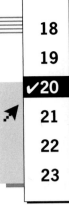

Corporate information systems strategy

All organisations have objectives which determine the way in which they function. For many organisations an objective may be to make a profit whilst for others providing a service may be the main objective. Breaking even to survive, growth or maximising sales are all possible objectives. Businesses need a long-term strategy concerned with defining what the business will do to become or remain successful. For example the strategy might plan areas of expansion and anticipate areas of growth.

Businesses are likely to have a *corporate plan* on how to implement *the corporate strategy*.

The plan is likely to include a marketing plan, an operations plan (including details of production), a human resources plan, an investment plan and an information technology strategy. An IT strategy is concerned with the planning, introduction and use of IT resources for the benefit of the entire organisation. The strategy must be closely linked to the other strategies/plans of the organisation.

It is important that there is a corporate approach to hardware and software purchase. Data files from one department may be used by another department. Staff using IT need continuity of both hardware and software. They get used to one type of keyboard and one word-processing package. To change to another package could cause unnecessary anxiety and require further training. In buying new hardware, it is essential that old software and data can be used easily.

Every organisation is unique and the information system developed within an organisation must meet its individual requirements.

The organisation structure and the management functions within the organisation will effect the information needs. For example, an organisation which is managed geographically will have regional managers who require reports summarising the performance of all functions within that region. In an organisation which is structured functionally, a production manager would require summaries of the performance at all factories in the organisation.

Figure 20.1 What went wrong?

The information flow within the organisation will also influence the details of the information system which is developed, as well as planning and decision making methods. Some organisations concentrate major decisions in a few individual managers whilst in others decision making is shared between several different management layers. Some organisations make extensive use of committees for decision making. Thus information needs will be different.

The hardware and software in use will also help to determine the information system. Mainframe based systems, WANs, LANs, the use of distributed systems; each of these will imply very different information systems.

Behavioural factors also have their effect. The personalities of the people who will use the system, their motivation and ability to adapt to change all have to be taken into account.

IT has an impact on other parts of the organisation and is in turn influenced by them. To develop the strategy, the organisation needs to identify its information requirements and provide the IT equipment to deliver this information. This means identifying information needs for decision making, planning, management organisation, information flow, behavioural factors (for example personalities, motivation, ability to adapt to change). Other issues that need to be considered at the strategy stage are health issues, avoiding accidental or deliberate corruption of data, compatibility of equipment from different sources, legislation and regulations.

Information management policy

All organisations store information, even if they don't use information technology. Traditionally, offices need to store information. In the past this information was kept in filing cabinets, stored in a particular order, usually alphabetical or chronological, so that information could be retrieved easily. Decisions concerning what information should be kept and for how long needed to be made.

For a number of years, microfilm has offered an alternative way of archiving old files. Paperwork such as application forms could be photographed and stored in a fraction of the space required before. However, the equipment is relatively expensive and retrieval may not be as quick as going to a filing

cabinet, so this makes the task of information management more complicated.

Decisions about the archiving of information forms part of an information management policy. Similarly the organisation may have a policy on which information should be stored on computer; whether it is stored on the network hard disk, a local hard disk

---

Case Study

A local council's Information Technology strategy

The excerpts on the following page are from the council's IT strategy. In recent years the council, like every other large business, has become increasingly dependent on computers for its administration.

Previously computers had been used simply as a means of handling large volumes of data and of carrying out calculations quickly. As such their use was mainly confined to the major financial systems, and consisted of the processing of 'batches' of data leading to the production of large volumes of printed output. Examples were the payroll, rates bills, and cheques for the payment of creditors.

More recently, computer technology and software have developed rapidly. The development of on-line systems and associated elimination of paper records mean that whole areas of operation are now totally dependent on computers for both storage of data and for calculations.

■ THE NEED FOR A STRATEGY

The whole installation has developed over a number of years as and when required. Generally, these developments have taken place either in response to a perceived need to increase efficiency, to provide an improved service to the public, or in response to legislation.

In spite of this fragmented approach to computer development, a high degree of compatibility between systems and departments has been maintained. However, a large part of the present installation is now technically obsolete and will need to be replaced within the next year. If the replacement is to be done in the most cost-effective way it needs to take place within an agreed strategy for information technology.

At present the Council has no such formal plan and it is now time to adopt a broad strategy within which future acquisition of equipment and development of systems should take place. □

A strategy For Information Technology

There follows an extract from the IT strategy of a local council.

1 The Council should continue to maintain automated office systems which permit the maximum possible transfer of data between users. All future system developments and acquisitions of equipment should be designed with this principle in mind.

2 The Policy and Resources Committee should be asked to approve each year a rolling programme for the development of new computer systems. This programme should be designed to take account of both legislative requirements and potential improvements in efficiency (whether internal initiatives or external audit recommendations).
The programme should indicate the reason for the proposals and the level (if any) of cost savings which will result.

3 Whenever appropriate new computer systems which are recommended in the rolling programme should be capable of exchanging data with existing systems. Data common to more than one system should be held so far as possible in one place only.

4 As a general rule the Council should aim to purchase commercially produced software whenever possible,

unless a suitable product is not available or it can clearly be shown that in-house development is more economical. This strategy recognises that in some cases the purchase of stand-alone Personal Computers in order to run specific software might be the most cost-effective solution to a problem.
Before such purchases are made, however, a proper technical appraisal of the options should be carried out by specialist computer staff within the Finance Department. This appraisal should take into account the compatibility of both the proposed hardware and software with that already being operated by the Council.

5 The Council's major financial systems should continue to operate on the existing mainframe computer until it is in need of replacement. At that time a full appraisal of the means by which these systems are to continue should be undertaken.

6 A full appraisal should be undertaken into the possible advantages of increased use of electronic data transfer, either by e-mail or alternative means. If increased use of this facility is recommended, formal procedures for such use should be drawn up.

7 This strategy should be reviewed periodically at intervals of not more than two years. □

or on a floppy disk. The policy may include how long is it kept for and who has access to it.

A well thought out policy is even more important if the data is of a personal nature because the organisation is bound by the Data Protection Act. This means that unnecessary, out-of-date information must be deleted and access must be restricted only to authorised personnel.

The organisation may need a policy on how to communicate information internally. In the past, memos, notices on notice boards, passing a note round, weekly information bulletins and internal company newsletters all may have been most appropriate. Today, internal e-mail provides a very easy way of informing staff. Mailing lists can be set up of say, directors, personnel, or sales staff. It is then easy to send a copy to everyone on the mailing list. The use of work-group software such as Lotus Notes allows users to work on the same documents and provides access to each others' diary.

Corporate IT security policy

As has been discussed in Chapters 16 and 17, adequate security is crucial to successful operations. Many companies have a formal IT security policy, which they publish and give to all staff. Staff may also receive a lecture on security when they join the organisation as part of the induction training programme for all new employees.

The introduction of one company's IT Security booklet states:

The company is in a highly competitive industry in which the loss or unauthorised disclosure of sensitive information could be extremely detrimental to the Company. These guidelines have been prepared to ensure that all staff understand the importance of safeguarding Company information and the protective measures that need to be taken.

Most companies' work involves some secrecy and thus there is a need for security. Postmen, for example, must not disclose where they have delivered letters and have to sign the **Official Secrets Act**. This has to be signed by all civil servants, whatever their job, saying that they will not disclose confidential details.

Companies storing personal data are obliged to abide by the Data Protection Act, which says that personal information will be kept secret. The company is responsible for ensuring that this data is not divulged and that company staff are aware of the legal requirements.

An IT security policy is established so that misuse can be prevented, with methods of detection and investigation being put in place. Staff responsibilities should be drawn up and disciplinary procedures agreed so that any misuse is dealt with. A staff IT security document is likely to specify who can use company computer systems and to set out the password policy. It will lay down the steps that should be taken to provide protection against viruses and the physical security of computer systems. Rules will be provided to ensure that all computer use is within the law.

A corporate security strategy will need to be frequently modified as new systems are introduced and old ones modified. For example, when insurance salesmen are issued with laptops for the first time, to take with them when they visit clients, the company's security policy will have to be extended to contain rules to protect the data, the hardware and the software.

If an employee is discovered to have broken the organisation's code of conduct, for example by installing unauthorised software on the organisation's network, he is likely to be subject to one of a number of sanctions.

He may be given a verbal or a formal written warning. If the offence were of a serious nature, it could lead to suspension or even termination of employment. The employee could face action under the Computer Misuse Act.

He is likely to have his network usage monitored very carefully and his access rights restricted.

Back-up strategies

An information management strategy must include a strategy for backing-up files. Computer files are very valuable to the organisation. Lost files will result in wasted time and money. It is essential that a company has a back-up strategy to cover all

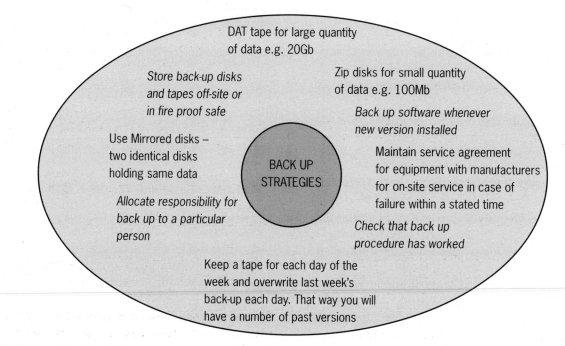

Figure 20.2 Examples of back up strategies

eventualities, including accidental damage, deliberate damage and damage due to equipment failure.

A formal strategy is needed to ensure that back-ups are taken in a consistent and appropriate manner. They need to be done in such a way that, if necessary, the system could be restored in a reasonable amount of time. With an on-line system, suitable back-up facilities must be in place to ensure that all lost transactions can be restored.

The strategy will need to specify which files need to be backed-up at what frequency. It is important to establish when back-up should take place and by whom. This is often at night as many systems only run during working hours. Arranging back-up for systems which are on-line for 24 hours a day is more complicated.

The place where back-up copies are to be stored must be specified, as well as the length of time and number of back copies that are kept. See page 196 for details on back-up planning.

Future proofing

Future proofing concerns finding ways of making sure that a system has a reasonable life and does not need to be totally replaced too soon.

Computers have developed so rapidly that machines that are four or five years old seem slow and cannot cope with recent software. It is not possible to predict the future other than to say it is unpredictable.

However it is sensible to take steps to prevent problems. It is often true that the data stored on computer is more valuable than the hardware itself. If a fire occurred it is possible to buy replacement hardware but the data can only be replaced if back-up copies have been stored safely.

When buying a new computer system, it is important to buy one that won't be out of date soon.

Old data must be able to be used. Programs must have backwards compatibility: the ability to read files from previous versions. High density disk drives must be able to read and drive to the old double density floppy disks. Faster CD-ROM drives must be able to cope with older CD-ROMs.

Hardware performance is constantly being improved by manufacturers in terms of both processing speed

and memory capacity. New versions of software include extra features and usually require extra main and backing storage memory. It is important that any computer purchased has a large enough RAM and hard disk to cope with likely future requirements, both from software and expanding files. It should be possible to expand memory at a later date.

It should be possible to add extra cards and peripherals if needs change.

Considering future needs is even more important when setting up a network. When establishing the cabling in and between buildings, care must be taken that future growth in network traffic is catered for. The network infrastructure of cables, switches, servers and so on are costly to purchase and install. Frequent changes to these basic, underlying services can be disruptive to work and need to be avoided through careful forward planning. There must be flexibility in the number and positioning of work stations so that changing future requirements can be catered for.

Upgrading hardware

After some years of use, a company may wish to upgrade their computers. This may be as a result of an increased volume of data or a desire to decrease processing time. Required changes in software might make the hardware upgrade necessary as new versions of software often have greater resource requirements.

The organisation may have a policy to upgrade hardware after a certain time in order to provide an up to date image for the company or to maintain good staff morale.

If computers are kept for a long time, they can become obsolete and spare parts become unavailable so that they cannot be repaired when they break down.

When an organisation upgrades a decision has to be made on the new hardware platform. Ideally it should be compatible with the old platform.

Compatible hardware

When different hardware manufacturers produce machines that all support the same software and data files the machines are said to be compatible.

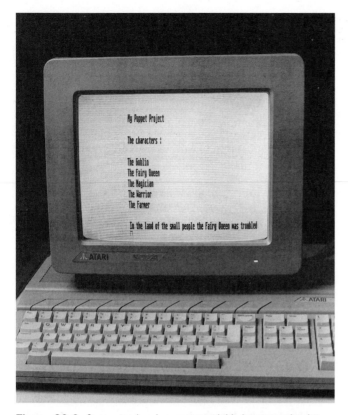

Figure 20.3 Computer hardware can quickly become obsolete

Some applications are dependant upon a particular hardware configuration (e.g. processor type, memory configuration, VDU configuration). For example, Microsoft Office 2000 will not run on a PC with a 486 chip. The term compatible hardware is often used to refer to those hardware systems that conform to a particular minimum hardware specification, having similar architecture and supporting the same peripheral devices.

Emulation

A problem occurs when changing to a new hardware platform if old systems are still required to be used. If a change to incompatible new hardware is made, it may be possible to run old software using a software emulator. This is systems software that acts as an interface between the hardware of a system and any applications running on that system in order that the application software can run on a hardware platform other than the one for which it was designed. Thus the original software can be run. The computer gives the appearance of being a different platform.

Emulation software

It is possible to turn your PC into a 1980s home computer like the Spectrum or Commodore 64. You might wonder why anyone might want to turn a powerful, modern machine into an obsolete museum piece but thousands of games from these old machines can be run on today's computers using **emulation software**.

A software emulator is a piece of software for one computer that simulates another computer so that programs written for the original machine can be run on the new one.

Eddie Kirk owns an Apple Macintosh computer at home, but at work he uses a PC. The specialist software he uses at work is not avialable for the Apple. When Eddie had a lot of work to do at home, he had two choices

* to purchase a PC

* to purchase an emulator for his Mac

Eddie decided to buy a PC emulator for his Mac for as little as £50. However emulators don't always provide a solution. To create an emulator, programmers have to translate microprocessor instructions written for one chip into ones understood by another. As a result emulators are complicated and often run too slowly to be of practical use.

There are at least two companies producing Macintosh emulators for PC users who want to run Mac software. There are about 60 Sinclair Spectrum emulators running on a variety of machines. There are 15 Nintendo emulators for the PC alone.

Using software emulation allows access to a greater range of applications that might not be available on the given hardware platform. The use of an emulator allows data to be transferred between platforms.

However, software emulation may not provide full functionality of the software and will not exploit all the facilities of the new hardware. As an extra layer of software is in place, the emulator can make heavy demands on system resources of memory and processor, often causing the application software to run slowly.

Summary

Organisations should have policies on:

☐ Information – What to throw away, what to keep and how to keep it

☐ IT – A corporate approach to buying equipment, ensuring current data can be used

☐ IT Security – How to keep your information secure

☐ Back-up – How often to back-up files

☐ Organisations should take account of likely future developments in their planning and purchasing.

☐ When upgrading, compatability needs to be considered.

☐ Emulation provides a way of running old software on a different hardware platform.

Corporate information systems strategy questions

1 Explain what is meant by the term 'future proofing'. (3)

2 A university provides staff and students with access to its computer network.

a) Activity on the university's networking system is monitored and an accounting log is automatically produced. Suggest what this log might include and explain why it is useful. (8)

b) Appropriate staff have access to personal and financial data. What steps should be taken to preserve the security of the data in such a system? (4) *NEAB 1997 Paper 4*

3 A hospital information system holds program files which are rarely changed and large database files which are constantly changing.

Describe a suitable backup strategy for this system, explaining what is backed-up and when, together with the media and hardware involved. (8) *NEAB 1997 Paper 4*

4 A computer user has bought a large number of packages for a NEAB PC computer. Due to increasing workload it is necessary to replace this model with a more powerful computer. The user has a choice of
either: buying an NEAB SUPERPC machine which is compatible with the NEAB PC,
or buying a MEGAMACHINE which is a completely different piece of hardware but provides the software emulation of the NEAB PC.

a) Why does the user need to relate the new machine to the NEAB PC? (2)

b) Explain the terms compatible and software emulation. (4)

c) Discuss the relative merits of adopting one choice of computer as compared to the other. (4) *NEAB Specimen Paper 4*

5 'If I need an IT system I buy whatever hardware and software I want without any regard to anyone.' This statement was made by a manager of a department in a company.

Why is this an inappropriate approach in a large organisation? (6) *NEAB Specimen Paper 4*

6 'I don't care which version of a word-processing package the rest of the company uses. As a senior company manager I intend to upgrade my department to the latest version.' Give **four** potential problems this attitude may cause for other IT users in the company. (4) *NEAB 1998 Paper 5*

7 A company has been running a large number of application packages on a personal computer. Although the computer works and has no hardware faults, the manager of the company now wishes to upgrade to a more powerful computer to run the same type of application packages.

a) Give **four** distinct reasons why the company may wish to upgrade their computer. (4)

b) The company could buy a computer which is 'compatible' with the current machine in use. An alternative is to purchase a different type of computer, with 'software emulation' of the current hardware. Explain the terms 'compatible' and 'software emulation'. (6)

c) Describe the advantages and limitations of adopting a 'software emulation' approach. (4) *NEAB 1998 Paper 5*

8 A hospital information system holds program files, which are rarely changed, and large database files, which are changing constantly. At present the backup strategy uses a tape storage device, and has the following characteristics:

Each evening the information system is taken off-line and a full backup is made of the entire system. Three sets of tapes are in use and are referred to as sets A, B and C.

Set A is used one evening,
Set B is used the next evening,
Set C is used the following evening.

This sequence is then repeated, starting the next evening, with Set A again.

An advisor has suggested a change is required to improve this strategy. Give, with reasons, four changes that could be made. (8) *NEAB IT05 1999*

Activity

■ TO UPGRADE OR NOT TO UPGRADE

It is not always easy to decide whether or not to buy software upgrades. There can be hidden costs and difficulties with upgrading such as incompatibilities with previous versions or other software installed on the computer, or a lack of sufficient system resources.

Users may need to be trained to use the new version of the software that is likely to include new features. Whilst such new features can be ignored until the user feels ready to investigate them, upgrades can also include changes to existing features which will immediately effect the use. Such changes can irritate and confuse. Users will need extra support when software upgrades have been made. If training is not provided, it will take users longer to perform familiar tasks at first.

In a large organisation, an upgrade can be trialled in one department before use throughout the whole organisation. In this way expertise can be built up and problems highlighted on a small scale. However, if this is done there are likely to be complications in file transfer between departments.

Very often, it is not necessary to purchase a full copy of a new version of software, just an 'upgrade' copy that will convert your current version to the new one. This is usually a cheaper option, but does have some disadvantages.

David Seek is a graphic designer who runs his own business that employs four other artists as well as himself. He owns four AppleMacs and Quark is one of the main software packages that they use.

Over the years David has bought the upgrades to the original version that he purchased. The updates were not really needed for their work as the earlier versions provided all the features they required, but their clients and publishers used the more advanced versions and would send files to them which could not be read by the older software.

Recently David's system was struck by a virus that resulted in all the software having to be re-installed. This took a considerable time as every upgrade had to be installed separately, in order.

1 Explain, in your own words, the difficulties and costs involved when upgrading software in a large organisation. (6)

2 Discuss the advantages and disadvantages of giving extensive training to users when upgrading takes place. (6)

3 Explain the phrase 'a lack of sufficient system resources' in the first paragraph. (4)

4 Outline the problems of running two versions of the same software package at the same time within an organisation. (4)

5 For a particular software package with which you are familiar highlight additional features that he been added on an upgrade, and features that have changed.

Software strategies

Software evaluation

Choosing appropriate software for using on a home computer is often relatively simple. There are many magazines on the market which compare different software packages of the same type, highlighting features and making recommendations. Software publishers sometimes provide a demonstration copy. These can be made available on CDs given free with magazines such as *Personal Computer World* or via the Internet.

Choosing the appropriate software to use within an organisation is a much more complex task. There might be hundreds of workstations throughout the organisation being used for a wide range of tasks. In such situations, software evaluation needs to be structured and planned since an inappropriate choice could have far-reaching consequences.

Finding the most appropriate software involves establishing the specific needs of the user and which aspects of a software type is most important to them. Available packages should be investigated to establish their capabilities with respect to the user's

needs so that the best match can be found. Of course, no suitable software may exist, in which case alternative solutions need to be sought. (See Software solutions page 246.)

> **ESTABLISH USER NEEDS**
>
> MATCH
>
> **SOFTWARE CAPABILITIES**

It is important that the criteria for evaluation should be established at the start. The problem specification should be agreed. Many aspects to be considered have been discussed in full in Chapter 9.

> A **benchmark is** a standard set of computer tasks, designed to allow measurements to be made of computer performance. These tasks can be used to compare the performance of different software or hardware. Examples of tasks are, how long it takes to re-format a 40 page word-processed document, how many pages can be printed in one minute, how long it takes to save 1000 database records to disk. (Taken from the BCS Glossary)

Figure 21.2 Benchmark

Evaluation criteria

Functionality – what features does the software offer? Does it actually **do** what it claims?

Functions of a word-processor might include: mail-merge, thesaurus, cut and paste editing, and so on.

Performance – how well does the software perform the chosen features? How does it perform to benchmark tests?

Usability and human-machine interface – is the software suitable for a beginner? How convenient would an experienced user find it? What are the help facilities? How complex is the software to learn? Are there shortcuts for the experienced user? Is the interface consistent?

Figure 21.1 Music software

Database	Microsoft Access 97	Lotus Approach 97
General	*Access 97* is part of *Microsoft Office 97*, and is the most widely used PC database. *Access* requires *Windows 95* or *NT*. *Access* may be very complicated for the typical user although it includes several wizards to ease use.	*Approach 97* is part of *Lotus SmartSuite 97* and is simpler than *Access*. *Approach* is designed to be easy to use, using flexible data formats.
Ease of use	*Access* uses a familiar interface to set up tables, queries, forms reports, macros and modules. It is possible to set up databases with simple relationships, queries, forms and reports. *Access* includes a Visual Basic package for Structured Query Language (SQL) programming. However Visual Basic programming is not easy.	*Approach* is quick to learn and quick to use as a database. *Approach* is not so easy to use if the user needs to use SQL programming. It was designed to avoid the need for any use of programming but this may be necessary in complex applications.
Data format	The database is stored in a single MBD file. However users are vulnerable if this file gets corrupted.	*Approach* uses the default format of DBF, compatible with many other applications.
Security	Files can be password-protected. Different permission levels can be set.	Files can be password-protected with four levels of privilege.
Validation	Rules can be defined to validate data as it is entered.	Validation can be set up in *Approach*.
Problems	Performance may suffer if too many users are trying to access the same database. *Access 97* needs more memory to run than any other *Office 97* application.	As a database needs to become more complicated, *Approach* becomes less suitable. The programming language, *LotusScript*, which is similar to Visual Basic, does not fit well into the rest of the package.
Data transfer	*Access* is linked by wizards to *Word* for mail-merge and to *Excel*.	*Approach* has very good portability to other applications.
Price:	£235 + VAT or bundled in *Office 97 Professional*	£40 + VAT or bundled with *SmartSuite 97*
Good Points	Well integrated with *Office 97*.	Simple to learn. Good price.
Bad Points	The advanced features of *Access* are hard to learn.	*Approach* is not friendly for programmers.
Contact	0345 002000 www.microsoft.com	01784 455445 www.lotus.com
Conclusion	The biggest seller, *Access 97* is the package to beat.	*Approach 97* is a good database if it is not to be used for complex applications.

Selection from a table produced by *Personal Computer World* showing characteristics of a number of database packages.

Compatibility – is this software compatible with the software currently in use? Is it similar to the existing software so little training is needed? Would its use involve extensive retraining or the replacement of some current software?

Transferability of data – is it easy to transfer data from current files into those for the new system? Will there need to be any major data conversion which will prove unwieldy or expensive? In the worst case, would data have to be re-entered?

Robustness – how well tested is the software? (.0 versions may contain bugs). Does it have a wide user base who will have tested the software more widely? Has it been used in circumstances similar to those required by the system? Are the advanced features that are needed fully developed? Can it handle the volume of data required?

User support – what kind of support is provided? Is it adequate for the needs of the users? How expensive will adequate support be? Are there books/videos/training courses available? Is there a telephone help-line?

Resource requirements – will the software run on the existing hardware platform, or will new equipment be required? If so, is the requirement within acceptable limits? Does it require a large amount of memory and disk space? Will it run under the current operating system? Will changes in other software be needed? Will further staffing be required

Dynamic Data Exchange. This is where data from one application can be copied into another application. If the data in one application is changed, then it is automatically up-dated in the other application.

Figure 21.3 DDE

Object, Linking and Embedding. This is where a file from another program can be embedded in or linked to a file in this program. If the first file is up-dated, so is the second file.

Figure 21.4 OLE

to run the new software? Can current staff be retrained? At what cost?

Upgradability – is it likely that upgrades of the software will become available in the future?

Portability – are adequate filters provided to allow data to be transferred to other systems in the way that the system demands? Can it export data to other software? Can you share data with other applications using DDE? (Dynamic Data Exchange). Does it support OLE (Object, Linking and Embedding)?

Financial issues – what are the financial implications of purchasing this software? Will it require additional hardware purchases? Will other, associated software need to be bought? What will the training and file conversion costs be? What will be the development cost of installing the new software? Will the use of the new software bring development opportunities?

Evaluation report

When an evaluation has been completed, a report is written for senior management who would then make a decision based on the information provided. The function of the report is to document how the software performed against the criteria set, to enable a decision to be made.

After explaining the purpose of the particular report by defining the user requirements, it should include a summary of the **methodology used**. This would include a description of how information was gathered together with a description of all benchmark tests undertaken.

The **actual evaluation** should then follow, where the appropriate evaluation criteria are discussed. This is often shown in the form of a grid, where the characteristics of different software packages are shown together.

Recommendations, with **justifications,** are then made on the basis of the evidence.

Software to support specialist applications

Some software is generic and can be used in a variety of applications. Examples of generic software include word-processors, databases and spreadsheets. These can be used in a wide range of circumstances and can, if desired, be customised to fit specific needs. Another class of software is that which is designed for a specific application. Some important categories of such software are now described.

Figure 21.5 An architect using specialist software

Music

Computers can easily be linked to electronic musical input devices such as a keyboard. They can also produce output, through an amplifier and loud speakers, which can be synthesised to mimic different musical instruments and other sounds.

Software allows music to be composed in two ways. The composer can write notes onto a conventional musical stave shown on the screen, using the mouse. Thus the method of composition is the same, but the software offers the same benefits as a word-processor offers a writer. As with word-processing, it is possible

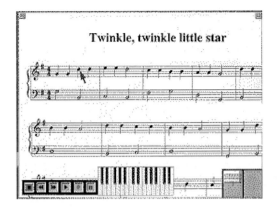

Figure 21.6 Music software

to make small alterations without having the re-write the whole score. The score can be played by the computer so that the composer can hear her work. The score can then be printed using conventional printers. A typical program can also compose intermediate passages in music and generate suitable accompaniment when the user is playing.

In the alternative method, the music is entered at a musical keyboard input device. The software converts this to standard sheet music notation.

MIDI (Musical Instrument Digital Interface) is a communications standard for exchanging information between computers, synthesisers and other electronic equipment. A sequencer is a form of software which allows the user to record music, play it back, and edit the performance from a range of MIDI instruments. While the composer plays at a keyboard, the sequencing software records the MIDI data that is produced and stores it in memory. Another part can be added that will be synchronised with the first track.

An example of music software is Sibelius 7. This is a very extensive package which includes hundreds of features including:
- editing: can add, delete or change notes, clefs, key signatures, rests etc.
- instruments: music can be written for up to 128 different instruments per score
- MIDI: 64 channel MIDI feedback; follows music while recording and rewinding
- rhythm: automatic grouping of rests according to beats
- text: menus of commonly used text; automatic bar numbering
- specialist notations for different types of music (for example jazz)

Mathematical and statistical software

Mathematics is a tool for people in many jobs, not just for mathematicians. There are a number of sophisticated packages on the market, such as Mathematica which provide a huge range of Mathematical facilities for the user. Like any such software, the user has to have a reasonable knowledge of the subject matter to be able to make good use of the package.

MathCad 7 is another example of a mathematics package that covers 18 distribution functions and can work on data imported from Excel.

A spreadsheet such as Excel offers a wide range of statistical and mathematical functions that can be used on data. As well as this the package includes a number of tools such as data analysis tools called the Analysis Toolpak that allow the user to save steps when developing complex analyses.

Statistical packages are used to obtain information from data. For many situations a spreadsheet such as Excel provides adequate statistical functions for a user's needs. However, specialist programs can be needed for more advanced manipulation.

Statistical packages can process the laborious work of statistical analysis. It is essential that the data entered in 100 per cent correct. These packages may be quite specific or they may be general purpose. A popular package is SPSS (Statistical Package for Social Sciences). This package can carry out a wide range of statistical analyses and includes a huge variety of charting and graphing facilities. It also has the ability to run simuations.

Certain situations demand prediction. If a predicted value is to be obtained with maximum certainty it is possible that a Neural Network package will be required.

Mathematica, developed by Wolfram Research, offers a fully integrated environment for technical computing. This tool is useful in all areas of science, technology and business where quantitative methods are used. Mathematica supports numerical, symbolic and graphical computation. It can be used as a simple calculator and as a full symbolic programming language with a host of powerful capabilities. It allows users to organise text, computations, graphics and animations for technical reports.

When choosing an appropriate package, there are factors other than functionality that should be taken into account. One is the screen layout as a screen of multiple statistical analyses can become cluttered. SPSS offers a 'session tree' which permits click-on access to previous analysis. How data is input is also important and a good package will integrate well with other software such as a spreadsheet or a word processor.

SPSS is a favourite package of UK local government and offers a wide range of extra, add-on modules. It is easy to learn and use and the modules allow it to be customised for a specific environment. The package is used by the Commission for Racial Equality to analyse data relating to discrimination in order to discover causes.

The direct marketing department of Abbey Life uses SPSS data analysis package together with their data mining module Chaid for market analysis. Its use ensures that customers only receive marketing information that is relevant to them. This minimises wasted mailing and reduces costs.

OLAP (On line analytical processing) has the ability to access and analyse multi-dimensional data very fast.

This analysis tool is combined with a querying facility that can bring together data from a variety of data sources and is used to find trends in the data that can be reported in any format that the user chooses. Such an activity is known as data mining.

Figure 21.7 Statistical software

Project management

Project management software is used to help a manager plan and monitor the progress of a project. Typical projects could be the implementation of a new information system, the building of a new section of motorway or the development of a new model of car. Such projects are usually so large that they cannot be implemented by just one person. A team of people working together normally undertakes them; each person having specific tasks to do.

Project management involves dividing the project up into smaller tasks which are allocated to teams. The project leader needs to use the resources available in the best possible way to get the job completed successfully and on time. Records of what is being done when must be kept.

Project management programs will organise the project into smaller tasks and then schedule these tasks. This can be a very complicated operation, as certain tasks cannot start until other tasks have been completed. For example, when building a road, the task of laying the tarmac must be completed before the task of painting white lines can start. The schedule can be displayed as a Gantt chart (a form of timetable) or a PERT chart which shows how the tasks relate (see Chapter 15). As the project progresses and jobs are completed or delayed, the charts can be up-dated to show the current position.

Project management software also deals with resource management, both physical and human. For

A company building houses needs to make sure that the jobs are done in the right order. The walls cannot be built before the foundations for example. Different craftsmen will be needed for different tasks, for example plumbers, bricklayers, electricians. Project management software will draw up a schedule so that

- the work is done in the right order

- the right craftsmen are there to do the work at the right time

- the craftsmen are not double booked at another job

- the materials these craftsmen need are delivered in time

- labour and material costs are monitored

Features of a project management software package

- ability to add or modify tasks, start or end date or resources available
- calculate project completion time
- schedule tasks to find critical path or best performance time
- assign resources to tasks
- print in graphical format (e.g. Gantt or PERT chart)
- calculate resource requirements

example, much expensive heavy plant is needed in road building; it is vital that the right machines are in the right place at the right time since idle machines waste money. It is also important not to schedule a person to carry out two tasks at the same time. The software also monitors costs, keeps records of work done and generates reports for analysis or presentation.

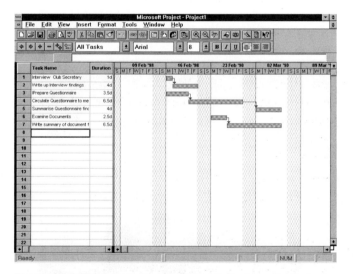

Figure 21.8 Project planning software displaying a Gantt chart

Computer Aided Design (CAD)

Using a CAD program such as AutoCAD, a designer can produce designs more quickly and accurately than using a conventional drawing board. Ideas can be sketched, stored, recalled and modified. A powerful computer is required to deal with the complex graphics and the need to re-draw designs regularly. Designs can be input using a mouse or, with greater precision, a graphics tablet, and output using a plotter.

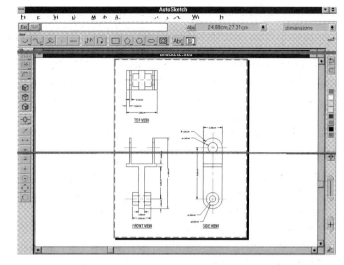

Figure 21.9 A CAD in use

Images are stored in vector graphics form. Lines and shapes are stored mathematically. A straight line can be represented by the co-ordinates of its start point, its length and direction. Similarly, a circle can be represented by the co-ordinates of its centre and its radius. This means a shape can be moved or altered. For example, the circle can be moved by changing the position of the centre point and made larger by increasing the radius.

As with word-processing, small alterations can be made easily without the need to start again. CAD packages offer a wide range of drawing aids. Lines can be drawn in different thicknesses or formats. Tools are available which serve the functions of the ruler, compass, protractor and arc: the basis of traditional technical drawing. Functions are included to allow the user to zoom in on part of the drawing, rotate or invert objects, lock points to a notional grid to ensure accuracy and allow automatic scaling.

USES OF CAD

- engineering drawing
- architecture
- interior design, for example kitchens
- mapping
- printed circuit and microprocessor design
- computer animation
- film special effects

Figure 21.10 Uses of CAD

Files can be stored on disk for possible future use. Cutting and pasting makes copying very easy. Standard components, such as a window, can be inserted into a drawing in many places without redrawing. Images can also be stored in layers. One layer might be the building outline. Another layer might be the electrical wiring. Another could the plumbing. Another might be the proposed new extension. Any combination of layers can be printed.

Designs may be in 2-dimensions or 3-dimensions and a CAD package allows the user to view from different angles.

Information can be exported into a spreadsheet or both programs can be linked by DDE (Dynamic Data Exchange). This allows costing to be worked out and stresses and strains calculated to ensure the safety of the design.

CAD software is used in a variety of application areas. For many fields, such as interior design and architecture, specialist CAD packages have been developed. These may be linked to customised graphics tablets as well as having libraries of pre-drawn objects.

Computer-Aided Manufacture (CAM)

Computer-aided design is often used together with computer-aided manufacture (CAD/CAM). CAM is a term used to describe the use of computers to control the manufacture of products. There are two main types of computer controlled machines used in manufacture: the CNC (Computer Numerically Controlled) device and robots. CNC devices carry out operations such turning materials, milling, drilling and cutting under the control of a computer program. Robots can be programmed to carry out specific tasks such as inserting screws or grabbing and moving objects. In simple terms, the CAM software converts the specification produced by the CAD package into instructions for the machines.

CAD/CAM can be used in the clothing industry for both design and manufacture of garments. The software works out the most efficient way of cutting the pattern from the material and the cloth can then be cut out automatically from the design.

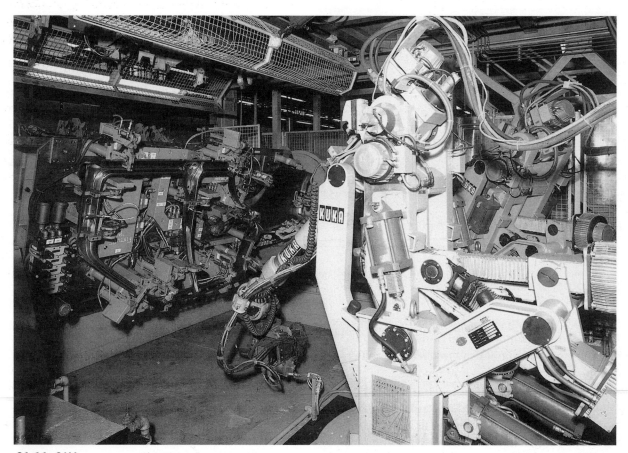

Figure 21.11 CAM

Geographical Information Systems (GIS)

Geographical Information System software has developed dramatically over the last few years. Such software combines data held in databases with graphics to display information geographically linked to maps. The information could be about the location of factories and customers, the routes for deliveries of goods or the boundaries for sales territories. Seeing information linked to a map provides an understandable user interface.

In the UK the Ordnance Survey maintains up to date maps of the whole country. These are held in digital form and stored on a computer. It is therefore possible to produce printed maps in a form that suits individual needs. The types of features, colours, scale and level of detail can be selected. The map data can be purchased in its digital form and can be used by companies in conjunction with data stored in a database. For example, details of sales could be linked to the map and displayed, so that sales territories can be studied and modified.

GIS are also used in marketing. Census data, including the age, occupation and income breakdown at a household level can be linked to maps. This data can be aggregated to produce profiles of areas which can be used to select locations suitable to receive direct marketing of a particular product.

> Vast amounts of geographical and meteorological data can be measured by computer and stored in a database. Meteorological data may be captured by observer satellites or by sensors, which can be left in remote areas, dangerous to humans. They may store details on weather, climate, land use, pollution, habitats, etc. and be available on-line to subscribers. Computer programs can also be used to draw weather charts and make forecasts. The computer does not need to be plugged into the sensors all the time. Data for several days can be stored and down-loaded when required.

> Telecom Eirann, Ireland's telecoms supplier, has a GIS to integrate digital maps of company equipment and services with customer records and billing systems.
>
> Cleveland Constabulary provides police services for 600 000 people in NE England. It uses a GIS to display intelligent maps of incident locations and for real-time monitoring of all available police resources.

> Packages such as AutoRoute Express are readily available to plan journeys on a desk-top computer and are easy to use. They allow the user to specify a range of constraints, such as maximum speed and favoured typed of road between selected start and finish points and will then provide a suggested route. Thus business trips can be planned more efficiently.

Use of specialist software

Software packages of the kinds mentioned above are for specialist users. Knowledge of the work concerned is required to use the programs successfully. For example, music software would be of no use to someone who lacks an understanding of musical principles. With a specialist package a computer can be used to calculate statistical information. This information still needs to be analysed and interpreted.

Software solutions

There are four main ways of acquiring application software. It may be possible to make use of a pre-written package which can be bought in shops or by mail order. The use of pre-written application software has advantages. There is no time delay before installation as it is ready for immediate use. Because the development cost is being spread over many sales, it is usually less expensive than an application software produced in house. Software from a reputable software house will already have been tested. It is usually possible to customise some features of the software to meet an individual user's requirements.

In certain situations, a suitable alternative for a user could be to buy a generic software package such as a spreadsheet or a database and customise it for his own use. This might involve writing macros or modules to act as a user interface which is automatically loaded.

However, if there is no pre-written application software on the market that fits the special needs of a user, then purpose written software will be needed. Such software will be tailored to the user's exact specifications. It will be costly to develop as skilled programmers are expensive. The process of establishing precise specifications can be time-

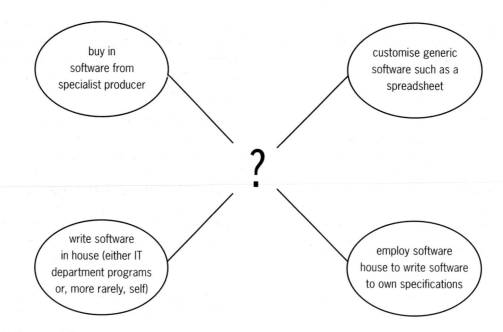

Figure 21.12 Software solutions

consuming. This software can be written in house, if a programming team is employed. In some cases it may be appropriate for the user to develop a solution for themselves.

If in-house production is not possible, due to lack of programmers with the necessary expertise and time, a software house can be contracted to write the programs. A software house will employ skilled programmers to develop customised software to a specification agreed with their clients. Obviously, customised software is an expensive option.

Computer services

Specialist computer companies offer a range of services to users. These companies are called **computer bureaux**, **software consultancies**, **software houses** or **systems houses**. The services offered may include:

- selling and installing standard, off-the-shelf software
- tailor-made software
- computer consultancy and advice
- networking advice
- Internet advice
- web-page design
- hardware and software packages

- hardware and software rental or leasing
- help in data preparation

Software reliability

For commercial software producers, marketing software that is virtually bug-free is crucial. The credibility of a product and the level of customer satisfaction will depend upon the reliability of the software.

Bugs

A bug is an error in a program. It often results from a trivial mistake in a line of program code. It could just be that one character is wrong, but this will mean that the program or one part of the program won't work. It is practically impossible to write software that is absolutely bug-free. A package might contain many thousands of lines of program code, leading to millions of different routes through. It would not be possible to test every route. It is very hard to write programs for an environment such as Windows, where the program must work perfectly with different hardware, different printer drivers, different screen resolutions and different versions of Windows.

Programming staff are not always as rigorous as they

should be in designing and testing programs. Software producers need to impose strict standards for the writing and testing of programs to minimise the likelihood of errors. They should be tested by colleagues who have not been involved with the actual production themselves.

However, commercial demands to keep software development costs within defined limits and to keep development time as brief as possible, as well as a need to ensure that the product beats competitors to the market, means that it is not feasible to test all software fully. Producers therefore usually provide help lines for users where they can record details of new bugs and receive help, often in the form of program 'patches', which will repair the error. Upgraded, .1, versions of software are often distributed free or at considerable discount to existing users of the .0 version.

Alpha and beta testing

New software needs rigorous testing. Alpha testing is in-house testing, undertaken by the company who wrote the software. It will test the software with its own test data and will check the software matches the design specification. This will involve testing with different hardware and old versions of the operating system.

Beta testing is carried out off-site by real users, perhaps companies or individuals who have bought previous versions of the software or who have expressed an interest. These valued customers enter into an agreement with the software house for special discount and support in return for the testing work done. The software will be tested using real data in a real situation, using different platforms (different processors, different memory sizes, and so on). This may detect errors not previously found.

Beta testing is a vital part of software development. The testers are independent of the producers and therefore impartial. The product is tested in the 'real world' and the customers also provide valuable feedback. A company needs enough beta testers to test the product fully but not too many so that the product becomes too public.

Only if both types of testing are successful will the software be released for sale. Windows 95 took several years to develop. Firstly, faults had to be ironed out in house. Eventually, six months before the program went on sale, beta copies were sent out

to testers. Several bugs still remained which had to be corrected before distribution. Of course, distribution outside the company reduces security. By the time Windows 95 was released, the press knew exactly what it would do.

However well they are tested, programs can still be sold with faults in them. These faults may not even be noticed until the program has been marketed. Software companies will try to provide a corrected version of the program when available and will supply a **bug fix** to upgrade the program to remove the bug. Later versions may appear similar but improve performance, speed or memory usage. The fix may be supplied free to all registered users.

Maintenance

The first version of new software that is released is usually closely followed by further maintenance releases which carry out changes. These changes could be:

perfective – ones that improve performance: speed, memory usage and so on

corrective – fixing bugs which only came to light after release

adaptive – making changes to fit differing needs, for example metrication, changing tax laws

The software producer must ensure that all purchasers are given details of maintenance changes. These can be dealt with by mail-shots to all licensed users, the dispatch of up-dates on floppy disk, the use of bulletin boards with details of patches, fixes and known errors. It is therefore important that all users register with the software producer by completing and returning licensing agreements.

HP9000 Trial Release Software
HP9000 is in Beta testing for Ingres and Sybase Database interface. Would you like to become a Beta Tester? Just fill out our Request Form and you can login to our secured Beta Pages to download the Installation keys to activate these features. Please make sure you mark the correct Beta Program that you would like to enter.

....

– Pages created solely for McDonnell Information Systems, PRO-IV Business Unit in Irvine.
Last up-dated on Thu June 19 11:02:27 PDT 1997 by WebMaster.

A company payroll system inputs the numbers of hours of overtime worked by an employee during the week. This number must be greater than or equal to zero and less than or equal to 100. New rules limit the maximum number of hours to be worked in a week to 48. The data entry check is adapted to make sure that the number is between zero and thirteen (assuming a 35 hour basic working week).	**SOME MAINTENANCE CHANGES ARE MADE TO IMPROVE PERFORMANCE,** **SOME FIX BUGS AND** **SOME MEET CHANGING USER NEEDS**	A supermarket chain introduces a new type of offer – if a combination of specific products is bought then an extra number of reward points are awarded. For example, if 2 punnets of strawberries, 1 carton of own brand double cream and 1 packet of wafer biscuits are all bought together, then 50 extra points are awarded. The billing software will need to be modified to cope with such an offer.
A particular package has a menu bar that appears at the bottom of the screen . Users have complained that it is hard to use. A new version is produced with the menu bar at the top of the screen.	Maintenance Examples	A game is produced that requires the player to work through a number of levels, building up points as they go. Users are frustrated that there is no function to allow them to save a play half way through. This extra function can be added.
A system developed for an estate agency allows photographs of properties to be scanned in and displayed for clients, both on screen and as a printed copy. The agency has bought a digital camera and wishes to have the system modified to allow photos to be added directly from the camera.	A program crashes whenever a certain combination of menu choices is followed by pressing the escape key. This was not picked up in testing.	Air traffic control software installed at Heathrow airport some years ago had been adapted from US software. Unfortunately, it did not take into account the fact that the 0 meridian runs through Britain – the software therefore did not differentiate between 1° East and 1° West. Planes flying over Norwich were plotted as being over Birmingham. The software needed changing fast.

Figure 21.13 Examples of software maintenance

Summary

- ☐ Software evaluation should address the following critertia:
  - functionality
  - performance
  - usability and HCI
  - compatibility
  - data transferability
  - robustness
  - user support
  - resource requirements
  - upgradability
  - portability
  - financial issues

- ☐ In choosing which software to buy, companies must establish their needs and find software that satisfies these needs.

- ☐ As well as generic software such as word-processors which can have many applications, there are many types of specialist software for specialist applications, for example music software, GIS, mathematical, CAD/CAM and project planning.

- ☐ It is possible to have software written especially for a particular task

- ☐ New software has to be thoroughly tested, if possible testing all branches of the program on different specification computers

Software strategies questions

1 The Ordnance Survey has maps of the whole of the UK stored on computer. Describe three ways in which this digital data could be used by a company manufacturing and selling white goods, for example fridges and cookers. (6)

2 A range of software packages can be described as 'Project Management Software.' What is project management software and what does it do? (2) *NEAB Specimen Paper 4*

3 You are asked to evaluate a software package and produce an evaluation report.

   a) Describe four criteria you would use to evaluate the package. (8)

   b) What is the function and content of an evaluation report? (4) *NEAB Specimen Paper 4*

4 A software company is preparing to release a new application program. Describe the two types of testing carried out before the final release of the software. Explain why both are needed. (6) *NEAB 1997 Paper 4*

5 A recording studio is considering providing a range of specialist software solutions to support its musical recording and composition work. List **six** features you would expect such software to include. (6) *NEAB 1997 Paper 4*

6 A particular organisation uses a computerised stock control system. On performing the half yearly stock check it is discovered that the actual stock levels of some of the items are below that shown on the system.

   a) Describe the functionality which should have been built into the software to minimise the possibility of this happening. (2)

   b) Explain why this functionality is required. (2) *NEAB 1997 Paper 3*

7 The management of a local college has decided to buy a computer-aided design (CAD) package to help to plan the best use of the available space in the college.

   a) Describe suitable hardware to support the use of the package. (4)

   b) Identify two features available in a CAD package that are not generally found in a simple drawing package. (2)

   c) Give, with reasons, three advantages of using a CAD package rather than manual methods for this application. (4)

   d) At various times the management is required to produce statistical reports using a spreadsheet with data currently held in the CAD package. Describe the functionality required by the CAD package to allow this to happen. (2) *NEAB Specimen Paper 4*

8 An examination board is considering developing a system which is to be used for maintaining and processing module test results of candidates.

   a) Describe the different ways in which the examination board may be able to provide a software solution.

   b) Discuss the issues the examination board should consider before choosing any particular solution. (20) *NEAB 1997 Paper 4*

9 Before releasing a new package the software company carries out alpha and beta testing.

   a) What are these two types of testing and why are they both needed? (6)

   b) Explain why, once the package has been released there may be a need for maintenance releases and how might these be dealt with. (6) *NEAB Specimen Paper 4*

10 Many businesses need to use relational database software. Two popular packages for *Windows* PCs are *Microsoft Access 97* and *Lotus Approach 97*. In deciding which package to buy, Ashford Recycling used the table on page 249. Study the table on page 238 and explain the major issues that should influence Ashford Recycling's decision. (12)

11 A range of software packages can be described as 'Project Management Software'. List **six** features that you would expect such packages to include. (6) *NEAB 1998 Paper 5*

12 You are the IT manager of a college. Your principal wishes to implement a computerised student identification card system. One way of providing the software for this system is to use a generic applications package, and to customise it to meet the project specification.

a) Describe **two** ways of providing the software other than using a 'generic applications package'. (4)

b) The college has a clearly set out IT strategy, however this project has not been included. Identify and describe **four** issues that should be considered when making a final choice from the above three methods (8) *NEAB 1998 Paper 5*

13 A company is about to change its accounting software. In order to evaluate the different packages available to them, they have drawn up a number of evaluation criteria.

a) Why are such evaluation criteria needed? (2)

b) Explain the issues involved with each of the **three** evaluation criteria given below:

- Functionality

- User Support

- Hardware Resource Requirements. (6)

c) Identify and describe **three** additional evaluation criteria that you might also expect the company to include. (6) *NEAB 1998 Paper 5*

14 A software company has notified customers of a maintenance release for its accounting package. The notification states that a programme of alpha and beta testing will be carried out to ensure that the maintenance release is reliable.

a) State **three** reasons why a maintenance release might be necessary. (3)

b) What is meant by the terms:

i) alpha testing?

ii) beta testing? (2) *NEAB IT05 1999*

15 A market research company wants to purchase a statistical package to analyse the results of surveys.

a) List features you would expect such a software package to include. (3)

b) Explain what other factor is critical for the effective use of this software, apart from an appropriate computer system. (2) *NEAB IT05 1999*

16 A university has decided to buy a new payroll package. They are considering several options and have drawn up a range of evaluation criteria to help them select the most appropriate one.

a) The criteria used by the university include:

- 'performance',

- 'robustness',

- 'user support',

For **each** of these criteria, describe **two** issues that you would expect the university to consider. (6)

Describe three other criteria you would expect the university to apply when comparing systems. (6) *NEAB IT05 1999*

Activity: Software Reliability

An NEAB Information Technology past paper question asked:

'Articles in the media referring to computer software which fails to work properly are commonplace. Discuss the difficulties facing software companies when testing and implementing complex software, and the measures that software providers could take to minimise these problems'.

The mark scheme listed the points given below. Explain and expand each point, giving examples wherever possible. Take care to explain the words which are underlined.

- Wide variety of different user <u>hardware platforms</u> is difficult to anticipate.

- Economic need to bring to market leads to inadequate testing.

- Competition from rivals forces reduced development time and testing.

- Many routes to test.

- Importance of well designed **test** <u>strategy</u>.

- Difficulties of providing a full range and variety of test data.

- Software may perform differently with different volumes of data.

- Full implication of changes is hard to predict.

- <u>Alpha testing</u>.

- <u>Beta testing</u>.

1 Explain why it is not possible to run software that is designed for a PC on an Apple Mac. (3)

2 Emulators are increasing in popularity.
   a) Why is this? (1)
   b) Explain why some emulators are not successful. (2)
   c) A company with Macs finds that the new software it requires is only available for PCs. It has to choose between buying new PCs or buying an emulator for the Mac. Explain the issues that are likely to influence their decision. (8)

Communication and information systems

Local Area Networks

A local area network (LAN) is a collection of computers and peripherals that is connected together by cables. It is located in one building or site. At its simplest, a LAN consists of a few computers sharing some resources. Very often the computers connected to the network in this way are called workstations or terminals.

A powerful, high performance computer with large disk capacity can act as a file server. Data can be stored in one place, allowing access from all terminals. It is common to have a number of hard disks, as well as a number of CD-ROM drives. Different databases can be stored on different disks. Such a network does not prevent an individual user from accessing resources locally at the workstation, such as hard disk, floppy disk, CD-ROM drive or printer. It is often possible to use an individual workstation with its own resources, independent of the network.

If a printer is to be shared by all workstations, then a printer server is required which manages the allocation of the printer to different jobs and maintains a queue of waiting jobs. A LAN may have several different printers for different purposes, for example an A4 laser printer for documents, a dot matrix for draft documents and an A3 laser for diagrams.

On larger sites it is common to have more than one LAN linked together with bridges (see page 256).

Wide Area Network (WAN)

A WAN is a collection of computers spread over a large geographical area, which can be as small as a few miles or as large as the whole world. Communication between computers is made in a variety of ways including microwave, satellite link, dedicated cables or the telephone network. The telephone link can either be made through the public

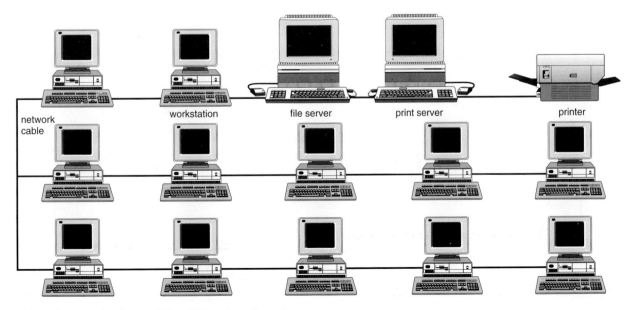

Figure 22.1 An example of a Local Area Network configuration

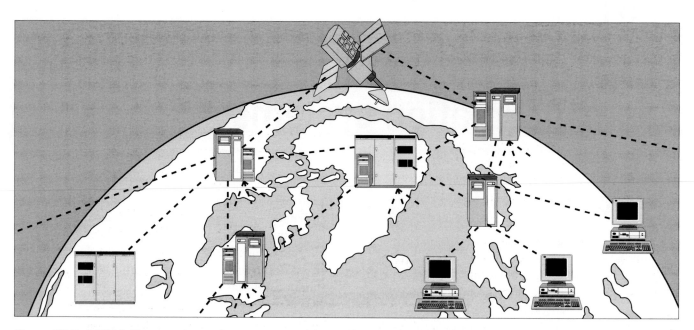

Figure 22.2 A WAN linking many types of computer around the world using a range of communication media

dial-up lines where the message is routed alongside others, or through a leased line which provides a permanent connection. Leased lines are practical when the volume of data being sent is large and communication is frequent.

Data sent between two computers on a global network might travel through a number of different communications media. The route is created via switching computers that create the necessary path from source to destination computers.

Public networks

A public network is a WAN that is available to all types of users. There have been a number of public networks developed over the years, but most have now been subsumed by the Internet.

The use of public networks has increased enormously over the past few years due to a number of factors. Many telephone networks have already been changed from old-style analogue to modern high-speed digital

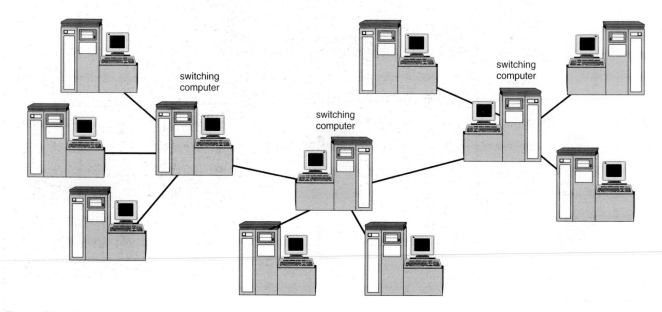

Figure 22.3 How computers can be linked via switching in a WAN

Case Study

Home worker

Paul lives in a beautiful, sleepy village in the heart of the Cotswolds where life seems to have been untouched by the computer age. This does not appear to be the most promising place for an IT professional to live.

However, from the converted stables at the bottom of his garden, Paul runs a business which relies on the latest technology. He designs and hosts web sites and builds intranets for clients as well as providing a range of other services.

For this work Paul has a high specification server with two client workstations. He has had a leased line installed to provide fast internet access for both his own and his clients' use. As well as hosting web sites (clients rent space on his server), he will design pages using FrontPage software by Microsoft. He trialled the beta version of FrontPage 98 on one computer, whilst running FrontPage 97 on the other. This provided security in case bugs arising on the new software which could lead to failure.

Paul has developed on-line forms and credit card purchasing facilities for clients. On-line credit card clearance can be provided by banks who then pass confirmation of the transaction to the vendor thus making it unnecessary for the vendor to be sent credit card details. This reduces the risk of fraud for the purchaser.

Paul has been working from home for three years now and he never intends to work in a conventional office again!

- Explain what is meant by the term leased line.
- What alternatives would Paul have had to having a leased line installed?
- Why do you think he chose to have a leased line?
- Describe a suitable back up strategy for Paul. ☐

Figure 22.4 Paul at work

technology and many more are scheduled to do so in the near future. The cost of connecting to and using networks has reduced. However, the speed in which data is transmitted has been a problem, particularly graphics and moving images. Improved compression techniques have reduced the amount of data that needs to be transmitted.

Intranets

An intranet is a web site that is accessible to a closed group of people within an organisation. It is accessed through the same browser software that is used for the Internet. Most intranets use web publishing, database and HTML to share information and create shared work group applications.

If an organisation wishes to set up an intranet, it must first establish a network – LAN or WAN. Web server software will need to be purchased and installed on a server so that the network will support the same browser software as used in the Internet. A firewall must be installed to stop access from the Internet to the intranet. A firewall consists of a set of programs working with bridges and routers to limit the flow of data between networks. It is used here to prevent external users from accessing the intranet.

Case Study

The *Daily Mirror*

Computers are involved in every stage of the production of the *Daily Mirror*. The main editorial work is carried out at head office which is located at Canary Wharf. Printing takes place in four locations around the country: in Watford, Oldham, Glasgow and Belfast. Editors work at Anderton Quay in Glasgow on a sister paper, the *Daily Record*. These six locations are linked to form a Wide Area Network. Watford also holds the *Mirror*'s picture library. Currently, between 500 000 and 600 000 photographs have been scanned in and held in digital form on disk. These images are available to editors for the inclusion in articles as appropriate and are catalogued according to subject.

At Canary Wharf, a Local Area Network links the Apple Macintosh computers used by journalists and editors. All articles are written using Microsoft Word; pages are built using the publishing package, Quark Express and art work is done using Adobe Photoshop. Quark Publishing System is used; this allows the chief sub editor to view a whole page and then assign individual articles to different sub editors via the local area network. The sub editors are then able to work on their assigned story while viewing the whole page, without having the rights to alter other stories. Internal e-mail is achieved using Lotus Notes.

The networks are supported by 21 servers in Canary Wharf and six in Glasgow. Back-ups are done at night by copying across vital files to a different location.

An on-line version of the *Daily Mirror* is published on the Internet by SuperNet, internet providers located in Jersey. □

Figure 22.5 The Mirror Building at Canary Wharf

An intranet is easy to use and relatively inexpensive to set up. It can be accessed by users on different hardware platforms and provides a standard appearance on all equipment. An intranet can provide users with the specific information they require at all times, both on their own desk tops or when away from the office. Information is centralised and, ideally, organised in a logical way.

Uses of networks

The use of a network can be found in most working environments. It is increasingly common to find networks used in schools and colleges, doctors and dentists surgeries, shops and offices of all kinds.

Distributed systems

In the early days of computing, before the development of microprocessors, mainframes were the only type of computer that could be used. Data processing was centralised and all data and information had to pass through the hands of the data processing department who were in control of all operations. All data had to be sent to this department for input. Information, in the form of reports would then be sent back to the appropriate user department. Any requests for non-standard information would require special programming and could take weeks or months to be met. When terminals became available, giving on-line access to the central mainframe, they were dumb terminals which had no processing power. At this time, control was firmly in the hands of the data processing department.

With the development of the microprocessors, control has shifted towards the local branch or department. LANs, which have a telecommunications link to the head office mainframe, now allow a high degree of local control. Less data needs to be sent to and from the mainframe computer. This is known as distributed processing.

The distribution of control of systems brings a number of advantages. Local branches or departments have greater autonomy. Processing capacity is spread around the organisation which allows the work of one branch to be covered by another in case of breakdown.

Ways in which a network can improve communication and productivity within an organisation

E-mail can increase the speed of communication by replacing written messages whilst avoiding the constant interruption that can result from the use of the telephone.

Video conferencing can reduce the time and cost of travel for meetings, bringing together participants from different locations.

An **intranet** allows the sharing of information throughout an organisation without the need for circulating, filing and retrieving physical documents.

The use of **work groups** allows employees to share documents, diaries and other computer files.

Telecommuting allows personnel to work at home thus reducing office overheads and absence due to family circumstances.

The use of the **Internet** provides all the facilities of the world wide web and opens up the opportunities of e-commerce.

The use of **Electronic data interchange** (EDI) allows one organisation to transfer electronic data from their computer system to that of another organisation thus removing the need for paper transactions.

A retail organisation can use **point of sale** (POS) terminals with **electronic funds transfer** (EFT).

Using a network allows **stock control** to be managed in real time. When goods are sold, via a POS terminal, the level of stock can automatically be decreased. Enquiries from other computers on the network can be provided with up to date stock levels.

Distributed databases

Databases can also be distributed. 'A distributed database is a database that consists of two or more data files located at different sites on a computer network. Because the database is distributed, different users can access it without interfering with one another. However, the DBMS must periodically synchronise the scattered databases to make sure that they all have consistent data'.

Traditionally, a distributed database was a collection of data which logically belonged to a system, but which was physically distributed to a number of locations that were connected by a communication network. Nowadays it is also likely to be a collection of possibly independent database systems with facilities for exchanging data.

A distributed database can either be on-line at the host computer in a central location but also available to remote locations, or part of the database at the host computer can be duplicated and placed in a remote computer or copies of the entire database and DBMS can be at each remote location.

■ ADVANTAGES OF DISTRIBUTED DATABASES

The distribution of a database can provide local autonomy: local data is managed independently of other sites. Data can be shared by different users or branches of the same organisation, or even different organisations. The data from one location can be available at another, even though the hardware platform or operating system may be different. New locations can be added to the database without requiring a complete re-writing of the entire database.

The opportunity of remote back-up to another site reduces the likelihood of data loss.

■ LIMITATIONS OF DISTRIBUTED DATABASES

A distributed system, which hides its distributed nature from the end user is more complex than a centralised system. This means that such a system is much more expensive to install and maintain. The DBMS has to co-ordinate messages and transactions between the different locations. This need to transfer data from location to another increases the security risks to the data.

As all the data is not stored in one location, if one station were to fail with inadequate back-up, other locations might suffer a loss of data.

The Internet

Internet structure

The Internet can be thought of as a network of networks. The Internet backbone is a high-speed network provided by telecommunications companies and used by Internet Service Providers (ISPS) to route information around the world. ISPs (also known as Internet Access Providers) give a user access to the provider's computer system, which is connected to the Internet. These providers have a Point of Presence (POP) in many sites around the world. Users link to the Internet via their nearest POP: this should only involve a local phone call. Well known Internet Service Providers include Compuserve and AOL.

The software needed to access an ISP from your computer is readily available. CD-ROMS can be picked up free in supermarkets, Oxfam shops or found on the cover of computing magazines, Internet Service Providers are keen to sign up new users and offer a range of different services. They can differ in the following ways:

- the number of hours per month of free access time;
- the number of e-mail accounts supported;
- the amount of disk space allocated for storing a web-site of your own;
- the browser software available;
- any restrictions in the use of facilities such as chat or newsgroups.

The occasional home user only needs a telephone line and modem to link their computer to the Internet.

The software provided on the CD-ROM has to dial up the nearest POP, organise re-dials if the line or the site is busy, and select the appropriate rate of data transfer for the modem. The user is allocated an identification number and password which the software can enter automatically.

Institutions with many users require an ISDN (Integrated Service Digital Network) connection. This gives a higher bandwidth and a telecommunications company such as BT can provide such a line. Data transfer is much faster. The line is considerable more expensive to rent than a standard line. There is no need for a modem as the signal is sent in digital form. An interface between the computer and the line is still required, called a Terminal Adapter (TA).

If 24-hour on-line access is needed, than a leased line is appropriate. A leased line is a permanently open connection. Many users can access the Internet concurrently using the same SDN line.

World Wide Web

The World Wide Web (www or Web) is a system that allows users to produce multimedia pages that are accessible by other users of the Internet. These pages can contain text, graphics, sound or video clips. The web is built on top of the Internet. Web clients and web servers communicate with each other using HTTP protocol. The Internet is the network structure that supports the transfer of information in the Web.

An Internet browser is needed to access information from the web. A browser is a user-friendly front end that allows the user an easy way of accessing and displaying documents without the need of any technical knowledge or extensive training. A web browser will contact a web site and access information from the site as a page in HTML format. Web pages are stored at a web site where the home page or main page is the first to be displayed.

The power of the Web as an infomation source is in the use of hypertext and hypermedia, an interactive navigation tool. A mouse click on a particular piece of text or graphical image moves the user to a different page, either at the same time or a different one.

Case Study

E-Commerce

E-commerce is a huge growth area in IT. E-commerce is not just about having a presence on the Internet: more interactivity is required. A formal definition is given as 'all commercial transactions taking place over the Internet where there is some exchange and money involved, in real time'. Perhaps a simpler definition could be: buying and selling through the Internet.

The US is the furthest ahead in web-based commerce with sales in 1998 accounting for 0.5% of all retail transactions. Purchases of computer equipment make up the greatest number of on-line sales, followed by financial brokerage and on-line travel and holiday sales.

Europe is seeing the greatest advances in developing techniques for access. Three techniques are currently being developed:

- multimedia PC/TV – the method used in the US
- interactive digital TV
- mobile phone – already used for sending limited text messages. In Sweden it is possible to trade shares using a mobile phone.

On the Internet, it is easy to display catalogues, take orders, deliver invoices, automate payments as well as carry out many other transactions. The golden rule for virtual shopping is to make the buying process as simple as possible. Intelligent responses to the client, using a profile of customer data can allow items of interest to the client to be highlighted.

However, all e-commerce growth depends upon building confidence, both in buyers and sellers. The most practical payment methods are debit and credit cards. Buyers need to be sure that their details are safe and not open to fraud and mis-use through hacking. The government, at the time of writing, has published a draft Safe Trading Bill which tackles both this problem and related ones. It proposes that:

- electronic signatures should be admissible in court so that paper documents can be replaced with electronic ones;
- a scheme is developed to regulate the enforcement of cryptography that ensures all personal information stored on-line cannot be read by others.

A vital requirement of e-commerce is that the vendor stores customer information electronically. All such data must comply with the Data Protection Act.

Abbey National is launching services on digital TV that allows customers to carry out transactions such as paying bills, obtaining financial advice and arranging insurance from their armchairs.

As well as retail sales, the Internet is also used for business-to-business transactions. This use builds on the EDI (electronic data interchange) networks that have been used by large organisations to transfer transactions, such as invoices, electronically. These networks are dedicated to their task and are highly secure. Progression to using the Internet is a natural one that allows the use of such data transfer to be carried out by small businesses as well as large ones as no expensive, dedicated network has to be set up. However the use of the Internet is less secure.

E-commerce, at the moment, only accounts for a small percentage of sales and other financial transactions. Once the public has confidence that the security risks are controlled, it is likely to increase enormously and become a normal part of daily life.

- Find 5 people who have ordered goods or services over the Internet. List the sites and the goods involved.
- What implications are there for society in the growth of e-commerce? ☐

Case Study

Peter Symonds' College

Peter Symonds' College uses a LAN throughout its site for curriculum and administrative use. Charles Parish, Head of Information Systems at the college, describes how the network has developed.

We purchased our first ten IBM PCs in 1990 (to replace ageing BBC Bs) and it was clear from the outset that we would have to network them in order to share programs, data and equipment such as printers. We decided to use ethernet rather than token ring as it was cheaper and the standard in most educational sites. The network operating software we purchased was Novell Netware (then version 3.0) running on a 386 PC with 600MB hard disk.

The College is accommodated in a number of buildings on a fairly large campus with an adult education site located about a mile away. Many colleges have physically separate curriculum and administration networks, but we decided that we would link them in order to give teaching staff access to MIS data and to share printers etc. This meant that we would have to span between buildings and fortunately we decided from the start to use fibre-optic cable between the buildings, while using coaxial cable indoors.

Over the years the network has grown in size and complexity. It now reaches all the teaching and administration areas as well as three boarding houses and the Adult Education site; 16 buildings in all. We have over 20 file servers for the curriculum, administration, intranet, e-mail and other uses. Two of the servers are in separate buildings and are dedicated to running backups of all other servers every night. All the buildings are linked via fibre optic cable except for the Adult Education site which uses a 2 Mbit per second leased line connection.

About 400 Pentium PCs running Windows 98 are now connected to the network and this, along with the increasing

size of applications and data files, means that we are always trying to increase the performance. We realised a number of years ago that configuring the whole college network as one segment where every packet transmitted by a computer is repeated to every other computer would lead to a very high rate of collisions and unacceptably poor performance.

One way that many organisations have tackled this problem is to segment their LANs so that each department has its own server which is separated from a backbone segment by a bridge. Users would log in to their local server and access their applications from it as well. The backbone would be used for occasional access to other servers.

Traffic from Design Department users destined for the Design server would not get on to the backbone cable. Only traffic destined for other servers would use this. This is fine so long as users have their own PC and mainly access their local server. This would not work in our college as students (and staff) have the annoying habit of moving around the site. They may log in from the Art department first thing in the morning, then from the Humanities Block after break and finally from the IT Centre in the afternoon. There can be no such thing as a fixed local server for such a user.

The introduction around 1994 of a device called a switch enabled us to segment the network and reduce traffic. We now have more than 20 switches, some running gigabit ethernet. All the computers on the network have access to the Internet via a firewall, router and leased line to our local university and from their to JANET (Joint Academic Network).

- Why did Charles decide to use fibre optic cable between buildings?
- List the similarities and differences between the network described and that in place in your school or college. ☐

Converging technologies

Until recently the use of various devices was straightforward. A television set was used to watch programmes, a mobile telephone was used to have conversations with others around the world, a computer could be used with a modem to access the Internet and a games console could be used to play games. Things are not so simple any more!

The Network Computer (NC) is a low priced computer designed only to work on a network (LAN or Internet). All resources are obtained from the network instead of being located in the computer. Digital televisions are now produced that are network ready: they contain the equivalent of a Network Computer. A user can access the Internet through the television via cable connection, and order goods and services. The volume of goods ordered in this way is expected to grow very fast.

Games consoles are now available with modem and e-mail facilities. These devices allow users to play games with other users over the Internet.

Dreamcast

Dreamcast is the revolutionary digital entertainment system for the new millennium. Thanks to cutting-edge technology Dreamcast's 128bit performance immerses you in incredibly realistic 3D environments and delivers the fastest gameplay we've ever created. Dreamcast is the world's first console to offer networking capabilities, which will include e-mail, Internet access and multiplayer gaming on your TV.
(From promotional material)

Voiceover IP (VOIP) is a protocol that allows voice messages to be sent over a data network. Long-distance phone conversation can take place using fast ISDN links to the Internet at a fraction of the cost of normal phone calls.

It has been possible to transmit text messages using mobile phones for some time. Recent developments have made it possible to access the Internet and transmit e-mails using a mobile phone. In fact there is a major shift towards mobility – as well as the mobile phone, the palm top organiser is being developed into an even more versatile tool. They can be used, with small keyboards, as very mobile word processors. Text can be entered on the move on trains or planes. On returning to their desktop computer, a user can transfer the text from organiser to computer with ease.

Blue Tooth wireless initiative was founded by IMB and Intel. The technology will allow devices such as lap top computers and mobile phones to exchange data wirelessly.

Microsoft death of PCs
(based on an article in *The Guardian*)

The death knell for the personal computer sounded when Microsoft announced a major investment in the next generation of mobile phones.

They have made an agreement with Ericsson, the world's largest manufacturer of mobile electronic devices to make Microsoft's e-mail system compatible with mobiles and personal organisers. Industry experts say that by 2003 more than 600m people will be accessing the internet from mobile devices compared to 550m using desktop computers.

Phones, walkmans, personal organisers and even digital cameras will provide access to the internet with users able to surf, read books, speak to friends, e-mail and shop from the same device.

Microsoft has already identified British computer company Psion, best known for its handheld organisers, as one of its biggest threats. Psion, together with mobile phone companies, are attempting to creates a global operating standard for handheld devices.

The new technology that will drive the growth is called Wireless Application Protocol (WAP). It is a piece of software that filters text information from web pages and displays the words on your phone. Nokia have launched the UK's first WAP phone offering access to e-mails, sports information and entertainment.

Summary

- A LAN is a collection of computers over a small geographical area.
- A WAN is a collection of computers spread over a large geographical area.
- A public network is a WAN that is available to all types of users.
- Intranets exist within an organisation and use Internet browse software.
- Computer systems can be distributed over a number of locations. Distribution can apply to both control and data.
- Internet Service Providers give users access to the internet.
- There is a growing convergence in communications technologies.

Network strategies questions

1 Describe in detail what a new home computer user would need to do to gain access to the internet.

2 Describe the latest developments in wireless computing. (8)

3 Explain the term e-commerce. (3)

4 Define the term distributed database. Describe the problems arise due to data being distributed. (6)

5 Describe the role of an Internet Service Provider (4)

6 A manager has upgraded his desktop computer to take advantage of his company network environment.

State **two** changes that you would expect him to see as a result of such an upgrade. (2) *NEAB IT05 1999*

Network strategies, standards and protocols

A network exists when two or more computers are connected to each other by cable, over telephone lines, or through wireless communication. When connected to a network, users can share resources on their computer, such as data files, software, printers and modems.

An organisation's network will be maintained by the network administrator. The duties include installing software, allocating accounts and assigning initial passwords for users, making regular backups of data and software stored on the network and optimising the performance to the user.

The following components are normally required for a network:

- One or more *file servers*. A file server is a powerful computer that is used to store and distribute files over the network. The file server identifies each user's file separately so that other users do not have access to them.

- A *network adapter* card for each computer that is part of the network. A network card or other device inside the computer that physically connects your computer to the network.

- *Cabling* is used to connect each network computer and printer to the file server. The cost of cabling is an important part of the cost of installing a network.

- *The Network Operating System* is an operating system that allows computers to connect to a network.

- *Network accounts* are required by users to access the network. Each user has their own account which is characterised by a user identity, password and access privileges. The network administrator creates network accounts. A user will have access to a range of centrally stored software and will be able to access their files stored in their own user area on a file server as well as those files to which they have access rights.

Network User Interface

A special network operating system is required for use on a network to enable a computer to communicate with other computers. The same network operating system software must be used on all computers on the network. Servers will require additional software. A file server will need software to allow it to provide secure data storage for users. Examples of network operating systems include Novell Netware and Windows for Workgroups.

When a computer that is attached to a network is switched on, it can 'boot up' from the network file server. The user will then be presented with a log in screen and will need to enter their id number and correct password before being able to access any files or software.

The user will have more drives available. Conventionally, on a PC based system, **a** to **e** are reserved for the local floppy, hard disk and CD drives while **f** to **z** are network drives. Typically, **f** drive might be the user's own private storage area on the file server and other drives could be used to store shared files and software. It is likely that a user will have different access rights for different drives.

Once a user has opened an application package, their view will be very similar to that of a user of the same package on a stand-alone machine. An awareness of being part of a network would only become evident when files are loaded or stored, or when some form of network maintenance was taking place. Maintenance activities could be backup procedures, virus checking or the transmission of messages from the network manager.

Certain activities are more complicated for a user on a network than on a stand-alone computer. The use of shared printers is one such example. All documents for printing would be sent to the appropriate job queue and printed when the printer was free. The user would need to ensure that they had selected the correct print queue from their software, otherwise they could find it hard to track down their document! The management of files and appropriate access to different drives require care and training.

Network security

When an individual uses a stand alone computer ensuring data is secure is a relatively simple matter. Use of a keyboard and disk lock can prevent other people from using the computer and accessing files. The use of a screen saver with a password can prevent casual prying when the user is away from their desk.

When an individual's computer is part of a network, then security becomes a more complex issue. It is normal for network users to have an individual and unique username. This allows a directory to be allocated to each user so that users cannot access another user's files. If the user uses a password it provides security for the user. A user is usually only able to log on to at one station at a time.

The use of a unique username also allows the network manager to keep a record of who has been on the network and when and at which station.

The user is allocated a username, but has to choose their own password. This should be chosen with care, and obvious choices such as 'SECRET', 'COMPUTER' and the user's name should be avoided. Ideally, the password should consist of a mixture of letters and numbers and should not spell out a meaningful word. Passwords should never be shared with others. Many systems are set up to require that a user changes their password on a regular basis, perhaps every month. Usually, the user is given a number (typically three) of grace logins, attempts at entering the correct password. If all attempts fail then the network manager is alerted and the user account is disabled for a period of time.

It is common for users to need access to shared files and databases. In a hospital, different doctors, nurses and clerks will need access to parts of the patients' database. In a college, students will need to access shared software and files containing assignments created by their teachers. There needs to be some means of allowing different users different levels of

access. Users can be assigned access rights to directories by the network manager. These can be full rights (allowing the user to read, alter or delete files), or limited to read only or access may be completely forbidden.

Access Rights (privileges)

Normally users do not have access to other users' areas. However more senior users or the network administrator may need access to all areas. It is now possible for the network administrator to 'steal' the screen of another user, something that is increasingly common in schools and colleges.

Access Rights

Some examples of access rights which can be allocated to a user for a particular file or directory:

- Supervisor (the most powerful)
- Read
- Write
- Create
- Erase
- Modify
- File Scan
- No rights (denies any access)

Figure 23.1 Access rights

Viruses

A security hazard that is a serious problem with a network comes from viruses. Some viruses are simply annoying, making a program run unexpectedly whilst others can have catastrophic results.

On a stand alone computer, the damage that can be done by a virus is limited, whilst on a network it can destroy the files of a whole business. Viruses are transferred to systems from floppy disks that have themselves been infected from another computer. Some CD-ROMs have been distributed containing a virus. Access to public networks such as the Internet also poses a threat as files downloaded can contain viruses.

A number of measures can be put in place to help prevent a network becoming infected by a virus. Special software can be bought which will check all

Case Study

▶ Problems with a virus

The publicity officer of a university was recenty caused considerable inconvenience by a virus. Over a period of several weeks, she had put together the proofs of the prospectus for post graduate courses. This involved considerable effort as entries had to be gathered from a range of contributors in different departments. A number of edits were made and the final versions were recalled from disk and visually checked on the screen with great care. The files were copied to floppy disk and transferred to the printers, who were located on the other side of town, by motorcycle courier. When the files were loaded into the computer at the printers they no longer incorporated the latest edits. The disks were returned. This process was repeated several time as the final deadline got closer and closer. Finally the problem was pinpointed to a virus which affected the copying of files.

- How could the virus have got on to the system?
- Outline precautions that the publicity officer should have taken to prevent this problem from occurring.
- Research some of the effects caused by the latest viruses. ☐

new files for viruses. This software is relatively expensive and needs to be upgraded on a regular basis as new viruses are being developed all the time.

Preventing viruses is better than trying to cure them. It is important to follow good practice to avoid infection. Users can be forbidden from using a floppy disk in a networked computer without first checking it for viruses.

Firewall software can be used with the Internet to filter and check files that are down loaded. A firewall is a system placed between an internal network and the outside world which ensures that all traffic passing from the inside to the outside, or the outside to the inside, must pass through it. Only traffic which is authorised by the security policy is allowed to pass. It is designed to protect a safe and trusted system from a risky and untrusted system.

Back-up

The need for back-up becomes crucial when a network is in use as the implications of breakdown and data loss could be enormous. Details of back-up procedures are discussed in Chapters 8, 17 and 20.

Again good practice is important. For example, backing-up should occur regularly and back-up copies should be checked as soon as they have been created to ensure that the process has been carried out correctly.

A crucial requirement of maintaining security is user awareness. Users should be made aware, in a variety of ways, the dangers of network use and the necessary precautions that should become an automatic part of their working habits.

Security and WANs

Data transferred on a network is vulnerable to misuse. Misuse can occur unintentionally when a user, who has perhaps not undertaken sufficient training, makes an incorrect transaction. Misuse can also be intentional when an unauthorised user attempts to gain illegal access to a network. The risk of authorised access increases when data is transferred over a WAN using public communication links.

People who attempt illegal access are known as hackers. There are two main categories of hacker: those who break into a system and those who act as imposters.

Although there is no absolutely secure system, without eliminating all possibility of outside connection, a number of measures can be taken to reduce risk.

With a network system each account holder is issued with a unique identification number and chooses a secret password. Strict rules need to be applied to password choice, as described earlier. Imposter hackers sometimes use 'packet sniffer' programs to intercept identification numbers and passwords, which they then store for later use. Systems can be designed to prevent the use of such programs.

A widely used way of countering the effectiveness of sniffer programs is the use of encryption, where data is sent in a coded form that has to be decrypted on receipt before it can be understood.

The danger of break-in can be guarded against by the use of a firewall.

Firewall

A firewall is an intelligent device that is used to prevent unauthorised access to an organisation's network. The firewall is placed between the network file server and the external network, often the Internet. The firewall checks all messages sent to the file server from outside and filters the contents. Access may be blocked to certain applications whilst being restricted to others.

An access log can be created that records detail of all attempts to access the site. This can be scrutinised by the network manager. The use of a firewall prevents external users from accessing internal computers directly.

Data Encryption

Data encryption is the process of 'scrambling' data so that it can only be recovered by people authorised to see it. It is used when data is transmitted or stored to ensure that the data does not fall into the wrong hands. When the data is to be used, it needs to be 'unscrambled' or decrypted.

The basic concept of encryption is simple, although the actual transformations carried out can be very complex.

Consider the kind of code used by children. The most common is the type where a letter in the alphabet is replaced by another a set number of places ahead of it, say 4. So A is replaced by E, B by F and so on. This simple example can be used to highlight two terms relating to encryption. The **key** in this case is 4, the number of places the letter is shifted and the **algorithm** is the idea of the shifting of the letters.

■ Network accounting

A user himself controls the files stored in a stand-alone computer and is able to make best use of the disk space available.

A networked system might have many gigabytes of disk space to be shared between users. It is important that an individual user is not allowed to use up more than his fair share of this resource. The network manager can allocate each user a maximum allowance; any requests for more space would need to be justified.

An operating system such as Novell Netware allows the network manager to set a wide range of parameters for an individual user's account. Login can be restricted to certain times of the day and an expiry date and time for access set. If necessary, an account can be disabled.

Restrictions can also be set for password use: a minimum length can be specified and the user can be forced to change the password within a timescale that can be set by the network manager.

Access rights can be given for different files and directories.

Users can be assigned to different user groups which can then be assigned access rights. The individual user would have all the rights that have been assigned to the group.

When a printer is shared on a network, it can be important to keep track of the amount it is used by individuals. In some environments, individuals might be charged for the number of sheets used, either individually or to a departmental account. Alternatively, each user might be allocated an allowance. The network operating system could perform this tracking function, but on the whole this currently is an area of weakness of many network operating systems. Utility software is available which can carry out this tracking function. Users or individual printers can be assigned a limit which, when reached, will prevent further printing until some action is taken.

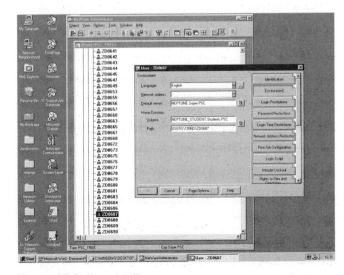

Figure 23.2 User details

Monitoring the use of a network

It is possible to monitor the use of a network either by the network operating system or by specially purchased utility programs. Tracking can take the form of a stored log file that is available for later inspection or can be on-line. Examples of activities that can be tracked include:

- Logins of users specifying workstation, time and number of attempts
- Software used
- Number of concurrent users of a piece of software
- Files accessed specifying time, user and work station
- Access to devices such as printers tracking user and number of pages printed

Accounting software

The increasing use of networks and the Internet by employees within organisations has led many managers to look for ways of keeping track of individual and departmental use.

Traditionally departments within an organisation have been billed for consumables such as paper and other stationary, secretarial support and the cost of phone calls. Until recently, similar costings were not undertaken for use of a network. The transfer of graphics uses considerable network time compared with text: as the transfer of voice and video become more and more common the growth of network traffic will be immense. Such potential growth is leading many organisations to begin charging for network time as the use of video conferencing and other techniques is becoming more widespread.

Network accounting software provides a means of collecting, storing and reporting on network use within an organisation. It can provide the IT manager with a range of information including:

- details of what services have been provided for users;
- band width use which can form the basis for billing individual departments;
- information on usage trends that can be used to plan improved service provision.

A growing problem in some organisations is the misuse of private Internet access by employees. Clear rules need to be laid out for employees in a code of practice. For example, excessive work time should not be used 'surfing' the Net, inappropriate sites should not be accessed and the use of private e-mail should be limited.

Network accounting software can provide information for the employer on:

- Who uses the internet?
- Where they log on and for how long?
- Which computer is used?
- Which sites are visited?

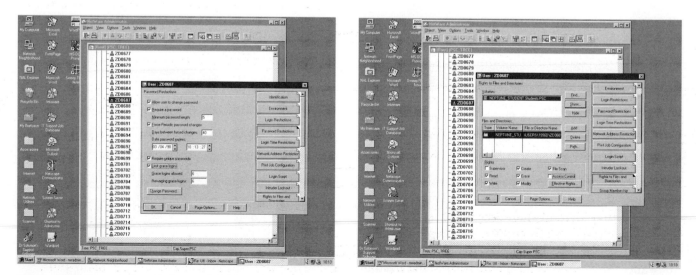

Figure 23.3 Password restrictions and rights to files and directories can be modified by the network manager

Maxtrack Systems (www.maxnet.com) use network accounting software to deliver reports on network usage by users, groups, departments or applications.

The benefits include improved security, assessment of network effectiveness and justification of future network investment. The software also provides information on network usage that enables costs to be allocated between departments or business units.

Audit software

Audit software enables the network administrator to keep track of network use: this will help in maintaining security. All logins and attempted logins are recorded so that illegal attempts at access at a particular work station can be highlighted. In a similar way attempts to access restricted files can also be recorded.

Protocols

Sharing data

It is vital that different applications can share data. This might mean two different pieces of software on the same PC, two different PCs sharing the same software, or even two different platforms running different software. For example, web pages can be accessed from different platforms with different browser software. Not only should text be transferable but graphics, sound and video files too.

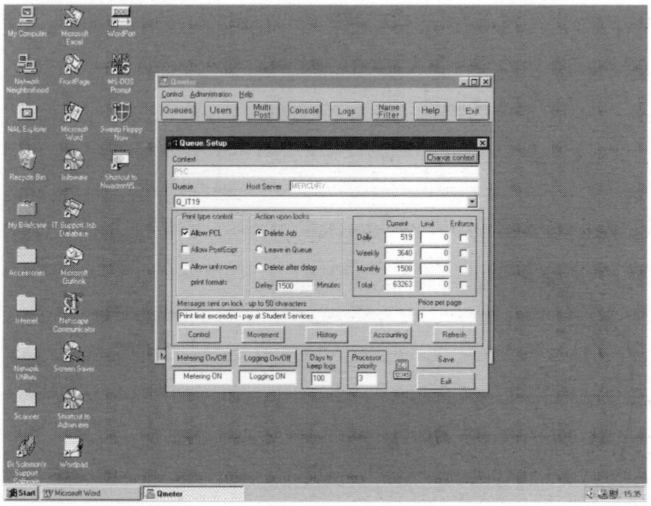

The ability to transfer data between applications is called *portability* and it was discussed in Chapter 9. To be able to transfer data between different computers, it is important to define *protocols* and *standards* first. Protocols are rules and codes for exchanging data between systems. TCP/IP (Transmission Control Protocol/Internet Protocol) is a protocol used to define how data is transmitted over the Internet.

Only computers conforming to specified standards can exchange data. Many standards for computers are laid down by the International Standards Organisation (ISO), for example, **jpg** images are compressed images conforming to an ISO standard.

Advantages of protocols

Protocols allow the use of *open systems* – computer systems that are independent of the manufacturer and the platform and can exchange data with other open systems.

The main advantage of protocols is that the user is not restricted to one manufacturer's equipment. Even if one company's computers are all the same make, they may wish to communicate with another company whose hardware is different e.g. for EDI. This would not be possible without protocols.

However as hardware technology develops, standards may become out of date. It is difficult and takes time to get universal agreement on the establishment of new standards. Open systems based on old standards may be unacceptably slow.

Development of protocols and standards

Some standards are formally introduced, often after considerable deliberation by a committee. ASCII, the American Standard Code for Information Interchange, is an example of such a standard. Others, known as de facto standards arise through historic precedence or as a result of marketing and sales success of a particular product.

Often de facto standards have evolved, not because they are technically the best but due to commercial or

Figure 23.4 Software allows for the monitoring of printer usage

other pressures. In the 1980s there were two types of video recorder, VHS and Betamax. Betamax was widely regarded as being the better quality but VHS became more popular due to better marketing. Betamax flopped while VHS became the standard.

A similar situation is happening in IT. PCs have a huge share of the market yet many people swear by the Apple Macintosh. Microsoft MS-DOS and Windows have such a dominant market share, they have become the standard operating system for a PC. It doesn't mean they are the best. However developers of new software are unlikely to be interested in producing new software for, say, the Linux operating system as the potential market is so small.

Internet protocols and standards

The Internet uses internationally agreed standards so that it can be accessed with a variety of hardware platforms. It is now possible to access the net with a mobile phone or a digital TV.

The protocols used include TCP/IP (Transmission Control Protocol/Internet Protocol) (see above), HTTP (Hyper Text Transfer Protocol) used to identify the address of a web page and FTP (File Transfer Protocol) used to upload files. Web pages are set up in an agreed language HTML (Hyper Text Markup Language).

Gif and jpg images are used on the Internet. They use compression techniques to store graphics files. This means that they are a fraction of the same picture in bitmap format. Utility programs like Paint Shop Pro can convert between different types of file.

In recent years, the Internet has become more and more commercial and is a very successful marketing tool. Advertisers want their image to be eye-catching – putting new demands on web page design. Animated gifs and Java script allow moving images, which are now common place. Once again commercial needs have driven technological developments.

ISO and the OSI Model

The Open Systems Interconnection (OSI) model provides a framework for data to be transferred between two networked computers, regardless of whether they use the same data formats and exchange conventions or are the same type of computer or are even on the same network.

The OSI model was developed by the International Standards Organisation (ISO) as a guideline for developing standards to allow the connection of different types of computers. The OSI model is not itself a communication standard, an agreed method that determines how data is sent and received, but merely a guideline for developing such standards. The OSI model plays a vital role in networked computing.

If a manufacturer's products obey a set of standards based on the OSI model, they can relatively easily be connected to another manufacture's products. Without the communication standards based on the model software development would be very difficult. It would be hard for a manufacturer to sell products that did not adhere to standards.

The OSI Model consists of seven layers, as the task of controlling communications across a computer network is too complex to be defined by one standard. Each layer of the OSI model contains a subset of the functions required to control network communications.

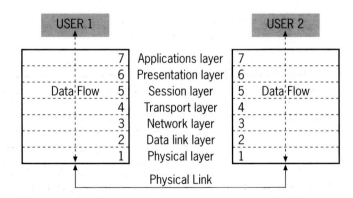

Figure 23.5 OSI seven layer model

Application Layer (7) This is the highest level of the OSI model. It provides service directly to the end users. Examples of its activities would include dealing with file handling, establishing passwords and determining if sufficient resources were present.

Presentation Layer (6) This layer carries out such things as data compression, encryption and formatting to provide a common interface.

Session layer (5) This layer ensures that a user can exchange information. It deals with synchronisation, grouping data and establishing communication as full or half duplex.

Transport Layer (4) This layer ensures that computers can communicate even if the data transmission systems are different. It provides error recovery between the two end points in the network connection.

Network layer (3) This layer routes information around the network. It is responsible for establishing, maintaining and terminating network connections. Standards define how data routing is handled.

Data Link Layer (2) Responsible for the reliability of the physical link established in the physical layer.

Physical Layer (1) This layer is concerned with the transmission of binary data over the transmission medium. It sets standards for electrical and mechanical aspects of interface devices.

An Internet protocol is a standard set of rules that are used to make sure that information is transferred correctly between computers. These protocols define:
- The way in which data is to be structured
- The control signals that are to be used
- The meaning of the control signals

File Transfer Protocol (FTP)	Allows a user to send a message that causes a copy of a file to be transferred from one computer to another. The FTP controls the sending of the file copy in blocks and checks for errors in the data when it is recieved.
HyperText Transfer Protocol (HTTP)	A standard for requesting and transferring a multi-media web page using HTML (HyperText Mark-up Language). *The URL (Uniform Resource Locator) is the address for the data on the Internet. Each file has a URL (its 'name') that specifies the file and the location where it is stored. The URL includes the transfer protocol to be used (e.g. **http**://psc.ac.uk) as well as the domain name.*
Internet Protocol (IP) address	Each computer linked to the Internet has a number, its IP address, which uniquely identifies it. The domain name (e.g. http://**psc.ac.uk**) corresponds to the IP address.
Transmission Control Protocol /Internet Protocol (TCP/IP)	Built on a set of agreed standards (part of the OSI model) this protocol allows providers and users to communicate with each other whatever hardware is being used. (see page 269)
Post Office Protocol 3 (POP3)	This protocol defines standards for the transfer of e-mail between computers
Point to Point Protocol (PPP)	This determines the communication between two directly connected computers. It is very often used between an individual user connected via telephone line to their Internet Service Provider.

Figure 23.6 Table describing some commonly used Internet Protocols

Summary

The use of Networks gives rise to increased problems of security.

A major role of a network manager is to maintain security.

Users of a network are assigned different access rights to data files.

Viruses pose a major hazard to network security.

WANS are particularly vulnerable to illegal misuse by both imposters and break in.

A Firewall can be used to reduce the risk of break in.

Use of Encryption can keep data secure.

Network accounting software is used to keep track of the use of network resources.

Network auditing software is used to monitor access to the network and highlight misuse.

Data is portable if it can be transferred from one computer platform to another.

Standards have developed for the interchange of text, numeric data and graphics.

Some standards are developed formally whilst de facto standards grow out of historic precedence.

The OSI model provides a framework for the development of standards to enable different types of computers to be connected.

Network Strategies, Standards and Protocols questions

1 Give three examples from everyday life that illustrate the use of standards (3)

2 Explain the term data encryption and describe the circumstances when it would be used. (6)

3 A GPs' practice has a networked system with a database of patient records. Doctors need to access details of a patient's medical history and drugs that have been prescribed. Nurses and receptionists need to access basic personal information. Describe the security problems associate with such a system, together with suggested ways of protecting against them. (12)

4 With the growth in computer systems being purchased for use on networks, there is a greater need for manufacturers to conform to standard protocols.

a) What are protocols and why are they required? (4)

b) The application layer is one of seven layers in the OSI model. Name **three** other layers. (3)

c) Briefly describe the role of the application layer in this model. (3) *NEAB 1997 Paper 4*

5 Give two examples of when a computer may use serial transmission and one where it would use parallel. (4)

6 An international company wants to set up a new computer network. Although many staff currently use stand-alone desktop systems the company has no experience of networking. As an IT consultant you have been asked to prepare a report for the company directors, outlining the issues, and the potential benefits, to communications and productivity that such a network could bring. Your report should include:

- a description of the various network components which would be involved;

- a description of the relative merits of different types of network which could be considered;

- a description of the security and accounting issues involved;

- an explanation of networked applications which could improve communications and productivity within the company.

Quality of language will be assessed in this question. (20) *NEAB 1998 Paper 5*

7 The IT manager of a large college is about to change the software that is used to record student attendance in classes. Given that this new software must provide different access permissions and types of report, what capabilities and restrictions should the IT manager allocate in order to satisfy the needs of each of the following groups of users?

- students

- teaching staff

- office staff

- senior managers (8) *NEAB IT05 1999*

8 A company has a computer network system.

a) Activity on the network system is monitored and an accounting log is automatically produced.

 i) State **four** items of information that this log might include. (4)

 ii) Give **four** reasons why such a log is useful. (4)

b) An IT consultant has suggested that the company changes from a peer-to-peer network to a server based network Give six features of these network environments which contrast the two different approaches. (6) *NEAB IT05 1999*

9 An IT consultant wrote in a trade journal:

'The growth in technologies such as personal computers (M), the 'world wide web' and 'wide-area networks' has only come about because manufacturers and suppliers of network hardware and software have adopted standard communications protocols. The OSI seven-layer model has been a key factor in this development . . .'

Discuss this statement. Particular attention should be given to:

- the meaning of 'communications protocols' and why they are required;

- the OSI model and a description of the role of three of its layers;

- the benefits and limitations of standards.

Illustrate your answer with specific examples. (20) *NEAB IT05 1999*

Activity: Extranets at Tesco

In January 1998 the Tesco supermarket chain launched the Tesco Information Exchange (TIE). TIE is a pilot project which allows some of Tesco's main suppliers to access Tesco's own intranet via an *extranet*.

This means the suppliers can get up-to-the-minute sales data for their products, enabling them to adjust production accordingly and avoid over or understocking. Suppliers only have access to their own data and not on other products.

The scheme involves five big suppliers; Britvic, CCSB, Nestlé, Proctor and Gamble and St. Ivel and two small ones; Kingcup mushrooms and St Merryn Meats.

Tesco expect to extend this service to all suppliers in the future. TIE can be accessed via the Internet but is a secure web-site. A combination of firewalls, passwords and security protocols provide security for the project.

1 Tesco have spent a lot of money to set up this system to provide more information to their suppliers. Describe the advantages that the new system would bring for are there for
   a) the suppliers
   b) for Tesco. (8)

2 Explain in your own words the terms:
   a) firewall
   b) protocol. (6)

3 Explain why security is such an important issue. (4)

For more on portability of data, see Chapter 9

Activity: Network accounting

It is likely that your school or college has a network and that you have a user account. If so, answer the following questions.

- What is your user ID/
- How are these IDs allocated?
- What rules govern your choice of password?
- How frequently do you have to change your password?
- Can you log in to more than one work station at the same time?
- How many grace logins are you allowed?
- What happens if you exceed these attempts?
- List the directories/folders/drives that you have full access rights to.
- List the directories/folders/drives that you have no access rights to.
- List the directories/folders/drives that you have read only rights to.
- What categories of people have different rights from you, and how do they differ?
- Is printing monitored?
- If so, explain how this is done.
- What disk space allocation do students have?

Human/computer interaction strategies

▶

Strategies for human computer interaction depend on physical factors, the use required and the user themselves. The latter two can be grouped and called psychological factors.

■ Physical factors

The European Commission Directive on 'Minimum Health and Safety Requirements for Work with Display Screen Equipment' EEC 90/270 became law in the UK on 1 January 1993. It lays down a minimum allowable set of physical conditions affecting computer interaction which must be adhered to. These relate to computing equipment and the working environment. There are set guidelines for the physical aspects of good computer interaction.

Physical factors affect the ease with which a computer is used by a human. Good lighting is essential for ease of use and to avoid eye strain. This means that there should be no reflections from lights or windows. It advisable to fit adjustable blinds to the windows. Liquid crystal display (LCD) screens on lap-tops can only be used with ease at a certain angle otherwise they appear grey. Screens need to provide a stable image with no flickering. They should have easily adjustable brightness and contrast controls, be tiltable and have a swivel base as well as be free from reflective glare. Special non reflective panels can be bought to fix over a screen.

Accommodation must be suitable; there needs to be plenty of space to work in. Rooms must be of adequate height and users need sufficient space around them to work comfortably. Keyboards should be separated from the screen and there must be sufficient space in front for the user to rest his hands and arms. This is essential if repetitive strain injury is to be avoided.

HUMAN COMPUTER INTERACTION

Physical factors	Psychological factors
Screen: • always visible • correct lighting • use of screen filter • not facing window • avoid shadows • adjustable blinds • colour schemes suited to the environment Space: • sufficient space around work station • uncluttered desk • safe cables • suitable ventilation and temperature • not too noisy Furniture: • ergonomically designed chair • adjustable chair • chair at right height to avoid RSI • footrest (legal entitlement)	User friendly: • meaningful icons • uncluttered screen layout • common look and feel between packages • easy to read text • intuitive feel Help to novices: • context sensitive help • wizards • tips • demos • clear error messages Short cuts for experts: • function keys • customised tool bars and menus • keyboard entry for menu choices instead of mouse Make use of long term memory: • standard function key – F1 for help • icons

Figure 24.1 Human Computer Interaction Factors

▶

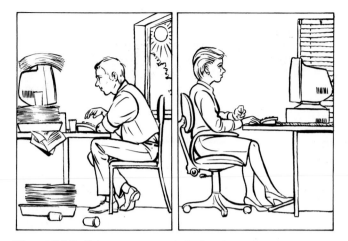

Figure 24.2 Using a computer (a bad and a good way) – spot the differences!

Figure 24.3 A Wizard

Thought needs to be given to the placing of documents being used alongside the computer. Document holders should be used which will keep the head movement of the user between screen and document to a minimum.

A work chair must be chosen carefully. It is important that a computer user sits at a height that allows them to work with their eyes level with the screen, so the seat height and tilt of the back must be adjustable. The chair must be stable and swivel to allow easy movement. A footrest must be available if the user requests one.

It is important to position noisy equipment far enough away from computer users so that it does not disturb concentration or distract attention. Care should be taken to minimise heat and radiation emissions from equipment. For example, laser printers emit ozone and should therefore be kept in a well ventilated position.

The choice of colour schemes on screen should be suited to the environment. Microsoft Windows allows the user to customise such features as text and background colour. Many users find it easier to work with white text on a dark blue background than black on white.

Psychological factors

Computer systems are used most effectively if they take into account the psychological factors which effect human/computer interaction. The interface should be user friendly. Screens should be clear and,

whenever possible, self explanatory so that a user has all the information he needs on the screen. It is very important that the screen is not too cluttered. The use of prompts, which guide the user through a dialogue, reduces the amount of prior learning that is required.

Adequate and consistent help should be given to novices. Many packages include wizards which provide a novice with prompts that take them through a particular task.

An expert who uses the same package very frequently can become frustrated if they are taken through a number of menus and prompts when they enter data. It is important that they are provided with short cuts which allow them to avoid time wasting dialogue. Packages can be set up to have hot keys, special key press combinations, which allow pre-set tasks to be carried out without having to make lengthy menu choices.

Many software packages such as Access, Word and Excel allow the user to customise software to his own requirements. In these packages buttons can be added to run macros, automatically load a customised front end screen interface and set up templates which provide a skeleton for types of documents.

It is good practice to make use of human long term memory to maximise efficiency. The use of icons is an example of this. In Windows programs the same icon is used to represent a particular action. This reduces the amount of learning needed when a particular Windows program is used for the first time.

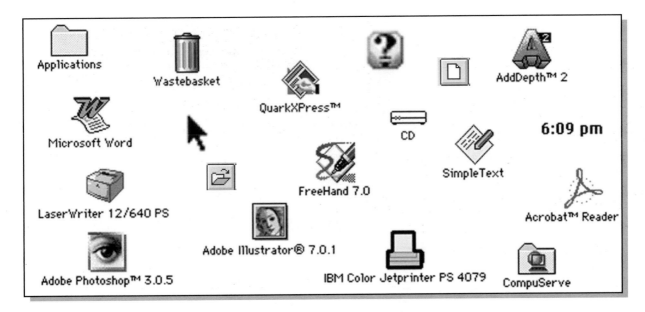

Figure 24.4 Icons

Human/computer interface

Command versus menu structures

A command based interface (where individual commands are typed in) is most suited to experienced, regular users who can take advantage of keyboard shortcuts. A command line interface is also more efficient in that one line can give as much information to the system as a number of menu choices. However, command formats need to be memorised.

Menu formats can be full screen, pull down and pop up. They provide an easy way for inexperienced and infrequent users to interface with a system. They provide the user with a range of choices and do not require any prior memorising of key strokes.

Screen design

It is important to choose a screen design that is appropriate for the likely users. Features such as the size, typeface and amount of text, the use of colour and the incorporation of graphics, need to be chosen carefully. Consistency of headings, menus and layout are very important.

Error messages

The error message is a key feature of an interface. Such messages alert the user whenever a possible mistake is being made e.g. closing a file without saving it. They should be displayed clearly and be consistent in form and positioning. Many modern packages can be configured to modify the number of error messages that appear. Warning sounds can be used to inform the user of an error.

Availability of help

Nearly all recent software includes help facilities for the user. This can be context sensitive. This means that when help is requested (via a key press or a menu choice), information is given which relates to the current function being displayed. It is standard for the F1 key to load the help screen and for users to be able to search for help on a key word.

User friendliness

There are a number of factors, other than those already mentioned, that make a package easy to use. In many packages the size of the standard font displayed on the screen can be adjusted to meet the needs of the user. An A4 page can be displayed at 75 per cent or 100 per cent of full size for normal work, 200 per cent when details are to be checked or whole page view for checking layout. Sounds can be incorporated to help the user; these are often used to bring errors to their attention.

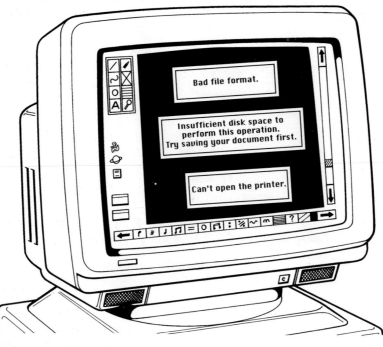

Figure 24.5 Error messages

Resource implications

The use of user-friendly HCIs may produce problems. Windows 3.1 and Windows 95, for example, are Graphical User Interfaces that help the user. They use Help screens, Icons, Menus and Pointing devices. The screen designs are similar and users can set up their own desk-top menu. Data can be taken from one application to another very easily. However running Windows takes up computer memory. It takes time to load and requires a lot of disc space. (About 10 Mb for Windows 3.1 but more if you want lots of fonts.). On-line help makes considerable demands on system resources including extra hard disk space to store help files.

Powerful computers with large memory (RAM) and fast processing speeds are necessary due to complex use of graphics. To run Windows 95, it is recommended that you have at least 16 Mb of RAM and a Pentium processor. See Figure 24.6, which explores the resources needed for a sophisticated HCI.

Customising software packages

Certain software packages can be customised to meet an individual user's specific needs. Word processing packages allow the user to choose the tool bars and icons to be used. The contents of menus can be altered to meet specific needs. A personal, supplementary dictionary can be set up so that commonly used names and words specific to the business can be added. A package such as a spreadsheet can be customised to hide unwanted functions from an inexperienced user and add a user friendly, task specific interface.

Biometrics

Biometrics is a technology that digitises human characteristics such as fingerprints and voices and compares them with a stored version to verify identification. Biometrics work because of the physical uniqueness of human bodies. Although several more obscure body parts have been considered, currently the face, the fingerprint, the shape of the hand, the eye and the voice are actually used.

The face is the easiest to measure but it is quickly changeable. It is therefore best for short term use. A face recognition system is used at an airport in Malaysia to identify travellers depositing luggage. A picture is taken of each traveller which is then converted by software to a digital template which is stored or a reusable smart card containing a computer chip. A similar card is placed on the luggage and the 2 are matched when they arrive at the departure gate.

RESOURCES	HCI FEATURES
Backing Store (e.g. disk) Needed to store large graphics files, large help files and complex programs.	**On line help** Context sensitive searching on different topics, tutorials and wizards all require considerable backing storage.
IAS (Main store – RAM) Complex graphics, high resolution screens with many colours require large amounts of memory to store bit map of pixels and carry out multi-tasking.	**Use of Colour** Enhanced user friendliness but high use made of IAS (More colours require more bits storage for each pixel).
Processor clock speed and word size Complex graphics require a fast processor to be displayed smoothly. Multi-tasking involves the swapping and sharing of the processor between tasks.	**Graphical user interface** Icons, scroll bars and all the other features of a GUI require disk space and IAS. The complex programming needed to run the interface requires a fast processor.
	Multi-tasking Several tasks can be run concurrently – for example a document can be repaginating whilst other tasks are being carried out. A spreadsheet can be re-calculating whilst a user is word processing a letter. Such activity requires a fast processor as well as adequate IAS.

Figure 24.6 Exploring the resource implications of a sophisticated HCI from view of resource and feature

There are 4 stages in a biometric system in use: data capture – a physical sample is captured; extraction – unique data is extracted and a template created; comparison – the template is compared with a new sample and a decision is made as to whether or not the two samples match.

In the past, biometric technology has been used to control access to buildings, but it is now being tested in applications such as ecommerce, banking transactions and data security.

Police have used the biometric of fingerprints to catch and convict criminals for years. Fingerprint technology has been so successful that in the US administrators of welfare benefits and regulators of border crossings have installed similar systems. Other biometric systems are also used. The US immigration service uses a system that measures the 3-dimensional configuration of an individual's hand to speed up entry for low risk travellers at certain airports. IBM has developed an experimental system, FastGate, which uses a hand, voice or fingerprint to ease business passengers through passport control. This system identifies travellers by comparing their fingerprints, voice patterns or palm prints with a digitised record stored on a central database.

In the UK, Nationwide Building Society has been evaluating speech verification software to give customers access to their accounts. The software could be used as an alternative to PIN numbers and code words which customers have to use to prove their identity.

A range of biometric techniques are shown in Figure 24.7.

Iris

Camera captures image of the iris. An iris never changes, is not susceptible to injury and contains a pattern that is unique to an individual.

Facial recognition

Used for verification and identification. A camera acquires an image of the face and the system analyses its geometry (e.g. the distance between the eyes and the nose). Can compensate for glasses, hats and beards.

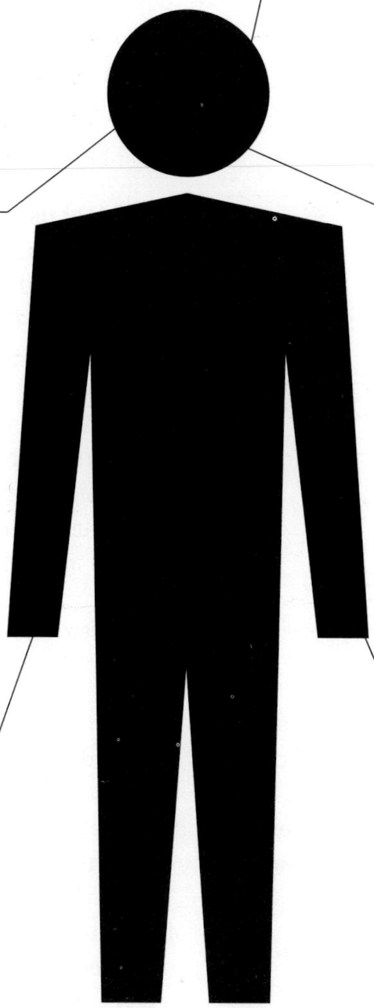

Speech verification

The user states a given phrase and the system creates a template based on a number of characteristics, including pitch, tone and shape of the larynx. Background noise can affect how well the system operates. The user can alter the voice – the technology is less accurate than some other biometric measures.

Hand geometry

This involves the user placing his hand on a plate where it is positioned by lining it up with 5 pegs. The system takes a picture of the hand and examines 90 characteristics, including 3-d shape, width of fingers and shape of knuckles.

Finger scanning

Used for verification systems. Scan can be captured optically or through ultra sound.

Figure 24.7 Biometric techniques

Summary

Physical factors effect human/computer interaction:

☐ screen should always visible and not facing a window

☐ user requires adequate space and adjustable chair

☐ colour schemes should be suited to the environment.

☐ Psychological factors effect human/computer interactions:

☐ user friendliness with help for novices

☐ short cuts for experts

☐ use long term memory to maximise efficiency.

Sophisticated HCI require large amounts of RAM to run. They take up a considerable amount of disk space. A computer with a fast processor is needed if the software is to run at speeds that are acceptable to the user.

Biometrics is a growing technology used to verify identification of individuals. Most common techniques: facial recognition, hand geometry, iris scanning, finger scanning and speech verification.

Human/computer interaction strategies questions

1 Explain the difference between a command driven and a menu based driven interface. For each, name the kind of user for which it would be appropriate. (4)

2 Describe, in your own words,

a) the physical

b) the psychological factors that effect human/computer interaction. (8)

3 With reference to a spreadsheet package that you have used, describe how software can be customised to meet the needs of an individual user. (6)

4 A technical author purchases a new word-processing package which he customises to fit his specific needs. Explain the term customise in this context. Describe what such customisation would involve. (6)

5 a) What are the factors you would need to take into account when designing a screen layout for a database application? (6)

b) What are the resource implications for providing a sophisticated human/computer interface? (4) *NEAB Specimen Paper 4*

6 a) Give **six** of the physical and psychological factors which govern how people interact with computer systems. (6)

b) Give **three** factors which should be considered when providing a sophisticated human-computer interface, explaining the impact of each one on the system's resources. (6) *NEAB 1997 Paper 4*

7 A university uses a complex CAD (computer aided design) package. The package has a sophisticated human-computer interface which also places considerable demands on the system's resources.

a) Give **two** examples of a system's resources that would be affected by such a package and explain the demands placed upon them. (4)

b) Describe **three** features you would expect to find in the human-computer interface which would merit the description 'sophisticated'. (6) *NEAB 1998 Paper 5*

8 A supermarket chain has recently implemented a new stock control system in each of its branches. This has affected those staff who have not used computer systems before. Many of the staff have described the system as being 'user friendly'. However, when the package was, implemented in one particular store, it was not well received by its staff.

a) Give four features of software packages, that would merit the description 'user-friendly'. (4)

b) Both physical and psychological factors can influence how people interact with computer systems. Both may have contributed to the poor reception of this system in that store.

i) Describe **two** such physical factors.

ii) Describe **two** such psychological factors. (8) *NEAB IT05 1999*

Activity: Resource Requirements

Draw up a table like the one below and enter details
for 8 major application packages used in your school
or college. Use manuals, the Internet and magazines
to gather the required data.

Package	Minimum Processor Speed	RAM needed	Disk space Required for full implementation	Disk space required for Help files

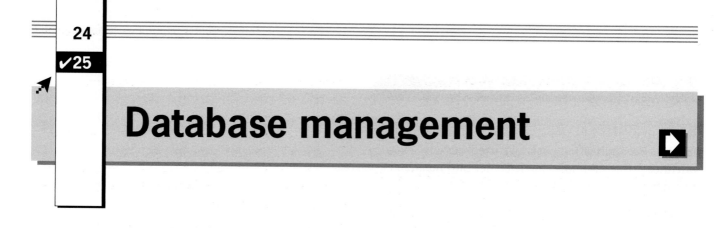

Database management

In Chapter 11 you were introduced to many aspects of database design. The importance of choosing the correct table structure was emphasised and the concepts and use of primary keys were explored.

The choice of the correct structure is not always obvious; a process called data normalisation has been developed to ensure that tables are set up that minimise data redundancy. The process is made up of a number of defined stages.

Database Normalisation

Before normalising, write out all the attributes required by the system.

Referring to the order form shown in Figure 25.1, and listing all the attributes:

Order Number
Date
Customer Number
Customer Surname
Customer Forename
Customer Title
Phone Number
Item Code ⎫
Description ⎬ *There can be any number (Minimum 1) of these fields in any orders. They are called repeated fields*
Quantity ⎪
Unit Cost ⎭
Total Cost

Leemingville Supplies ☐

31 Waverley Road
Midchester

Order Number
99/0005643

Date
23/4/1999

Customer Number
654983
Name
Mrs Penelope Higgins   **Telephone Number**   01962 875439

Item Code	Description	Quantity	Unit Cost
H1786	Yellow Duster (pack of 3)	5	1.75
H3350	Tea Towel (pack of 2)	3	2.50
H4562	Scouring pads (pack of 12)	2	1.25

Total Cost   £18.75

Payment should be received within a month of the date shown on the order. Thank you for your custom.

Figure 25.1 Order form showing attributes

These can be written in **Zero Normal Form (0NF)** in standard notation.

> - Standard Notation for Listing Fields in a Table
> - The table name is written in UPPER CASE letters
> - The table name is a singular noun – VIDEO not VIDEOS
> - Fields (attributes) are written in parentheses after the table name
> - Primary key field(s) are underlined,
> - Repeated fields have a line over them

ORDER (<u>Order Number</u>, Date, Customer Number, Surname, Forename, Title, Phone Number, $\overline{\text{Item Code}}$, $\overline{\text{Description}}$, $\overline{\text{Quantity}}$, $\overline{\text{Unit Cost}}$, Total Cost)

Order Number has been chosen as the primary key as it is a unique number for each order. A line is drawn over the four fields to indicate repeated fields. There can be many entries for these fields in one order.

First Normal Form (1NF) is the next stage. 1NF occurs only when there are repeated fields in the 0NF. In the Order example, the fields Item Code, Description, Quantity and Unit Cost are repeated. These fields are removed from ORDER and a new table is created. This will be called ITEMORDER as it consists of the details of the order for one item. The two tables can be listed:

ORDER (<u>Order Number</u>, Date, Customer Number, Surname, Forename, Title, Phone Number, Total Cost)

ITEMORDER (<u>Order Number</u>, <u>Item Code</u>, Description, Quantity, Unit Cost)

Note that the key field Order Number has been copied to ITEMORDER. If this were not done, there would be no link between the items ordered and the main order. However, Order Number by itself is not correct as a primary key for ITEMORDER as there may well be many entries in ITEMORDER for the same order. The **compound key** of the 2 fields Order Number and Item Code together form the primary key for this table.

Second Normal Form 2NF exists only if 1NF produces a table with a compound key. All attributes in this table should be examined to see if they are dependent upon just one part of the compound or composite key. Examining the table ITEMORDER we find

Order Number is part of the key
Item Code is part of the key
Description is always the same for a given Item Code,

regardless of Order Number so it is dependent on Item Code.
Unit Cost is always the same for a given Item Code, regardless of Order Number so it is dependent on Item Code
Quantity relates to the Item Code but will vary for different orders – thus it is dependent on <u>both</u> Order Number and Item Code.

We can create a table, ITEM that holds all data about a particular item. The attribute Item Code is also left in the ITEMORDER table to provide a link.

ORDER (<u>Order Number</u>, Date, Customer Number, Surname, Forename, Title, Phone Number, Total Cost)
ITEMORDER (<u>Order Number</u>, <u>Item Code</u>, Quantity)
ITEM (<u>Item Code</u>, Description, Unit Cost)

Third Normal Form (3NF) is a 'tidying up' stage where all tables are examined for dependencies (links between attributes). In other words, in 3NF no non-key field should be a fact about another non-key field.

Such links can be found in the ORDER table where Customer Number, Surname, Forename, Title and Phone Number are all linked, Surname, Forename, Title and Phone Number can be said to be dependant on Customer Number. These attributes can be taken out to form a new table – CUSTOMER. Once again, Customer Number is also left in order to provide a link.

The final design of tables is as follows:

ORDER (<u>Order Number</u>, Date, Customer Number, Total Cost)
CUSTOMER (<u>Customer Number</u>, Surname, Forename, Title, Phone Number)
ITEMORDER (<u>Order Number</u>, <u>Item Code</u>, Quantity)
ITEM (<u>Item Code</u>, Description, Unit Cost)

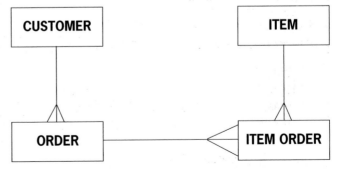

Figure 25.2 Entity relationship diagram for Order database

Note that some data duplication remains, but this is necessary redundancy as the fields are used to create links.

If the CUSTOMER NUMBER had not been present in the original data, a suitable field would have had to be created to provide a primary key for the customer table.

TOTAL COST should not strictly be included in the process of normalisation as it is a derived field: the value is calculated from other attributes and therefore does not need to be stored.

The Database Management System (DBMS)

A DBMS is software that accesses data in a database. It can find data, change existing data and add new data. The DBMS works directly with the data on behalf of the user who does not himself access the data. The DBMS acts as an extra layer between the user program and the database. It deals with all search and update requests from users and carries out other tasks such as maintaining indexes.

The DBMS may provide its own, user friendly interface or may deal directly with user programs that request and use the data available in the database. The DBMS will support a number of different user applications. User queries are made via the DBMS using either SQL (structured query language) or QBE (query by example) to communicate with the DBMS.

The functions of a DBMS are summarised as follows:

1 **Data storage, retrieval and up-date.** The DBMS allows users to store, retrieve and up-date information as easily as possible. These users are not necessarily computer experts and do not need to be aware of the internal structure of the database or how to set it up.

2 **Creation and maintenance of the data dictionary.**

3 **Managing the facilities for sharing the database.** Many databases need a multi-access facility. Two or more people must be able to access the database simultaneously and to up-date records without a problem.

4 **Back-up and recovery.** Information in the database must not be lost in the event of system failure

5 **Security.** The DBMS must check user passwords and allow appropriate privileges.

Database administration

The BCS gives the following definition for a database: '*A database is a collection of data items and links between them, structured in a way that allows it to be accessed by a number of different applications programs*'. In other words, a database consists of data. The database is accessed using a database management system. This is a software package that works with the data for the user. A database administrator is employed to manage large database systems with many users.

Database Administrator (DBA)

The database administrator is the person in an organisation who is responsible for the structure and control of the data in the organisation's database. She is usually involved in the design of the database and will carry out any changes requested by users. This will be followed up by informing other users of changes affecting them. An example of such a change could be the addition of a new field, or the change in size or type of an existing field.

Monitoring the performance of the database is an important task. If the number of access increases, the end user response time might become unacceptably low. If such problems arise, the DBA will need to make appropriate changes to the database structure.

Many early databases became overcomplicated and failed to fulfil their function – no one knew the whole picture of the data stored. It is crucial that the DBA maintains a data dictionary, a file containing descriptions of, and other information about, the structure of the data held on the database.

The DBA will deal with queries for the user and provide help and training appropriate to their own use. She will allocate passwords to users and protect confidential information by implementing and maintaining appropriate access privileges that restrict what data an individual user can access and/or change.

Backup procedures will be managed by the DBA on a regular basis. She will also archive data and carry out any year end other special housekeeping tasks.

The Data Dictionary

The data dictionary is information about the database, such as:

- what tables are included and fields in these tables
- name and description of each data item
- the characteristics of each data item, such as its length and data type
- any restrictions on the value of certain fields
- the relationships between tables
- control information such as who is allowed access to certain data
- whether users can change data or only read it

Database security

As we have already seen data stored in a database is very valuable. Good security to prevent loss, theft or corruption of data is vital. Relational databases such as Microsoft Access and Paradox are multi-access databases. This means that on a network, more than one user may access the same database at the same time.

Relational databases provide different methods of database security:

1 The simplest method is to set a **password** for opening the database. Once set, a password must be entered whenever the database is opened. Only users who type the correct password will be allowed to open the database. The password will be encrypted so that it can't be accessed simply by reading the database file. Once a database is open, all the features are available to the user. For a database on a stand alone computer, setting a password is normally sufficient protection.

2 A more flexible method of database security is called **user-level security**, which is similar to the sort of security found on networks. Users must type a username and password when they load the DBMS. The database administrator will allocate users to a group. For example, in Microsoft Access their are two default groups: Admin (administrators) and Users. Additional groups can be defined. Which group a user is in will determine what level of access they have. For example, some users may be able to see some fields such as name and address but not others such as financial details.

User-level security is essential where users can legitimately access some parts of a database but not those parts which contain sensitive data.

Advantages of the database approach

Organising data in a well structured relational database brings a number of advantages over older, traditional individual file-based systems.

1 **Data independence.** Any changes to the structure of a database, for example adding a field or a table, will not affect any of the programs that access the data. In a file-based system, a minor change in a file structure may require a considerable amount of reprogramming to all the programs that access this file.

2 **Data consistency.** Each data item is stored only once, however many applications it is used for. There is no danger of an item, such as an employee's address, being up-dated on one system and not on another. If this happened the data would not be consistent.

3 **No data redundancy.** Redundancy occurs when data is duplicated unnecessarily. In a file-based system, the same information may be held on several different files, wasting space and making up-dating more difficult.

4 **More information available to users.** In a database system, all information is stored together centrally. Authorised users have access to all this information. In a file-based system data is held in separate files in different departments, sometimes on incompatible systems.

5 **Ease of use.** The DBMS provides easy-to-use queries that enable users to obtain instant answers. In a file-based system a query would have to be specially written by a programmer.

6 **Greater security and integrity of data.** The DBMS will ensure that only authorised users are allowed access to the data. Different users can have different access privileges, depending on their needs. In a file-based system using a number of files it is difficult to control access.

Disadvantages of the database approach

As well as advantages, the use of a database can bring some disadvantages. It is important that these are recognised and addressed when a new database is planned.

1 **Bigger computers and greater risk.** A DBMS is a large program which may require a larger disk and a bigger, more powerful computer than a file-based system. Storing all the information in one database can be risky if the database system fails. All departments of the organisation will be affected, not just a single department. Complex procedures are required to ensure that lost data can be recovered.

2 **Greater complexity.** As databases have more uses, they get more complex and their design is more difficult. This requires considerable expertise and if not done well, the new system may fail to satisfy the user's needs.

3 **Possible inefficiency or poor performance.** A tailor-made file-based system may be more efficient than a database.

Database Management Questions

1 What is the purpose of the data dictionary in a DBMS? (3)

2 The video shop could use a paper-based record-keeping system, a conventional computer file system or a database management system. Give four advantages of using a DBMS in this application. (4)

3 What is the purpose of the DBMS? (4)

4 A college library uses a relational database management system to operate a membership and loans system. Staff and students can borrow as many books as they wish at any given time.

   a) Name **three** database tables that you would expect to find in this system. In each case, identify the columns and keys required to enable this system to be maintained with minimum redundancy. (6)

   b) Draw an entity relationship diagram to show the links between the database tables named in part (a). (3)

   c) Describe the capabilities of the relational database management system that might be used to identify and output details of overdue loans. (6) *NEAB 1997 Paper 2*

5 A theatre has two performances a day (matinee and evening) for six days a week. It has 20 rows, lettered A to T. Each row has 16 seats numbered 1 to 16. Seats in rows A to J cost £12. Seats in the other rows cost £8.50

   a) Define the tables of the theatre booking database

   b) List the fields required for each table. Use standard notation. Identify the primary key of each table by underlining it

   c) List any relationship pairs and draw a relationship diagram

6 A company sports centre use a database management system to operate a membership and fixture system. Normally members register for at least three sports, although they can play any of the sports offered by the centre. Fixtures against many other organisations are arranged in a wide range of sports involving a large number of teams.

   a) Name three database files you would expect to find in this system. (3)

   b) For each of the database files you have named, list the fields required to enable this system to be maintained with minimum redundancy. (6)

   c) Draw a diagram to show the relationship between the database files named in part (a) (3)

   d) Describe three reports that the system might be required to produce. (3)

   e) The manager of the centre intends to send out personalised letters to each of the members. This is to be done using the mail-merge facility offered by a word-processor in conjunction with the database. Explain how this is achieved. (4) *AEB Computing Specimen Paper*

7 A car leasing company stores information about cars and who they have been leased to. Part of the CUSTOMER table is given in the table below:

   a) Suggest why this way of storing data is not efficient. (2)

   b) Suggest three tables which could be used in a relational database to store this data more efficiently. (3)

   c) Draw diagrams showing an entity relationship and how the tables are related. (3)

Customer code	Customer name	Comp name	Town	Car reg	Make	Model	Lease date	Return date
017312	Johnson, M	CDR	Stoke	N877THJ	Peugeot	406	010895	310796
017312	Johnson, M	CDR	Stoke	P981ESD	Peugeot	406	010896	
013442	Brazil, P	CDR	Stoke	P982ESD	Peugeot	406	010896	
009865	Smith, L	Cooks	Derby	N723KLJ	Volvo	440	010895	
016613	Brooks, M	AVP	Crewe	N623TYU	Ford	Mondeo	010196	
016613	Brooks, M	AVP	Crewe	P109TYT	Rover	214	01	

8 An estate agency handling property within one town decides that computerisation of its operations will bring a number of advantages to the business. Clients are of two types, buyers and sellers. The information on properties for sale is provided by the sellers. Buyers are seeking properties within a certain price range and of a specific size. They are also interested in additional features such as location, size of garden and age of property.

a) Show how a database package could be used to store records on individual properties for sale. Include the design of each record and describe the data type for each field. (6)

b) Explain how the database could be used to give information on all properties suitable to a buyer who wants to buy a three bedroomed house for less than £70,000. (4)

c) Describe briefly two processes which involve altering the data in the database and say how they would be carried out. (4)

d) Describe two other applications made possible by computerising the operations of the estate agency. In each case indicate the appropriate hardware and software necessary to support your example. (6)
Oxford Computing Paper 2 1995

9 A company makes use of a computerised flat file information storage and retrieval system. The company is experiencing problems due to the use of this flat file system.

a) Describe **three** benefits that the company would gain by using a relational database as opposed to a flat file system. (6)

b) The company currently has three files in use; customer, stock and orders. During conversion to a relational database system these files would need to be normalised. Explain clearly what you understand by the term normalisation. (2)

c) Examples from the three files are shown below. Normalise these files explaining any assumptions or additions you make to the files. (5)

Activity 1

For each of the following 5 tables, write out the attributes in standard notation. Normalise, showing first, second and third normal form. Then draw an entity relationship diagram.

1 A college library system: each record contains data about a student and the books they have borrowed

Stud#	Name	Address	TG	Tutor	Book#	Title	ISBN	Return Date
3124	Mary Smith	32 Hill Street Bluedale	D22	H Jones	2234	'PCs for Fun'	1 877556122	13-6-97
					3356	'Geography is good for you'	1 655536327	21-6-97
5464	John Bloggs	27 Mill Road Grenham	D23	J Pell	5883	'Biology for Brilliance'	1 223334554	18-6-97
3369	Peter Kelly	119 Leigh Hill Grenham	D22	H Jones	4343	'French for Frivolity'	2 346798443	16-6-97
					2112	'History is Hilarious'	3 444576661	28-6-97
7765	Gary Tolam	The Heights, Blogdale	D44	M Kelly	3412	'Sociology for Slackers'	4 558787432	14-6-97

2 Cub scout award badges: each record contains data about a cub scout and the badges they have been awarded

Cub#	Cub Name	Phone	Six	Sixer	Badge#	Description	Level	Date
95-22	Frank Blooms	866755	Blue	Jacob Giles	33	Washing up	1	15-4-97
					36	Advanced Washing up	2	28-5-97
95-36	Jerry Taylor	874532	Blue	Jacob Giles	33	Washing up	1	15-4-97
95-28	Martin Adams	802844	Green	Josh White	22	Swimming 500 metres	2	15-3-97
					33	Washing up	1	12-4-97
					41	Throwing a cricket ball	2	16-5-97
					07	Abseiling	3	2-6-97
95-07	Harry Thomas	899677	Red	Adrian Flint	33	Washing up	1	15-4-97
					09	Sailing	3	23-5-97

3 Scouting goods suppliers: each record contains data about different scouting goods and the available suppliers, together with the price charged

Product#	Description	Stock level	Supplier num	Supplier name	Phone	Selling Price
78780098	Large woggle	67	3345	Wogglicity	0131-338675	1.25
			5678	Best Woggles	0181-777-8654	1.39
			1123	Scouting Bits	01962-874532	1.09
65654333	Cub scarf: green	33	1123	Scouting Bits	01962-874532	4
78012544	Cub scarf: red	12	1123	Scouting Bits	01962-874532	
35422123	4 man tent	1	1123	Scouting Bits	01962-8745	
			4659	Tent city	0171-765	

4 Employee training records: each record contains employee details and the courses they have attended

Employee#	Name	Dept Code	Department	Course	Description	Venue Code	Venue	Date
321-93	Hilary Barr	HR	Human Resources	G1	Getting on with people	IH	In house	23-5-94
566-92	Penny Blyth	F	Finance	EX-A	Advanced Excel	PT	Pembly Tech	15-6-95
				M2	Financial Modelling	HH	Highham House	02-11-96
				AP2	Appraisal level 2	IH	In house	1-2-97
321-92	Neil Flynn	HR	Human Resources	G1	Getting on with people	IH	In house	23-5-94
				AP2	Appraisal level 2	IH	In house	1-2-97

5 Fine arts sales: each record contains details of a work of art being sold in auction

Catalogue#	Title	Painter	Seller Code	Name	Phone	Reserve	Medium Code	Medium
M0134	Hill Sheep	Conrad Dale	AT1	Andy Tomkins	0151-878787	£250	O	Oil
P9009	Abstract	Sally Saeter	CD4	Colin Drake	0121-844444	£1200	I	Bronze
M9843	Abstract	Conrad Dale	PT1	Primrose Tilly	877666	£700	O	Oil
G6677	Tempest	Anonymous	CD4	Colin Drake	0121-844444	£12000	W	Watercolour

Activity 2

1 Write a definition for each of the following terms:

- Data consistency

- Program – data independence

- role of DBMS

- Entity

- Data redundancy

- Role of database administrator

- Data di~~ction~~

~~...ite down the~~
~~...any, many to~~

~~...ecord~~

- **Context: Car hire company**
  Hire Car   Customer

- **Context: Library**
  Borrower   Library Book

- **Context: Football team**
  Footballer   Team   League

- **Context: School**
  Teacher   Pupil   Tutor

- **Context: Cruise holidays**
  Passenger   Boat   Cruise   Port

- **Context: Newspaper delivery round**
  Newspaper   Round   Customer
  Paperboy

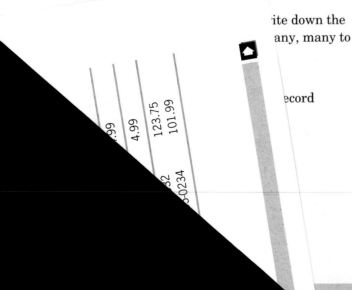

Activity 3

For each of the scenarios below carry out the following tasks:

- List out the data fields in ONF using standard notation.
- Underline the key field and draw a line over all repeating fields.
- Work through the process of normalisation, showing the data in 1st, 2nd and 3rd normal forms.

Somewhere College

Admission Number 3472   **Date of admission** 01/9/99

Surname Burgiss   **Forename** Mary   **Initial** I

Date of Birth 27/4/84   **Previous school** Hillfield
School address Manor Road
Burgham
PD32 3RE

Tutor Group 12PKW   **Tutor** Peter K Williams

Agreed courses:

Title	Board	Start date
IT A level	AQA	12/9/1999
Geography A level	OCR	12/9/1999
Swimming Activity	Internal	1/11/1999
French A level	OCR	12/9/1999

Malletshire Schools Sports

JULY 1999

Certificate presented to:

Simon Finden **date of birth** 23/5/90

of Kingsmead School
Cedar Road
Midchester

who took part in the following races:

Egg and spoon race (under 10) 3rd place
100 yards (under 10)
200 yards (under 10)
Sack race (under 11)   1st place

Presented by Mary Tomalin, Mayor.

Revision questions

1 In the life cycle of a system, one stage of testing new software is to run it together with the manual system, *in parallel*, using the same, live data. Explain why such a system would be used. (5)

2 When investigating a manual system, the first step is to carry out a feasibility study. Outline three aims of the process. (3)

3 Describe three daily reports a computerised till system can provide as part of an MIS to help the store's management. (6)

4 For computer-to-device communication there exists two forms of transmission – **serial** and **parallel**. Describe and illustrate these two forms of transmission.

5 A small printing firm has for the last year used a Local Area Network. Through expansion a new office has been purchased next door to their current location and a new network installed.

The new network works on a more advanced and different topology. Explain what extra hardware is required to connect the different networks on the same premises? (2)

Both networks use an OSI protocol model for communication purposes. Explain the term **protocol** and describe clearly what the **OSI model** is. (6)

6 Many businesses use distributed databases to control and store vital information. What is a distributed database? (2)

Explain the problems businesses might encounter when installing a distributed database.

7 A firm's personnel manager is considering transferring personal information about employees from paper to computer files. She is concerned about the integrity and security of computer information. She knows there are legal requirements and decides to talk to the DP manager. The DP manager believes there are no real causes for concern and agrees to produce a report underlining the strategies involved to avoid problems. Draft the DP manager's report. Include security, integrity and legal requirements. (20)

8 Explain the term software emulation. (2)

9 'Information is only as good as its timing'. By using examples show the relative importance of speed and accuracy? (6)

10 Describe the strategy that a manager could use to reduce resentment of staff over the introduction of new computer systems. (5)

11 What is an audit trail? (4)

12 Draw labelled diagrams to illustrate three different network topologies. (6)

13 Explain why software providers need to undertake both alpha and beta testing. (4)

14 Explain why a company needs

a) an IT security policy; (2)

b) a code of conduct for its employees (2)

15 A customer of a travel agent wishes to book a holiday. The travel agent uses a computer to do this. Describe the information flow. (6)

16 Explain what is meant by detailed, aggregated and sampled data giving an example of each. (6)

17 Computer data is very valuable and must not be lost. (6)

   a) Describe four possible threats to this data;

   b) Describe suitable ways of combating these threats that might be found in a disaster avoidance plan. (8)

18 Give four ways in which a company may train its staff. Comment on the suitablility of each. (8)

19 'I like my computer,' said a senior manager in a large company to the IT director, 'and I am happy using my software. Why do you insist I change to your computer and your software?' Why should she change? (6)

20 A small company has a pyramid organisational structure. Each manager has a maximum of three people directly answerable to him/her. The managing director is considering reducing the number of managers to cut costs.

   a) How is this likely to affect the organisational structure? (1)

   b) Give two advantages and two disadvantages of this approach. (4)

   c) How might the change in structure affect internal communication? (2)

21 An expert system, used in diagnosing a problem with an industrial machine, has a knowledge base.

   a) What is meant by the knowledge base? (2)

   b) Explain how the knowledge base is built up (2)

22 A leisure centre uses a computer to store information on booking of its facilities and to record financial details based on these bookings. The manager has been told that an MIS may be able to help him.

   a) Explain the term MIS (2)

   b) Give examples of the categories of information that an MIS might provide for the manager. (4)

23 Explain why it might be better for a software user support team to receive a problem by e-mail, rather than by phone. (3)

24 Teamwork is important in many businesses. Suggest some ways in which information technology has assisted teamwork. (4)

25 Stephen's computer space at work is too cramped for him to sit comfortably and use the mouse. Give four other physical factors that effect human/computer interaction. (4)

26 List the hardware that might be required by a specialist music program. (2)

27 A quantity surveying company is thinking of using some new software to help with their calculations.

   a) Explain the options that are open to them in obtaining the software. (3)

   b) Which option would you recommend? (2)

28 Describe the steps that should be taken by a company to ensure that new staff are aware of legal implications of the storage of personal data. (3)

29 A school is considering introducing a computer system to produce reports on pupils. Some staff are unhappy about this idea. Suggest six possible concerns that may exist and arguments that may be used to persuade the staff to accept the new system. (10)

30 Explain, with examples, the difference between physical and psychological factors in human/computer interaction. (6)

31 a) Explain what is meant by the term Geographical Information System (GIS). (2)

   b) Describe how a company that sells greenhouse heaters might make use of GIS. (3)

32 Describe 4 features you would expect to find in a mathematical software package. (8)

33 GettaJob is a successful employment agency which operates throughout the south of England. It has a small IT department and all its main functions are computerised. The company has recently implemented flexi-time, and after a successful trial period, have decided to continue with it. At present the recording system is a manual form-based one. The management wish to introduce a computer-based system. Describe **four** ways of providing a suitable software solution. (8)

34 'The development of sophisticated human/computer interfaces and the ability to customise generic software have played a major part in the huge increase of computer usage over the last few years.'

Discuss this statement, including in your answer:

- features of HCI

- the reasons why such sophisticated interfaces were not available ten years ago

- the ways in which generic software can be customised (15)

35 Describe the possible criteria for selection of software solutions to specialist applications and their place within the corporate strategy. (6)

36 Describe the advantages of a common user interface between different generic application packages. (2)

37 Computers can play a major part in musical composition. Descirbe two ways in which software can be used to aid a composer. (4)

38 Describe how specialist software can be used in the planning and monitoring of the implementation phase of an IT project. (6)

39 In the context of user interfaces, explain the following terms, giving an appropriate use for each:

  a) command language

  b) forms dialogue

  c) free form dialogue

  d) dedicated keys

  e) soft key. (10)

40 Explain what is meant by the term 'facilities management'. (3)

41 Describe the characteristics of a good team. (4)

42 Explain the term 'risk analysis'. (3)

43 A board member of an advertising company, Satchel & Satchel, reads with horror of a company which has been fined for using computer software illegally. He is concerned that the same thing could happen in his business.

  a) What legislation would have brought about this prosecution? (1)

  b) Outline the steps that Satchel & Satchel should take to ensure that they would not be liable to prosecution. (4)

44 A local council supports a LAN and most employees have a workstation on their desktop. Management have decided to replace their current DOS based word processor with a more powerful Windows based one for all employees at every level. Discuss the possible training methods that could be used to enable staff to use the new software and explain how each is appropriate. (9)

45 In a sixth form college, it is necessary to audit MIS student systems.

  Why is such an audit necessary? (2)

  Explain how an audit trail is used for student tracking. (4)

46 A large book publishers has built up an extensive use of IT over the years. IT is used in all aspects of the business, but systems have developed in an ad hoc way so that no one possesses a complete overalll view of all the software in use.
  The board has decided that the time has come to develop a corporate IT policy.

  a) Give reasons why such a policy might be needed. (4)

  b) Describe what should be included in such a policy. (4)

  c) Two months later, the board approves the policy. The personnel director asks the rest of the board how they are going to ensure that the policy makes any difference. Describe the methods that you would advise the board to take. (4)

47 What is a benchmark? (2)

48 a) Explain **two** reasons why an auditor might be employed by a business. (4)

  b) It is necessary for an auditor to be knowledgeable about IT? Explain why this is so and list **three** specific aspects of the IT with which they should be familiar. (4)

  c) Describe **two** problems which arise for auditors of IT based systems which did not arise in traditional ledger based systems. (4)

Glossary

Alpha testing	Primary software testing, in house by the software producers
Alphanumeric characters	Letters, numbers or other characters, for example punctuation marks
ASCII	(American Standard Code for Information Interchange). The binary code used in computers to store alphanumeric characters.
ATM	(Automatic Teller Machine). The official name for cash machines outside banks.
Back-up	To make an extra copy of stored data in case the original is lost or corrupted
Bandwidth	physical limitations of a communication system (usually bits/sec)
Batch processing	A form of processing where all the information is batched together before being processed
Beta testing	Secondary software testing, by independent external users before publication
Bit	(Binary digit). A binary number which can only have the value 0 or 1.
Browser	A program that allows the user to access a database (typically the Internet)
Buffer	Memory where data is stored while waiting to be processed, typically in a printer
Bugs	Errors in computer programs
Byte	A group of eight bits, normally storing one alphanumeric character
Cache	A very fast expensive memory
CAD	Computer Aided Design
CD-ROM	(Compact Disc-Read Only Memory). A small plastic disc used to store data.
Compression	A method of reducing the size of a file, typically to use less disk space
Configure	To set up a computer system for the appropriate hardware and software. A system will need to be configured for the printers, sound cards and so on.
Cyber-	A prefix alluding to computer communication often with reference to the Internet as in *cybershopping*, shopping by computer, *cyberspace*, everything accessible by computer communications.
Database	A structured set of data stored on a computer
Data integrity	The reliability of data, that is ensuring it is accurate
Data security	Keeping data safe from loss
DBA	Database administrator
DBMS	A set of programs allowing the user to access data in a database
DDE	(Dynamic Data Exchange). Shared data in two packages is linked so that when it is updated in one program, it is automatically updated in the other program.
Debug	Remove bugs from a program
Digital	Something that is represented in numerical form typically in binary numbers
Direct-mail	Advertising a product by sending details directly to potential buyers through the post
Directory	An area (usually of a disk) where files are stored. A disk may have several directories and sub-directories to make finding files easier and to aid security
Dongle	A piece of hardware, for example a lead that has to be plugged in to the computer before software will run. Usually used to protect copyright
DOS	Disk Operating System
dpi	Dots per inch – describes the performance of a printer
DSS	Decision Support System

e-banking	The use of the Internet to communicate with your bank.
e-commerce	The use of computers and electronic communications in business transactions, including web-sites, EDI, on-line databases and EFTPOS systems.
EDI	Electronic Data Interchange. Transferring information such as orders and invoices electronically between two organisations.
e-tailors	Retailers who do business on the Internet.
e-shopping	Using the Internet to purchase goods and services.
EFTPOS	Electronic Funds Transfer at Point of Sale. The system where customers can pay by debit (Switch) card and the money is taken electronically from their bank account.
EIS	Executive Information System. Software that provides management information about the organisation's performance.
Embedding	Including one file (such as an image or a document) in another file. See OLE (Object Linking and Embedding).
Encryption	To scramble data into a secure code to prevent it being read by unauthorised users
Extranet	The linking of two intranets usually to assist business transactions, for example linking a customer and a supplier
FAQ	Frequently Asked Questions. A file containing answers to common questions, for example about using a program.
Fax modem	A modem that enables a computer to send and receive faxes
Fibre optic	A cable made out of glass fibre and used in communications
Filters	An option in a program enabling the user to import files from or export files to another program
Flatbed scanner	A scanner in which the item to be scanned is placed on a flat piece of glass
Floppy disk	A small removable disk in a hard plastic case, used to store data
Gantt chart	Chart which shows progress through a project in horizontal bars
Gigabyte (GB)	A measure of memory capacity equal to roughly 1 000 000 000 bytes (it is exactly 2 to the power 30 or 1 073 741 824)
GIS	Geographical Information System
GUI	Graphical User Interface, for example Windows. It is sometimes pronounced 'gooey'.
Hacking	Unauthorised access to a computer system, possibly for criminal purposes
Hand scanner	A small device, held in the hand and dragged over the item to be scanned
Hard disk	A magnetic disk inside a computer that can store much more data than a floppy disk. Usually it cannot be removed but removable hard disks are becoming more common.
Hardware	The physical parts of the computer, such as the processor, keyboard and printer
HTTP (Hypertext Transfer Protocol)	The Internet protocol used to identify web addresses
Integrated package	A package which combines several different applications such as a word-processor, a graphics package, database, communications software and spreadsheet
Interactive	A system where there is communication between the user and the computer
Internet	An international WAN providing information pages and e-mail facilities for millions of users
Intranet	A private internal network using Internet software, that can be used for internal e-mail and information
IRC	Internet Relay Chat. A function of the Internet allowing users to send and receive real-time text messages.
ISDN	Integrated Services Digital Network. A telecommunications digital network which is faster than an analogue network using a modem.
ISO	International Standards Organisation. An organisation which defines standards for computer architecture and data storage.
ISP	Internet Service Provider. A company that offers a connection to the Internet.

Java	A programming language used for utilities on web pages
JPG or JPEG	Joint Photographic Expert Group. An ISO standard for storing images in compressed form. Pronounced jay-peg.
Kilobyte (KB)	A measure of memory capacity equal to 1024 bytes
Licence agreement	The document which defines how software can be used, particularly how many people can use it
Macro	A small program routine usually defined by the user
Magnetic disk	A small disk coated with magnetic material on which data is stored. It can be a floppy disk or a hard disk.
Magnetic tape	A long plastic tape coated with magnetic material on which data is stored
Mail-merge	A feature of a word-processing program that combines details from a file of names and addresses into personal letters
Master file	The file where the master data is stored. Data from this file is combined with data from the transaction file.
Megabyte (MB)	A measure of memory capacity equal to 1 000 000 bytes (it is exactly 2 to the power 20 or 1 048 576)
MICR	Magnetic Ink Character recognition. The input method used to read cheques.
MIS	Management Information System
Modem	Modulator/demodulator. The device that converts digital computer data into a form that can be sent over the telephone network.
MS-DOS	Microsoft Disk Operating System. The operating system developed for the PC.
Multi-access	A computer system allowing more than one user to access the system at the same time
Multimedia	A computer system combining text, graphics, sound and video, typically using data stored on CD-ROM
Multi-tasking	A computer system that can run more than one program simultaneously
Network topology	The way in which a network is connected.
Network	A number of computers connected together
OLE	(Object Linking and Embedding). A method of taking data from one file (the source file) and placing it in another file (the destination file). Linked data is stored in the source file and updated if you modify the source file. On the other hand, embedded files are part of the destination file.
On-line Processing	Processing while the user is in contact with the computer
Operating system	The software that controls the hardware of a computer
OSI	Open Systems Interconnection. A set of standards defining protocols for communications.
Package	A program or programs for a specific purpose
Peer-to-peer	A type of network where there is no server, with each station sharing the tasks
Pentium ™	A processor developed by the Intel Corporation™ for the PC
Peripheral	Any hardware item that is connected to a computer such as printers, mice or keyboards
PIN	Personal Information Number, used to check that the user is the person they claim to be, for example at an ATM
Platform	Used to describe a hardware or software environment
Port	A socket usually at the back of the computer.
Portability	The ability to use software, hardware or datafiles on different systems
Primary key	A unique identifier in a record in a database
Protocol	A set of rules for communication between different devices
QBE	Query By Example. Simple language used to search a database
RAM	Random Access Memory. The computer's internal memory used to store the program and data in use. The contents are lost when the power is turned off.
Redundant data	Data that is repeated unnecessarily (in a database)

ROM	Read Only Memory. Part of the computer's memory that is retained even when the power is turned off. Used to store start up program and settings.
Serial access	Accessing data items one after the other until the required one is found. Associated with magnetic tape.
Server	A dedicated computer that controls a network
Shareware	Software that can legally be distributed freely but users are expected to register with and pay a fee to the copyright holder.
Smart card	A plastic card, like a credit card, with an embedded microchip. The information in the chip can be updated, for example when cash has been withdrawn from an ATM.
Software copyright	Laws restricting copying of software
Software	A computer program or programs
Systems analyst	A person whose job involves analysis whether a task could be carried out more efficiently by computer
Toggle switch	A switch or button which if pressed once turns a feature on. If pressed again it turns the feature off. The Caps Lock button is an example.
Transaction file	A file containing new transaction details or changes to old data, which is merged with the master file
WIMP	Windows, Icon, Mouse, Pointer.
Windows ™	A GUI for the PC produced by Microsoft
WWW	The World-Wide Web
WYSIWYG	What You See Is What You Get

Answers

Chapter 1

1 This enables them to target these people who are likely to listen to their station based on information about their address, age, interests, and so on.

2 All our customers think our service is good or better – Not necessarily true
Our average score is good to excellent – True
Our customers think we are consistently good – Not necessarily true
The average score of 1.8 has no meaning in the questionnaire
The average is an aggregated score and gives no details about individual scores. For example, are all the scores around 1 and 2 or are the scores spread throughout the range?

3 Only up-dating once a day may mean that the supermarket orders too much to cater for extra demand. This leads to overstocking which ties up cash in stock, needs extra storage and could result in perishable stock going off. The opposite is when the supermarket under-estimates the days sales and does not have enough stock. Items may run out and sales will be lost.

4 Garbage In, Garbage Out, for example a wrong bar code being entered at a supermarket till will result in the wrong price being charged.

10 The direct-mail industry, like all industries is concerned about its public image. Many people dislike direct-mail, seeing it as a waste of time and paper. Registering with MPS enables you to choose not to receive this mail and saves paper and postage. Secondly, if the person receiving the direct-mail has no intention of opening or reading it, there is not much point in sending it out!

Many people would not want to register with MPS as it is possible that they may get some direct-mail that is useful. The fear of missing out on a real bargain is likely to discourage most people from registering.

Junk-mail or direct-mail is a cost-effective way of targeting individuals who may buy your product or service.

Chapter 2

1 Betta Biscuit plc could improve their security by ensuring the computer room is locked when unoccupied, by ensuring that all staff use suitable passwords, by ensuring that to access the payroll file you need further privileges, i.e. extra passwords, by telling staff to challenge unauthorised personnel in the room and by telling all new staff that they should not go into unauthorised areas and risk immediate dismissal.

Both Danny and Betta Biscuit have broken the law. Betta Biscuit have not ensured that personal data is surrounded by the proper security. Danny has committed offences under the first and third categories of the Computer Misuse Act.

2 Keep back-up tapes or disks in a fire-proof, water-proof safe. Keep these tapes or disks off site. Do not install computer system on the ground floor if flooding is a possibility. Have a proper fire prevention system – sprinklers, extinguishers, and so on.

3 Security means making sure you don't lose it. Privacy means stopping unauthorised people seeing it.

4 It is possible to write a computer program to try a large number of passwords. It would be possible to try every word in a dictionary for example.

6 This is based on a true story. The bank do not appear to have broken the Data Protection Act.

The bank can make up their own rules about who they lend money to.

11 The registrar felt that using information, stored on computer, that had been obtained for the purpose of sending out bills for the purpose of sending out advertising was either not being used for the purpose described in the register entry, or had not been obtained fairly.

13 a) It seems inconsistent that the DPA does not include data stored on paper.
b) No, as question 11 shows.
c) The law offers some protection but also the registrar acts as an ombudsman and an adviser on good practice.
d) The law should apply equally to everyone.

Chapter 3

3 Agree: Price of new technologies is reducing, society is more IT literate, on-line systems are more up-to-date, on-line systems already exist, the user can choose the pages they wish to see.

Disagree: paper is very convenient, easy to use, portable, no need for expensive equipment, teletext graphics poor, teletext stories are restricted by the size of the screen, teletext has not affected sales of newspapers.

4 Visa payments processed by computer, debit cards deduct money from your bank account by computer, payment for services can be by direct debit or other electronic transfer.

5 It is unlikely that cash will be completely replaced, for example for small transactions. Not all customers will have access to payment cards and not all establishments, for example school tuck shops will accept them.

6 Reduces doctor's work-load. May not be accurate. Can be programmed to suit pupil's needs. Pupil may get a friend to answer.
Ensures consistency. May not take account of special cases.

7 Lack of knowledge of IT. Increasing use of IT – greater dependence. Wide area networks may allow remote access. Too many books/films highlighting problems.

Chapter 4

1 LAN – small area, dedicated cables. WAN – wide area, could use telephone lines.

2 Quick processing of data
Accurate processing of data
Reliable processing of data
Automatic processing of data
Cannot be rigged
Can be processed overnight when phone and electricity costs are cheaper
Information is already in the computer – no need to re-type it

3 a) WAN b) LAN

4 Can share peripherals, for example printers. Can share software. It's cheaper to have a network software licence. Pupils don't need to sit at the same station. Security is better.

Chapter 5

1 Fraud, theft of information, theft of goods, malicious damage.

2 Checking the paying in forms at the bank, checking that the MICR numbers match the written numbers, for example by checking a sample.

6 Good security measures, following proper security procedures, using a firewall to prevent illegal access.

7 The Internet presents dangers of pornography, racist literature and using the Web for unauthorised access to information.

Chapter 6

1 Injury caused by repeated use of some machine or tool. It affects many parts of the body particularly hands, arms and legs. The symptoms include pains and stiffness.

4 Obviously the answer depends on your position. Employers thought they went too far. Trade unions not far enough. They could be extended to all computer users, not just in offices.

Chapter 7

3 Computers faster and more powerful, faster input than keyboard, now very reliable, good for people with learning difficulties.

4 Most new computers are multimedia, using text, graphic, video and sound. Loud speakers are part of this.

5 The first character must be a capital letter.
The last three characters must be letters.
The length must be between 5 and 7.
There is a maximum of three numbers.
The first number cannot be 0.

The real number may be R123ABC but R123ABV has been typed in. It is still valid.

6 Length must be 6.
First two numbers must be from 01 to 31.
Third and fourth numbers must be between 01 and 12.

7 The council in the newspaper cutting should have used a range check to prevent sending out bills of (say) less than a pound.

8 Keyboard. Data will include names, addresses and so on. OCR may be acceptable but it is unlikely that customers will fill in the form sufficiently clearly. Validation checks could take place, for example on date of birth – see above.

12 See question 5.

Chapter 8

2 a) Batch
b) Real-time

3 a) In a batch processing system all the data is batched in a transaction file and the whole file processed. In a multi user pseudo real-time system, the information is constantly being entered and processed. A multi user pseudo real-time system requires a fast processor, with a large memory and disk storage. These are not needed in a batch processing system.
b) An example of a multi user pseudo real-time system is access to a travel agents computer. An example of a batch processing system is the processing of bank statements.

c) Batch processing is likely to be cheaper with a good back-up system (generation of files). Multi user pseudo real-times system enables many users to access it, the files are up-to-date and you have the facility to search a file.
d) Batch processing maintains security of data by keeping generations of files as back-up. Real time systems need to back-up as well but more frequently.

4 For example; printing statements, up-dating accounts from the transaction file, adding records to the file for new accounts.

6 a) up-to-date files, search facility
b) old files cannot be transferred to new equipment, the need to train staff in using the new equipment, problems if designed badly, for example cannot cope with the amount of data.

8 Keeps data if lost or damaged, won't forget to back-up, back-up is off site, back-up is encrypted – normal back-up isn't – increased security, the right files are backed-up, the system is working and reliable, you can choose the time it is backed-up.

9 Lots of similar data, immediate up-date of file not required, no need to search data, good security required.

Chapter 9

1 Systems software helps run the computers smoothly, for example controlling input and output. Applications software performs a task which has to be done even without a computer, for example accounts or payroll.

2 A program that performs a routine activity on the computer, for example getting a directory of a disk, deleting a file, copying a file to disk.

3 Users of Word 6/Word 95 cannot load files written with the later version Word 97. (It works the other way round.) Users of Windows 3.1 may not be able to up-grade to the newer Windows 95 if the specification of their PC is not good enough. Therefore they cannot run a lot of software unless they buy a new PC. (This helps PC sales of course!)

5 A desk-top publishing package like PageMaker or Publisher would be suitable but many word-processing packages today like Word have enough features to be suitable. Columns, import graphics, different fonts, print preview, can choose different paper sizes, import from other (WP) software, boxes, shading, clip-art, different coloured text and many more.

Chapter 10

5 Demand from users for easy and fast input, demand from people unable to use a standard keyboard, bigger RAM, faster processing speeds, more reliable hardware.

8 a) Screen in wall of bank asking customer to choose from a list.
   b) A joystick and pointer on the screen.
   c) A simple menu screen so that the user knows the 3-digit number to choose or the colour with Fastext.
   d) Keyboard.

Chapter 11

1 Key field (record number), house number, street, town, post code, council tax band, parish (for parish council tax), discount (given for single person living alone), payment details, how bill is paid.

2 Manufacturer number, Name of company, Address, Phone number, fax number, Name of contact.

3 Type of relationships
   a) one-to-one
   b) many-to-many
   c) one-to-many
   d) one-to-many
   e) one-to-one
   f) one-to-many
   g) one-to-many

4 An attribute is a property of a table (a field)
   A relationship is a link between two tables

6 PRODUCT, <u>Number</u>, Name, Supplier code, Manufacturer code, Price, Cost, Delivery time, Minimum stock level.
   MANUFACTURER, <u>Code</u>, Name of company, Street, Town, Post code, Phone no, Fax number, Name of contact.

The Manufacturer code is linked to the code in the MANUFACTURER table.
A SUPPLIER table would be linked to the PRODUCT table using the Supplier code. Both these are one-to-many relationships.
A PAYMENT table could store records of payments to manufacturers.

Note: The types of these relationships are not necessarily clear. In question 3, can a racehorse have more than one owner? Sometimes several people to have 'shares' in the horse.

When you are designing a database, it can often be quite hard to decide what is an entity and what is an attribute. The database designer would have to decide if racehorse is an attribute of owner, or is owner an attribute of racehorse. Or is neither the case, since both are entities in their own right.

Questions 3

1 Can validate entries, can specify the size of data to save time/space, can sort in numerical/data order as well as alphabetical.

2 Avoid repeated data in the database, wasting time/space and leading to possible inconsistencies.

3 A field (not the primary key) that is indexed, that is the database is stored in the order of this field to speed up sorting.

4 Advantage – speeds up sorting. Disadvantage – slows data entry.

5 QBE – Query by example. Easy to set up using the functions of the database management system. SQL – Structured query language. You need to know the programming language to do this. They probably want to use QBE.

Questions 4

1 To record management information about the database, for example tables, fields, data types.

2 Can't get lost
   Can have password security
   Avoids duplication
   Easy to up-date
   Quick and simple to use

3 The DBMS is the program that provides an interface between the database and the user in

order to allow them to access data easily. Users are not computer experts in the internal structure of the database or how to set it up.

5 a) If entering details of a new customer, the customer number (key field) must be unique.

b) Verification means double entry of data to ensure it is correct. Two different clerks may be asked to type in the same data which is compared.

c) Depending on the nature of the company's products, its financial procedures, some information may be confidential or only entered by authorised staff.

Questions 5

1 A library database will work in exactly the same way as the video database in this section.

3 This question is a bit harder than the others and there are different ways of tackling it. A lot of information has to be stored. If bookings are being taken over the phone, there will need to be a CUSTOMER table storing names and addresses. This saves having to write them down when someone phones up.
CUSTOMER table, <u>Customer no</u>, surname, title, initials, address, town, postcode, phone number.
SEAT table, <u>Seat no</u>, price.
BOOKING table, <u>Booking no</u>, Customer no, Seat no, Date of performance, Time of performance.
The BOOKING table is linked to the other two tables.

Chapter 12

4 **School/College:** strategic: moving from single-sex to co-educational; tactical: planning when mock A Level examinations should take place; operational: deciding which students to enter for Foundation and which for Higher level at GCSE.
Shoe shop chain: strategic: opening another branch; tactical: selecting new lines to sell next season; operational: staffing rota for next month.
Multinational Bank: strategic: whether or not to diversify into Life Assurance; tactical: select a new, up to date ATM for branches; operational: decide whether or not to give a loan to a customer.
Car manufacturer: strategic: whether or not to make a takeover bid for a rival company; tactical:

pricing strategy for new range; operational: deliveries to make for next week.

5 Information flow across structure may undermine current chains of command; some layers may become redundant; due to lack of IT skills, senior management may become out of touch with state of business.

7 **Functional – advantages:** subject specialisation leads to higher levels of expertise and the sharing of ideas; good career prospects; **disadvantages:** hard to share ideas across functions, isolation, rivalry leading to lack of cooperation between functions.
product – advantages: all the skills needed working together, integrated, identity in product; **disadvantages:** unnecessary duplication of many functions; little sharing of ideas between experts, could lead to inappropriate rivalry;
Geographic – advantages: identify with local customers which should lead to a high level of commitment and service; can take advantage of regional differences; **disadvantages:** unnecessary duplication of many functions.

Chapter 13

3 Simulation packages, spreadsheets, using what if . . . and similar forecasting features; on-line queries in databases (eg QBE); expert systems.

4 **Include in answers:** inadequate analysis, lack of management involvement in design, emphasis on computer system, concentration on low level data processing, lack of management knowledge of IT systems and their capabilities, inappropriate/ excessive management demands, lack of team work, lack of professional standards.

6 **Managing director** – strategic. Decisions: whether to close a factory (information needed – productivity figures, transportation costs, predicted sales figures); diversify into comb production (information: potential set up and production costs, current and future state of the comb market).
Sales Manager – tactical. Decision: pricing of buttons (information needed – sales/price analysis over past 3 years; pricing strategy of competitors); whether to increase level of 'regular customer' discounts (information needed – customers purchases histories, analysis of lost customers).
Sales representatives – operational. Decision:

whether to offer a customer credit (credit rating of customer, cost); itinerary for day's visits (mileage, customers' availability).

Factory manager – tactical. Decision: whether to re-instate a night shift in the Autumn (staffing costs, product demand); whether to computerise a particular process (costs, effectiveness of new system used elsewhere).

Shift leader – operational. Decision: job assignment (staff skills level, production plan); adjustment of machinery (quality analysis of previous batch).

Personnnel manager – tactical/operational.

7 The managing director could use an EIS to help him in his decision over closing a factory. After examining a chart showing end of year figures in comparison with previous years, and finding that production costs were soaring, he could 'drill down' to examine the costs and production levels at each factory. He could explore further to find out details of transportation and local staffing costs.

9 a) **Purpose:** to convert data from internal and external sources into information.
   b) **Why required:** to enable managers to make effective decisions.
   c) Eg In a nation-wide distribution company the use of a MIS to monitor the movement of vehicles and revise strategic planning of the location of warehouses.

Chapter 14

3 Prototyping allows a systems designer to involve the user in all stages of design. For example, mock-ups of input and output screens, together with appropriate validation checks can be produced. The user can indicate where data fields are missing, layout is inappropriate, validation checks are inadequate. Modifications can easily be made at this stage and lengthy re-writes of a full system are avoided. Very often a user will think of extra features when viewing the mock-up which can easily be incorporated into the final design.

4 Possible areas from: Inadequate analysis eg failure to establish existing meeting patterns, volumes of changes; lack of management involvement in the design eg failure to consult on types of meeting; Emphasis on hardware or software rather than the 'solution': eg software does not reflect management

usage {data entry, reports produced etc.} Data processing orientated eg does not cross-schedule or allow attachment of documents; Management not IT literate eg failure of some to use it destroys objective.

Lack of teamwork eg as above; Lack of professional standards eg in the software or the approach to use by the management; Lack of proper evaluation of the potential products {covers technical inadequacies}; Frequency of changes; Readiness of managers to keep electronic system up to date.

5 Retail chain: pilot; traffic lights: direct (possibly phased); library: parallel; cinema: direct; time keeping: pilot; diary: direct.

6 Perfective (adding extra features), corrective (correcting errors), adaptive (meeting new circumstances). Payroll – perfective: add new monthly exception report listing all employees working more than 30 hours overtime; adaptive: dealing with new way of calculating National Insurance payments.

7 Interview librarian: current practices, problems, IT skill of users, volume, constraints; examine documents (inputs, outputs and data stores): details of data; data flow; observation: procedures, exceptions, data flow.

Chapter 15

2 **Diary** – internal, formal, future; **car owner** – internal, formal, historic; **delivery note** – external, formal, current; **illness** – external, informal, current; **widgets** – internal, current, formal.

4 Data aging during the improvement stage. Overcome by setting up the model first with test data, do export at final stage.

5 **Strategic:** to decide where to locate warehouse need detailed geographical information on the region including potential sites and transportation details. **Tactical:** information regarding progress of stages in building; operational: information regarding materials required.

7 a) Factors influencing information flow: Organisation structure: the number of levels through which information must flow;

geographical structure of the organisation: distributed; how data originates within an organisation; where data originates within an organisation; the validity of data (re-collection affects quality of information); the preparation and input of data (including timing); the volume of data to be collected and input; the processing cycle; the specification of reports; the report distribution cycle; the report timing cycle; Formal versus informal requests and responses; quality of data; the techniques/structure for monitoring and organising the information flow.

b) Techniques:
Inspection of current I/O sub-systems; observation of current I/O sub-systems; tracking of documents for input; tracking of documents for output; inspection of development requests; inspection of report/information requests; interviews with end-users; questionnaires (at any level as above).

8 a) **OMR:** attendance marks recorded on pre-printed class lists, batch return to central point for input; **Manual entry:** attendance marks recorded on pre-printed class lists, batch return to central point for input; **Radio transmission:** attendance marks recorded on 'folder' which holds existing class lists, real-time to return to central point; **Bar code/Swipe card:** bar code captured at reader, central system reconciles attendance at reader with known class list, real-time return to central point.

b) **class teacher:** eg class list showing attendance at each session over a period of time; operational information to resolve individual reasons for absence (ie a log of excuses) eg class pattern of absences; **pastoral manager:** eg individual student attendance profile reconciled against all classes for student; tactical information to address patterns of absence; **senior management:** eg monthly percentage attendance for 'faculty' or school; strategic information inform management decisions.

9 E-mail will help reduce problems as there will no longer be a need to physically transport memos etc.; users will not be interrupted by phone calls, e-mail messages can be dealt with when appropriate. Information will not get 'lost in the system'. E-mail offers the facility to send the same message to a group of people quickly and easily.

However, there may be a resistance to e-mail from some employees which could result in important messages being left unread; e-mail might place an unacceptably high traffic load on the network; if not all employees have a desk top computer the effectiveness of an e-mail system is greatly reduced. The communication might be a structural problem within the organisation; using e-mail would not alter this.

10 a) All levels of management require information on which to base decisions, to organise, to plan and to control.
Whilst timing is important, other factors such as completeness, accuracy and relevance are equally important in assessing the value of information to an organisation.
Clearly the quality of management is directly related to its timing, but this in itself is linked to the particular situation giving rise to the need for such information.
To illustrate, the following examples are given:
   i) **provision of historical information** – into this category come annual accounts where there is no conflict between speed and accuracy, as time is taken to produce the information required; most companies, however, have well-defined timetable.
   ii) **provision of information for control purposes** as, for example, in production or quality control. Speed and accuracy are important to avoid costly delays or bad production.
   iii) **provision of information for planning purposes:** here the time scale may be years and thus there is less pressure on time and no need for a fine degree of accuracy.
It is well to understand that by accuracy is meant an acceptable level dependent on the circumstances. The cost/benefit ratio is something not to be overlooked.

b) A well-designed management information system will provide the various levels of management with appropriate information to enable them to manage. Thus the starting point must always be the uses to which information is to be put.

Delays occur:

i) the organisation structure – the number of levels through which the information must flow. ii) the data processing cycle – starting with how and where data originates; the preparation and input of data; processing and output of resultant information.

why: human failure; software unsuitability; lack of planning; volume of data to be entered; information to be printed.

Chapter 16

1 Legal requirements: refer to DPA; inform staff through initial training, handbook, appointment of DP officer to oversee use and register. Conditions of staff: healthy and safety legislation.

2 Computers have altered the work of auditors (on-line systems; 'auditing round the computer', need auditor input into system design and testing; audit packages).

4 Time and date of access; terminal id; person id; person password; files accessed; activity eg menu route; log off time.

5 a) The college must apply for registration under the Data Protection Act. Adequate protection should be applied to the system and data. The registration will specify the data use and data can be held. Only authorised users should have access to the data as specified under the registration. The college must supply details of the data held to the data subject on request.

b) There should be an in-house policy to inform staff of the college terms of registration. This may include a list of 'good and bad practice' points for staff. Appropriate examples may include: handling data and disks, access levels, password changes. File security measures, log-on/off procedures, physical security measures etc. Back up copies kept off site; roll back of failed transactions. This may include any contractual matters or disciplinary measures for staff who fail to comply.

6 a) Mailshot targeting on specified criteria or socio-economic grouping. Customer purchasing analysis to improve purchasing or manufacturing quotas. Customer purchasing profiling to predict specific likely product needs of customer and thus target.

b) Improvements in analysis of predicted demand means that stock is always available via distribution. Improvement in analysis of predicted demand means that stock is always available via manufacture. 'Customer profiling' is the only way to 'know the customer personally' and his/her needs as in the days of the local shop.

7 Necessary to meet formal audit requirements and ensure protection of the system from fraud or the use of the system from the accusation of fraud. An audit trail is the software functionality to produce a selective record of what has been happening on the system, who has been using it, when, how long for and what this person did with the data. There must be several ways of analysing the record relating to different levels of task.

Chapter 17

1 Risk analysis should highlight: how likely breakdown is to occur, what effects of such a breakdown would be, and the associated costs. This information should be used with the councillors.

2 Disasters can bankrupt businesses. Threats include: physical security, document security, personnel security, hardware security, communications security and software security. Contingency plans include backup strategy, employing specialist disaster recovery company, fault tolerant hardware, duplication of hardware.

3 Information systems are more secure to threats such as fire and flood as it is possible to duplicate all data off site. To combat unauthorised access some measures (locking doors, identifying authorised personnel etc.) are common to both. However electronic data is vulnerable when it is being transmitted over public networks, large amounts of data can be copied and transported in a small space compared to paper. On the other hand, access to information can be restricted to authorised users by the use of individual identification, passwords and access rights.

Chapter 18

1 Novice users: basic features; text entry and modification, saving documents, page formatting and printing. User's converting from a different package: main differences, help features, new features of this package. Advanced users who wish to customise package for specific purpose: macros, mail-merge, form entry.

2 Training course at local college: relatively cheap, would have dedicated time to learn, undistracted by normal work interruptions. Self-teaching from the manual and on-line help; this would be suitable for users with experience in a wide range of packages. It is very cheap, the trainee can work at his own pace and only learn the features of the package that are appropriate to his work. Buying a training video. The trainee would be able to work on those features which were appropriate to his needs; the tape could be used by further employees at a later date at no extra cost.

3 a) Different users require different training. Describe the difference between skill based and task based training. Operator: eg inputting figures into accounts package; Supervisor level: eg producing standard reports, restricted routine operations eg adding cost centres. Management level: eg fundamental revision of structures, closing down financial year.

b) Reasons: routine but infrequent procedure eg end of year type procedure require refreshing; upgrades/new functionality; users progress in their understanding of software; power users develop new needs, continual need for refresh.

4 Skill based training: the user is taught what the package can do in general terms. An evening class where the basics of a spreadsheet package are taught would provide skills based training. Being taught to use an electronic till in a supermarket or an electronic registration system in a classroom are both examples of task based training.

5 a) Use a computerised call logging system. Each user given a call reference number. A datafile of Registered Users is maintained. A datafile (or knowledge base) of Known Errors is maintained.

b) Hardware base (or configuration); netware base (= *operating system*); software versions (or configuration); problem description; error message shown; number of users on system.

c) Calls logged per hour; response time to initial call; resolution time from initial call; how well the problem was resolved; number of 'repeat' calls on the same problem.

6 a) package version number, nature of problem, call reference number, date & time.

b) knowledge base software which allows enquiry in a variety of ways including keyword, reference code, enquiry type.

c) user avoids engaged and 'music' problem; may not be time critical; message must be acknowledged. Support line: no chasing absent users, not time critical, can smooth out demand.

Chapter 19

4 a) The surveyors will be able to use e-mail and a shared diary system. They will be able to access reports and documents directly. The drawing packages can be used to produce plans which can be speedily amended. The use of financial packages will speed up accounting process. If a secretary is on holiday or sick, another secretary will be able to access the files from her own work station. The filing process will be simplified with a suitable computer based system. The office junior will not need to spend so long acting as a runner as e-mail will reduce the flow of documents. The filing job will be reduced. As a receptionist, the junior will be able to access the diaries of the partners.

b) The junior may fear that there is no longer a real job for him; the secretaries may fear a loss of autonomy and a decrease in their personal contact with their Partner. The Partners may fear that their lack of IT skills will be shown up and that they will lose job satisfaction.

c) Need to examine current procedures and work out new ones as appropriate, involving all staff. All staff will need to undergo appropriate training. The Partners must agree to embark on the new system with enthusiasm.

6 Factors: attitude of management & workforce; skill levels and re-skilling; structure of organisation and key roles; conditions of service; internal procedures for operations; external image; culture of organisation.

7   Complete contracting out of IT support to a third party. Initially may involve the transfer of hardware/software assets to the FM supplier. This can be any or all aspects of IT support within the company: software development, hardware support & maintenance, help desk, software evaluation & maintenance. Existing personnel may transfer to FM supplier.
Advantages: reduced cost due to economy of scale; flexibility for future development; access to latest hardware & software; greater expertise in support available.
Disadvantages: lack of responsiveness; business is unpredictable.

10   The company has not purchased a license; the company has purchased a fixed number of licenses however the particular user has not been allocated access rights (or had loaded onto local disk); the software is share-ware but not authorised by the network manager; authorised source code has been modified without authorisation; personally owned software has been installed; software may introduce non-standardisation; software may facilitate unauthorised data changes eg by-pass audit log; software may compromise network security.

14   The Data Protection Act says that personal data must only be used or disclosed only for the purposes registered. If this data relates to an individual then the DPA will have been broken. If the information is part of secret government information, for example details of military installations, other laws like the Official Secrets Act will have been broken. If it is not personal or an Official Secret, but say confidential to the company, the employee should still not disclose it to a third party. They might not have broken the law but have broken Clause 8 of the BCS Code of Conduct. In all cases, they will have broken their company rules and would be disciplined, probably dismissed, if discovered.

16 a) Bill has not given impartial advice. He has broken the BCS Code of Conduct, Clauses 10 and 12.
b) Bill does not appear to have broken the law. Bill's clients do not have to take his advice!

Chapter 20

2 a) The log might contain: a record of facilities used by each person including processor time; number of pages printed or disk space used; details of systems failures/crashes; details files stored/updated/deleted; details of e-mail usage/storage; time and duration of log-in; ID of logged in users; network address/station ID; failed log on attempts.
Why it is useful: provide systems administration with information about network load; enable administrators to deal with network performance problems; facilitate sensible distribution of resources to users; to limit use of scarce resources, possibly through a charging system; inform decisions about any upgrade or systems enhancement; dealing with network misuse.

b) allocation of hierarchical password to all; different access rights for different users eg read only, read write; restricted physical access to hardware; restrict sensitive applications to certain terminals; organisational codes of practice; staff training to raise awareness of security procedures/issues; existence of appropriate security procedures; audit of security procedures; auto-log off; restrict access to hard copies/printouts.

3   Database files:
Strategy: recognising need for rapid recovery; mention of incremental dumping; use of bypass systems so that processing can continue if main computer fails – through use of intelligent terminals with local hard drives; possibility of transaction tracking where all transactions are logged due to possible loss of alterations between incremental dumps generation systems;
When: need to shut down to back up; during a terminal session all updated files are marked and when user logs out these are dumped to disk; may be dumped more frequently while user is working; Media & Hardware: mirrored disks on servers; DAT/tape/exchangeable disk packing; program files:
Strategy: periodic dump to tape; generations system; may need to shut down systems to backup/use bypass facility; backup process may be time consuming.
When: backup prior to systems maintenance; backup prior to upgrade.
Media Hardware: tape, changeable disk packs.

4 existing software base; continuity for users

   a) software that runs on NEAB PC runs on NEAB SUPERPC software that decodes and executes the machine code of another machine hence 'pretends' to be that machine).

   b) Any two relevant comments but must be one for each approach eg upward compatibility – software emulation – use existing software base reduce user training yet can move on to new software emulators tend to be slow development for the future.

5 Compatibility with existing systems. Compatibility with existing software and the extent to which new hardware platform supports existing software. The company is likely to have a data interchange and compatibility policy. The company is likely to have a training policy. The company is likely to have an environment/operating system policy. User support within organisation and ability to cope with new systems. Site licenses for software. Purchasing and leasing contracts may exist on hardware/software. Maintenance contracts and support for new systems. The company is likely to have a security policy.

Chapter 21

1 Planning deliveries, siting a new retail outlet, siting a new factory.

2 Software which provides tools to help the user schedule tasks, manage resources, monitor costs and generate reports for analysis and presentation.

3 a) Agreed problem specification. Functionality Performance – use of benchmarks. Usability and human-machine interfaces. Compatibility with existing software base. Transferability of data. Robustness. User support. Resource requirements including hardware, software and human. Upgradability. Portability. Financial issues: development cost, development opportunities.

   b) function: to document how the software performed against the criteria set to enable a decision to be made. Content: purpose, structure, methodology, recommendations, justifications.

4 Alpha testing – testing in-house with data provided by software house needed to test implementation against design specification. Beta testing – off-site/real user testing, using live data, needed to detect errors not detected at alpha stage. Beta testing involves a wider audience & different environments.

5 Features: storage/retrieval of compositions; playback facility; playback in different modes/keys etc.; output to a range of devices e.g speakers, synthesisers, CD, tape, etc.
edit/update of compositions; complex or multi-featured user interface designed to take account of expert musician/engineer skills; support for range of input devices inc. keyboard, mouse, musical instruments etc., ability to accept data in a variety of formats eg live recordings, CD, tape, other packages; auto repetition of sounds; graphical representation of sounds; mix sounds from several sources; simulation of different instruments; library of sound effects; print musical scores.

6 a) Functionality: an audit trail **OR** to produce a selective record of what has happened on the system; who has been using it, when, for how long and what this person did with the data theft or damaged stock – enter the adjustment and a reason for adjustment (eg code).

   b) Required: to meet internal procedures; to meet legal audit requirements; to protect the staff from accusations of fraud.

7 a) Top-range PC with high processor speed, disk capacity and screen resolution. Digitising tablet, plotter, printer.

   b) Any appropriate example: eg rotation of a vector image, layers etc.

   c) Advantages: library of diagrams, cut & paste sections, updating facilities, rotation and views.

   d) Dynamic link or produce an export file to a spreadsheet to allow detailed analysis on space eg total room utilisation.

8 *Note from the examiner:*

Mark allocations: maximum of 8 marks for discussion of alternative strategies for providing a software solution; maximum of 10 marks for a consideration of the issues which would influence the decision about which strategy is the most appropriate.
Maximum of 16 marks for the two sections combined. Up to 4 marks are available for the quality and coherence of the candidate's argument.
Ways of providing a solution: user written/internal development team/department; external software

house to examination board specifications; use of Generic package(s) customised to meet specific needs of the examination board; specific ie purchased from a company that specialises in software for examination boards.

Issues that should be considered before solution is selected: 2 marks available for each issue

Cost of alternative solutions; for generic large numbers sold so prices are low, less so with alternatives; development & testing time for alternative solutions; generic is thoroughly tested – reducing time not so elsewhere.

Ease of use of alternative approaches: extensive user base of generic suggests better user interface; quality/reliability existence of documentation provided; generic provided with extensive documentation others may not be so appropriateness of solution generic may need considerable work in customising to suit specific requirements.

Configurability: generic may require in depth knowledge of the package to configure the application/bespoke should already match requirements; Upgrade Paths provided by alternative approaches; new versions of product/ dependence on small company.

User Support overheads incurred by alternative solutions; wide user base/size of supplier organisation & ability to cope with support overheads.

Compatibility of alternatives with existing hardware base; need for upgrades, additional memory, faster processors, etc.

Compatibility with existing software; transferability of existing data files, interface with other generic packages, etc.

How do alternatives relate to Corporate strategies; for hardware/software licensing/purchase.

9 a) alpha testing: Testing in-house, with test data provided by software house, needed to test implementation against design spec;
    beta testing off-site of software house, with 'live data', by real users, needed to detect errors not identified at alpha stage.

   b) Need for maintenance: Perfective – improving performance, speed, memory usage, etc.
    Corrective: fixing bugs which only come to light after release.

Dealt with by mail-shot to all licensed users, dispatch of update disk/floppy, computer bulletin board with detail of patches, fixes, known errors, etc.

Chapter 22

2 (see page 260)
3 (see page 259)
4 (see page 258)
5 (see page 258)

Chapter 23

1 Wiring a plug, video tapes, photographic film.

3 Potential loss of data due to hardware failure etc.: backup strategy. Non-medical personnel could access patients' medical history: id. numbers with passwords linked to appropriate access privileges. Patients may be able to access data when a work station is left unattended. Use locked screen savers, require staff to log out whenever they leave the work station. Potential loss of functionality due to a problem (fire etc.) in one part of the system. Duplicate data and processors. Power failure: backup generator.

4 a) What are protocols: set of rules; covering standards for physical connection, cabling, mode of transmission, speed, data format, error detection, error correction.
    Why protocols: to allow equipment from different suppliers to be connected, to encourage development of more open systems.

   b) Presentation; session; network; data-link; physical transport.

   c) Application layer: highest/closest to user, deals with transfer of information between end-users, applications programs and devices; hides physical network from the user; giving a user oriented view instead; deals with entry control accounting, user id. eg: user need not know a database is stored on several computers.

5 Serial transmission: via a modem; parallel transmission: to a printer.

Chapter 24

1 Command driven: user keys in command, usually as a code. Appropriate for very experienced users carrying out familiar and frequently used tasks. Menu driven: user selects, with cursor keys or mouse, one of a selection of options. Appropriate for novices and occasional users.

3 Macros can be assigned to buttons to provide a front end for the user which insulates him from the detailed workings of the package. Data can be entered via dialog box and a macro used to update the sheet in the appropriate manner. All sorting and output of reports can be automated in a similar way.

5 a) Volume of text to be displayed on screen use of colour to indicate special keys/special areas, commands etc. position of pop-up menus, help messages, etc. level of user and vocabulary/user of terminology, etc. sequence of data entry related to sequence of data collection frequency of use of data validation rules of input data method of interface/windows/keyboard, EDI, etc. Speed of processing, appreciation of the overhead of increased processing time due to complex use of graphics and dynamic objects/ windows, etc. need for increased memory resource, RAM + hard disk as virtual memory.

6 a) Physical factors: position of screen, lighting conditions, seating conditions, choice of colour schemes, etc., ergonomics/design of mouse/ keyboard ventilation/room temperature. Psychological factors: user friendly interface (qualified), help available for novice users, short cuts for expert users, make use of human long term memory to maximise efficiency, functionality, technophobia.

b) Resource implications: a greater demand for memory/IAS/backing store and processor functionality and time/speed, might apply to the same factor of the H.C.I.
On-line help availability – increased need for backing store. Complexity of interface/ multiplicity of menu routes adds to size of resultant code thus increased IAS demands. Use of GUI – increased IAS demands; need for multi-tasking/ability to switch between applications/tasks – processor functionality overhead; faster searching of help file – processor speed overhead.

AS Revision questions – Notes for answers

1 Computer data has a commercial value if it can be useful to an organisation in carrying out its work. For example, the data files of a gas company are very valuable as they are needed to send out bills. If the data was lost, the company would be in deep trouble. This data is unlikely to be valuable to any other organisation. Sales figures and customers details may be useful to rival companies who wish to expand their business.

2 Jim might really be overdrawn, the cash machine may only be up-dated nightly and hasn't details of Jim's pay yet, Jim might not get paid until the afternoon.

3 Information must have a context or meaning, for example you owe me £5.50. Data has no context, for example 5.5.

4 Additional facilities include the ability to up-date files as stock is sold instead of every night, the ability to search for information. This means that the company's stock control should be more efficient. This will save money tied up in extra stock and stock perishing.

5 Back-up procedures may include: keeping original disks of network software, backing-up pupil files to tape every day, keeping disks of applications software, running virus checks to avoid viruses and using sensible security procedures to prevent malicious damage.

6 Weighted check digits would be a sensible way of validating this data.

7 Bar code/magnetic swipe card. Reader required in every room connected to the main computer. Both systems are reliable and quick if they are not abused, for example by getting someone else to swipe your card.

9 See the Glossary.

10 Advantages: local call charge, can send more quickly, can import into your word-processor. Disadvantages: they might not look in their mail-box, a computer is more expensive than a fax machine.

11 Check digits are used to validate data as follows. A calculation is performed on the first digits of the code. This should result in the last one or two digits. If the result agrees the code is accepted. If not, it has to be re-entered. Check digits are only a good method of validation for codes like account numbers or bar codes. They cannot be used to check a date of birth for example, where a range check can be performed.

12 OCR is used to read electricity meter reading slips. It is fast and reliable but it can only be used for limited uses. Even here it may result in confusion of, say, 4 and 9. OMR would be an acceptable alternative that would also be fast and reliable.

13 They need to examine what benefits they think they will get in the first place, considering what sort of information they expect to find on the WWW. They then need to consider the costs, the subscription to the ISP, the cost of the phone call, any on-line charges and the cost of any extra hardware – for example a fast modem. There is much valuable information but the school should be aware of the cost before it starts.

14 Designs can be sent quickly by e-mail or fax. Either would not need to be answered immediately like the phone so the difference of time zones is not important. Designs sent by fax are not likely to be not such good quality (they come out a bit fuzzy), restricted to A4 size and only in black and white. In contrast, a computer design stored as a Bitmap, a JPG or any other image can be sent as a file attachment to an e-mail. This might be a different size (bigger or smaller) and could be in colour. The receiver could load the design directly into their PC and amend it and possibly send it back. An e-mail can be sent for the price of a local phone call.

15 A LAN is a local area network restricted to one building or nearby buildings using dedicated cables. It is ideal for school use as disks are not required (these can lead to viruses) yet users can access software and files at any station.

16 An electronic mail system is virtually immediate and provides traders with a record of all transactions and are accepted as legal documents. In sending papers by courier is slow, particularly if the transaction involves a long distance, for example on the Tokyo stock market. Of course, the deal may be within the same building and e-mail not really necessary. If the

system is less than 100 per cent secure, there is a risk of fraud. (6)

17 The public phone network is suitable for much communication, for example a home user accessing the WWW or sending an e-mail. It is possible to use public ISDN lines for faster access which is useful for loading graphic intensive web pages. A private line is really only necessary when security is important. For example, banks use private lines to send details of financial transactions. CCTV cameras will send pictures to the control room by private lines as security and speed are important.

18 Most humans are social animals; we like face-to-face contact. We find out more face-to-face, for example by body language. Sending an e-mail next door makes us lazy. Walking to the next door office gives us a break from the computer (see Health problems), sending an e-mail doesn't mean that it has been read (or understood).

20 Many people like the idea of working at home; not having to commute and being near their family. Using a computer they may be able to do (nearly) all the work they did in the office at a time to suit them. Their employer is happy as it saves office space and so is cheaper. The employer can use the computer to check up on their staff by seeing how many keystrokes they have entered. However this may leave the employee with the feeling that they are being watched. They may feel they never can get away from the job nor have the opportunity to meet friends at work.

21 See Chapter 4.

22 The dairy farmer could make use of the Internet as much as a more high-tech industry. She may get information from web pages on the weather, milk prices, livestock prices, latest developments from the Ministry of Agriculture, Fisheries and Food or the European Commission. She may wish to send e-mail.

23 Teletext. A home user may find out football results and a business user may find out the weather forecast. Viewdata. A home user may use it for home banking. A business may use it for interactive advertising of products which users can order.

24 Information can be accessed at may workstations in wards, consulting rooms and offices, rather

than being stored in one place. E-mails can be used to send messages quickly between wards, offices, and so on. Details from file can easily be printed out or sent to the patient's doctor. The main problems are in security, preventing the information getting into the wrong hands and integrity, making sure that the information is correct. The hospital must register under and follow the requirements of the DPA.

25 Voice recognition systems, touch screens, special keyboards can be used for input. The output devices could include actuators to switch on lights, television, and so on.

26 EFTPOS (Switch cards) deducts money directly from our bank accounts to pay for goods and services. The magnetic strip on the back of the card can be read by store's computer. It automatically contacts the bank's computer to transfer the money. Smart cards (electronic purses) can be 'charged' with a certain amount of money. As you pay for goods in a shop, the money is deducted from the amount stored by the smart card. Both systems mean that we do not need to carry much cash. Electronic transfer is used to pay wages and salaries directly into an employee's bank account. This saves the employer from dealing with huge amounts of cash and is therefore much safer as well as being much easier. **Note:** credit cards are not strictly electronic money.

27 See the Glossary. The memory in the microprocessor embedded in the plastic of the card can easily be read and up-dated. They are suitable for use in storing medical records as the patient can keep their own records on the smart card, which they take with them to the doctor, hospital, dentist, and so on. The records do not need to be stored at the doctor, hospital or dentist so the records should be secure. However the patient may lose or damage the card or forget to take it with them to an appointment.

28 Staff can protect files by backing-up on a regular basis and storing back-up copies off site. Files can be rebuilt by restoring the last global back-up. (This should be sometime in the last month.) Then incremental back-ups can be restored in order to recreate the lost files.

29 Security means keeping data secret. Integrity means making sure it is correct.

30 Data stored electronically can be best protected by using passwords and encryption. This should be more secure than that stored on paper if security procedures are followed. However if these procedures are ignored, it may be possible to copy the data easily on to a floppy disk – this is easier than photocopying and carrying away hundreds of pieces of paper. It may also be easier to hack in from a remote site to avoid detection. (Both types of stored data should be protected by physical means – locks, guards, and so on.)

32 Data users are the people who store the data. The data subjects are the people whose personal details are stored by the data users.

33 A spreadsheet is ideal for examining the effects of changes in wage rates, prices of products and numbers of units sold. Buying a spreadsheet package is likely to be much cheaper than having the program specially written. It is also available immediately whereas developing a new piece of software may take years. Even then, it may not work as new software may be littered with bugs.

34 A spell checker is not perfect. The dictionary does not include proper names like place names but these can be added to a customised dictionary. The dictionary does not distinguish between different spellings of words that sound the same like to, too and two. A grammar check may sort this out. The dictionary may be in a different language, for example US English.

35 Peripherals are devices that plug into the computer, for example printer, keyboard, mouse. The operating system controls the running of input and output devices, booting up the computer, transferring programs and data between disk and RAM, carrying out the management of files, managing memory and allocating to users, logging of errors, checking and controlling user access.

37 Possible answers include

   a) Keyboard, mouse, scanner.
   b) Bar-code reader, swipe card reader, joystick, speech recognition device.
   c) Monitor, loudspeaker, printers.
   b) Plotters, LCD projectors.

38 An integrated package performs a variety of tasks combined in one easy to use package. An integrated package would normally offer several generic programs, probably a word-processor, a graphics package would normally offer several generic programs, probably a word-processor, a graphics package, database, communications (e-mail) software and spreadsheet. All the component parts have the same 'look and feel'; it is very easy to transfer data from one function to another, the integrated package may need less disk space and RAM. An integrated package is likely to be much cheaper but may not include all the features required and lack versatility.

40 The appropriate software will depend on the speed of the computer's processor, the amount of RAM available and the user interface used. An example might be Excel 95. This won't work unless Windows 95 is available. If an older computer is being used, say a 486 running Windows 95, the software may run too slowly. If a colleague is running Excel 5, they cannot read your files.

42 Speech recognition may eventually be able to replace the keyboard, preventing RSI and enabling users with disabilities to use IT. Disadvantages include the lack of 100 per cent accuracy of the system which means the input has to be thoroughly checked, background noise increases errors, users may disturb other people in their office and users may go hoarse or lose their voice.

44 Computers have become much faster, with more memory and cheaper. As a result the software has got much more user friendly. More computer communication as computers/modems have increased in speed. More computer control and robotics in manufacture.

46 Graphical user interfaces make it easier to run programs (for example icons help us), on-line help makes it easier for us to find out what to do if we are stuck, programs look similar so that we are more confident about using them.

47 The club will need stations in all the booths connected by a multi-access system to the database storing details of tickets sold. Each station will need a printer to print tickets. The multi-access system will prevent access to the same seat twice at the same time. The club will save money on printing the tickets as it only needs to print the tickets when they are actually sold. It will be able to answer enquiries more quickly and give a faster service to customers and probably saving staff.

49 The system would save having to employ security staff. Staff who arrived for work early – before the security staff could get in. The new system would lead to several security problems. A member of staff might tell the code to a friend who could gain access or they might bring their friend in with them, which they couldn't do under the old system. It might be possible for an intruder to sneak in as a member of staff is leaving and not be spotted.

A2 Revision questions: notes for answers

1 To provide back up in case of failure in new system; compare outputs from old with new to highlight any discrepancies; training; using in a 'live' way can highlight new errors which can be dealt with.

2 See Chapter 14.

3 Breakdown of sales by product for selected items on special promotions; total sales value for day; details of stock re-ordering.

4 See Chapter 23

5 If buildings are close enough together, a fibre optic cable can be used to link the 2 buildings. A bridge (see Chapter 22) will also be needed.

See Chapter 23.

6 Several computers on a network each hold part of the total data. They need to work together to make the data available to the user.

7 See Chapter 2 for details of DPA and Chapter 16 for what an organisation should do. Security: identification codes, passwords and access rights; rules for choices of passwords, regular changing; physical access to offices monitored – use of bar-coded badges etc. Integrity: verification and validation; back-up.

8 A program which allows one computer to behave as if it were another computer.

9 See Chapter 15.

10 Involve staff and, if appropriate, trade unions in project from start. Plan well ahead. Set up appropriate training/retraining schemes in good time.

11 Notes kept on a file of any changes made to the file and where relevant transaction documents are located.

12 See Chapter 4.

13 Alpha testing carried out in-house to remove main bugs before released to public. Beta testing carried out by selected users can reveal errors under real life conditions that were missed before.

14 IT Security policy: corporate, standard approach to security issues. Code of conduct: lays down acceptable behaviour for employees – provides boundaries.

16 See Chapter 15.

17 See Figure 17.3. Back-up; security measures; contingency planning: see Chapter 17.

18 On the job; in-house training; external course; computer based training.

19 Easy interchange of data with other users; maintenance; if she uses different software the technical support staff will be unable to provide support; cost: later versions of her software package are likely to be more expensive.

20 a) Produce a flatter structure. b) Each manager will have a wider span of control which might overburden them and reduce their effectiveness while the chain of command could be reduced which c) would allow for faster and more effective communication.

21 a) Facts and rules. Such as if patient is pregnant and is suffering from headaches then avoid certain drug. Many are 'rules of thumb' which are obtained from human experience. b) A knowledge engineer will gather information for knowledge base through talking to experts and translating findings into facts and rules.

22 a) Provides information to help managers in their role as decision makers. b) Summaries of information; exception reports. Often presented in graphical form.

23 Can prioritise jobs; can complete one task without interruption; no need to phone back – again and again.

24 Documents can be shared. E-mail and video conferencing can make communications easier.

25 See Chapter 24.

26 Sound card, MIDI (Musical Instrument Digital Interface) if connecting to instrument such as keyboard or guitar.

27 Buying off-the-shelf; employing software house to write tailor made programs or customising a generic software package to meet specific need. Suggest off-the-shelf as there are many potentially suitable packages on the market.

28 Induction training. Procedures manual.

29 See Chapter 19.

30 Physical: the environmental considerations e.g. lighting and furniture layout. Psychological: how the human brain processes information. E.g., short/long term memory.

31 a) Geographical Information System: A system which can store huge amounts of geographical data, modify and integrate the data and display it in map form. b) Link data on housing (gardens) and demographic data to produce maps of target areas.

32 Manipulation of formulae; graph plotting; ability to carry out simulations; data analysis tools.

33 See Figure 21.12.

34 Features: on screen Help; windows, icons, use of mouse or other similar pointing device; not available 10 years ago as memory (RAM) capacity and processor speeds were insufficient to support such interfaces. Generic software can be customised to create own menus, tool bars, buttons and so on that fit the specific needs of an application. (NB such customisation is carried out in your Minor and Major projects).

35 See Chapter 21.

36 Users can move between different packages with same look and feel without need for retraining.

37 Compose from keyboard: the software will 'correct' the rhythm and put in appropriate notes; can edit notation in the same ways as words can be edited in a word processor.

38 Using PERT software job tasks can be entered together with expected timings. The critical path (sequence of tasks most likely to cause bottleneck) can be worked out. Progress on completed tasks can be fed in. Such software can

also monitor and schedule use of resources both human and physical.

39 Command language: user types in commands to initiate computer tasks. Appropriate for experienced users. Data entry via a form where whole form is displayed on the screen before data is entered. Appropriate when entering orders for goods made over the telephone. Free form dialogue is when the user apparently has a conversation with the computer, responding to questions asked. Dedicated keys are keyboard key presses, either singly or in combination with keys such as Alt or Ctrl that have a specific assigned function. Useful for frequent, experienced users to shortcut menu choices.

40 Facilities management is the contracting out of all or part of computer operations to an outside organisation.

41 A good team needs effective leadership, who allocates tasks appropriately, should adhere to standards and monitor costs.

42 Determining the threats to a system together with their seriousness, likeliness to happen and cost to prevent.

43 Copyright Designs and Patent Act. Document details all software installed on system. Ensure that all software is installed via central service. Spot-checks on individual computers. Inform all staff of legal position and sanctions that will be taken if any breaches found.

44 See Chapter 18.

45 To ensure that claims for student activity made to the FEFC (Further Education Funding Council) are correct. Details of occurrences such as a student course change should be trackable.

46 a) Reduce costs by ensuring no 'white elephant' purchases; built up appropriate maintenance; ensure legal obligations are fulfilled. b) See Chapter 20. c) Circulate all staff – meetings, notices, handbook. Ensure that it is a senior manager's responsibility to ensure that policy is implemented.

47 See Figure 21.2.

48 a) Check that systems function correctly by delivering the expected results. S/he must ensure that no fraudulent activity is taking place. b) How computer stores data; how vulnerabilities to dishonest programming. c) On-line systems may not record all transactions so no record of when data changed and by whom. d) e) f) See Chapter 16.

49 See Chapter 16.

50 See Chapter 17.

Projects

As candidates for AQA AS level Information Technology you have to undertake one project, AS Module 3. This coursework module uses generic application software for a task solution. It is not within the spirit of the syllabus to use a programming language.

As candidates for AQA A level Information Technology you have also to undertake a second project, A2 Module 6. This coursework module uses information systems for problem solving.

Each project will take up most of one year, the coursework running in parallel with the theory work covered in this book.

Project timetables

With projects taking a long time, it is essential that they are broken down into sub-tasks, each with an appropriate deadline for completion. You need a realistic activity schedule to ensure that the work-load is spread evenly throughout the project period and allow for other factors such as mock exams, work-loads in other subjects, holidays and half-terms, etc.

You should not be tempted to start implementation before the design stage is completed. Nor should you fail to allow plenty of time at the end for the testing, the user guide and the evaluation stages, which constitute over half the marks available for the first project.

Those of you who do not have a good schedule of activities invariably find that too much is left to the last minute. Deadlines cannot be met and the testing, the user guide and the evaluation are weak leading to poor marks.

First project (AS)

This project aims to use information technology to provide a solution to a task. The task may be a genuine task with a real end-user or a realistic task with the teacher as end-user. It is important that there is an end-user so that you can gain feedback from the end-user at each stage.

Which software should be used?

The project must be implemented using generic application software. This offers a wide choice of possible programs but you will probably choose to do one of the following:

- a word-processing project;
- a spreadsheet project;
- a database project;
- a web page project;
- a DTP project;
- a presentation project (e.g. slideshow for a conference or multimedia presentation).

You may use more than one package. For example a mail-merge in a word-processing package may take data stored in a spreadsheet.

You are expected to demonstrate full knowledge of the software, including use of advanced features. Using every single feature of a software package would almost certainly lead to a very contrived project. It is better to choose a real problem with a real end-user, the solution of which involves some of the advanced features.

In general, most DTP and presentation software does not include many advanced features and it may be difficult for you to achieve the highest marks. Below are suggested some project ideas, showing the sort of features that could be included and what might be interpreted as advanced features.

Wizards should not be regarded as the solution of the problem but the starting for a possible customised solution. All projects must be thoroughly tested. Output should be annotated and cross-referenced to test documentation.

You are reminded that only a third of the marks are given for implementation of the project. The rest of the marks are for specifying the problem, the testing, the evaluation and the user documentation.

1 Word Processing Project

To obtain the highest marks, you would be expected to use most of the following features:

- recording macros for common tasks;
- customising toolbars to suit the user;
- adding icons to run macros;
- using selective mail merge, e.g. to send out agenda for a business meeting;
- importing graphics in files (possibly designed and scanned by the candidate);
- creation of templates, e.g. a business fax template;
- use of tables and graphs;
- customising of the spell check and AutoCorrect to include common words/errors.

The following may be regarded as advanced features:

- dialogue boxes, e.g. to choose and load different templates;
- form fields and protected documents;
- selective mail merge, e.g. to send out different agendas to different people for a business meeting;
- a customised front-end to hide the package from the user;
- drop-down boxes.

Project example 1

A local business wants to use a word-processing package to improve its administration. For example it may wish to:

- selective mail merge to inform committee members of meetings;
- templates for memo, fax, letter and invoice;
- customised front-end, menu bar and dialogue boxes to help the user.

Project example 2

You have been asked to help in the running of a conference. Amongst the things you might need to do are:

- print the name labels for all delegates;
- write to delegates with details;
- selective mail merge to print lists of names for each study group at the conference.

Other possible ideas:

Running a fan club/sports club/other club.
Designing a school magazine.
Organising a sports match.
Producing a football match programme or fanzine.

2 Spreadsheet project

Candidates would be expected to use some of the following features:

- a good range of charts;
- formulae and functions;
- macros;
- adding buttons to run macros;
- customising icons and tool-bars;
- including pictures/graphics;
- linking pages;
- absolute/relative cell references;
- validation.

The following may be regarded as advanced features:

- a customised front-end to hide the package from the user;
- dialogue boxes;
- look-ups;
- nested if statements;
- count functions;
- drop-down boxes;
- spinners.

Example project 1

You have been asked to present the accounts of an organisation (e.g. income of club, conference, etc.).

Example project 2

You have been asked to design a car insurance look-up and calculator. By entering details of the car (or choosing it from a list) the software will calculate the cost of the insurance.

Other possible ideas:

- automatic questionnaire analysis;
- how much does this PC cost?;
- dietary analysis;
- sports league table;
- costs of a school magazine, tuck-shop, school play;
- invoice production;
- heating costs calculation (from dimensions of house, etc.);
- accounts for a department in school or college;
- accounts for a fanzine;
- foreign exchange;
- stocks and shares records.

3 Web site project

This type of project is easier to evaluate as large files take longer to load. Speed of loading is important for a good web site. You can specify the maximum loading time before a user would log out of the site.

The project should not just be one page but several linked pages. As the project should be a repeatable system, it should include instructions on how to update the site.

Many HTML packages now exist such as *Microsoft Front Page Express*. You would be expected to use some of the following features:

- different size and colour text;
- hypertext links to other parts of the same page;
- hypertext links to other pages (including those not set up by the candidate);
- hyperlinked graphics;
- graphics, including thumb-nail pictures linking to full size pictures;
- interlaced pictures;
- background graphics or watermarks;
- tables;
- links to e-mail addresses.

The following may be regarded as advanced features:

- using frames so that only part of the screen needs to refresh when a link is chosen;
- using short Java script applets to improve the image of the page, e.g. dynamic HTMLs. Examples can be found at http://www.insidedhtml.com such as changing the colour of a link/graphic as you move the mouse over it;
- opening new windows containing graphics to avoid having to reload the previous page;
- counting number of visitors. There are many web-sites offering free counters. Look at http://www.thefreesite.com. It is also possible to include guest books;
- using animated gifs, which can be set up with *Microsoft Gif Animator*. However animated gifs will take longer to load;

using feedback forms allowing people who access your page to send you feedback to your e-mail address. These forms can include text boxes, drop down boxes, radio buttons, check boxes, etc.

Example

You have been asked to produce a web site for an organisation (e.g. school, business, conference, etc.).

Finished pages can really be uploaded on to the Net for testing. Companies like *Geocities/Yahoo* and *Angelfire* offer free web space, if required. Testing could be with different computers and different browsers. You should consider likely loading time in their design.

4 Database project

You would be expected to use most of the following features:

- related tables, identifying fields, primary keys and types of relationship;
- queries;
- forms to enter data;
- reports;
- macros;
- add buttons to run macros;
- customise icons and tool-bars.

The following may be regarded as advanced features:

- a customised front-end to hide the package from the user;
- parameter queries;
- validation and input masks;
- sub-forms;
- grouped data in reports;
- list boxes.

Example project 1

You have been asked to set up a database of students in your school/college and their computer network user-numbers.

For more examples, see the second project examples.

What should the project include?

The project should include the following sections:

- List of contents (pages should be numbered), name, candidate number.

Specification (9 marks)

- The definition of the problem in information technology terms.
- The requirements specification. What inputs, processing and outputs are required to meet the needs of the user? This may include interviews with the user to find out exactly what is required.
- Test plan. How will the system be tested? What test data will be used and expected outcomes.
- A project timetable.
- Performance indicators. What criteria will be used to assess the success or otherwise of the completed system?

Implementation (20 marks)

- Designs for files, screen layouts, print outs, data capture forms.
- Description of hardware and software available, its limitations and capabilities and how it will be used.
- Implementation of project, complete with annotated print-outs from the system, listings of any macros, and screen shots of the system in action.
- Implementation report of how the system was implemented, including advanced functions of the software used and a project log of how you have tackled the problem.

Testing (15 marks)

- Test results, expected outcome and actual outcome.

e.g.

Test	Expected outcome	Actual outcome

- Full report on testing following the test plan, including use of typical erroneous and extreme data.

- Annotated print outs and screen shots of testing cross-referenced to the report.

Evaluation (6 marks)

- Evaluation of the system based on the original requirements and pre-determined performance indicators and limitations of the solution.

User documentation (10 marks)

- User guide, including screen shots and installation details.

Test strategy

This is the plan of how the system is to be tested. One possible strategy is functional testing – testing every function of the system, for example all macros or all links to other web pages.

Test data should include typical data, erroneous data that should be rejected and extreme (boundary) data. Test data for a mail-merge may include different names and addresses to check that they fit all right into a mail-merge letter or an address label. Test data for a spreadsheet may include different values to check the calculations work.

The test strategy should also include user acceptance testing to ensure that the user is happy with the system.

Possible evaluation questions/criteria

Some of the following may be appropriate:

- Does the system do everything originally required?

- How long does it take to load (e.g. when a macro loads a template)? How does it compare with the old method?

- How much space does each file take on the disc? (Suggest a maximum.)

- Does the end-user think it does everything expected?

- Does the end-user think it does everything well?

- Is there a common theme to all aspects (e.g. a logo)?

- Does the end-user or someone independent think the icons are clear?

- Does the end-user or someone independent think it is easy to use?

- Does it print satisfactorily (on which printers)?

- If it is a colour graphic, does it look all right in black and white?

- Does output photocopy clearly (e.g. if using a watermark)?

- Does the mail merge work successfully and do the merged documents look correct?

- How long does it take to print?

- Is there any security and does it work?

Screen shots

As no disks can be submitted with the finished project, it is up to you to demonstrate that the new system has been implemented successfully. The best way to do this is to include screen shots in the project report. Use the Print Screen key to copy the screen into the computer's clipboard. Graphics software can be used to crop the image to the right size to include in the report.

Second project (A2)

As candidates for AQA A level Information Technology you must also undertake a second project, taking much of the second year of the course.

For this project, much more emphasis is placed on the quality of the analysis of the problem and the design of the solution. You will be expected to demonstrate knowledge of the system information flow and to use Dynamic Data Exchange between packages. The project should be of a cyclic nature, for example it may need to be updated every year. Old data must be cleaned down and archived.

The project should be a real project with a real user. Although it is possible to use other software, a relational database like *Microsoft Access* is the ideal applications package for this substantial project. It can include information flow and data dynamics and will also cover the theory of relational databases required in A2 Module 5.

You are again expected to demonstrate full use of the software, including advanced features. Below are suggested some project ideas and the sort of features that should be included.

You should be reminded that only a sixth of the marks are given for implementation of the project. Over one third of the marks are available for analysing the problem and designing the solution. All projects must be thoroughly tested and evaluated. The project should be presented as a report with pages numbered and an index.

Again, you should not be tempted to start implementation before the design stage is completed. Nor should you fail to allow plenty of time at the end for the testing, the user guide, the evaluation and the report stages, which constitute nearly half the marks available.

Good candidates would be expected to use many of the following features:

- different tables with data normalised;
- look-up fields in tables;
- validation and input masks;
- relationships between fields in these tables;
- queries including linking more than one table;
- parameter queries;
- customised forms to enter data e.g. list boxes, combo boxes;
- sub-forms;
- splashscreens;
- reports;
- grouped data in reports;
- macros;
- push buttons to run macros;
- a customised front-end (switchboard) to hide the package from the user;
- data dynamic exchange e.g. exporting data to a mail-merge or a spreadsheet.

Ideas:

- booking facilities at a leisure centre;
- finding details from a pupil's school timetable;
- running a video shop/library;
- car hire database;
- tickets for the school play;
- stock control in a school department;
- stock control in a shop;
- rail ticket price calculator;
- equipment repairs database;
- newspaper shop delivery database;
- bank statements.

What should the second project include?

The second project should include the following sections:

- List of contents (pages should be numbered), name, candidate number.

Analysis (18 marks)

- Full description of the problem (probably including transcript of interview with the user).
- The problem broken down into sub-tasks.
- Requirements specification for system and each sub-task. What inputs, processing and outputs are required to meet the needs of the user?
- Description of information flow and information flow diagram.
- Hardware and software available to you and how it could be used.
- Current IT knowledge and likely training needs for the user.
- Evaluation criteria for the system.

Design (16 marks)

- Relevant possible alternative IT solutions to the problem related to hardware, software available and evaluation of each solution.
- Chosen solution and reasons for choice.

- Solution broken down into IT sub-tasks.

- Input/processing/output requirements for each sub-task.

- Designs including data capture sheets, record structures, description of tables, data dictionary, relationships between tables, validation procedures, screen layouts and printed output.

- Test plan. How will the system be tested? What test data will be used and expected outcomes.

- Schedule of activities.

Implementation (15 marks)

- Full implementation exploiting all the features of the software, complete with annotated print-outs from the system, listings of any macros, and screen shots of the system in action.

Testing (15 marks)

- Test results, expected outcome and actual outcome.

e.g.

Test	Expected outcome	Actual outcome

- Full report on testing following the test plan, including use of typical, erroneous and extreme data.

- Annotated print outs and screen shots of testing cross-referenced to the report.

- Evidence of end-user testing.

User guide (8 marks)

- Full, illustrated user guide (including installation, general use, back-up procedures and trouble shooting).

Evaluation (10 marks)

- Evaluation of the system based on the original requirements and pre-determined performance indicators.

- End-user comments.

Report (8 marks)

- Illustrated implementation report of how the system was implemented, including advanced functions of the software used and a project log of how you have tackled the problem.

Syllabus Matching Chart

Topic		Chapter
10.1	Knowledge, Information and Data	1
10.2	Value and Importance of Information	1
10.3	Control of Information	1, 2
10.4	Capabilities and Limitations of information and communications technology	8
10.5	The social impact of information and communications technology	3
10.6	Role of communications systems	4
10.7	Information and the professionsl	6
10.8	Information systems malpractice and crime	5
10.9	The legal framework Software and data misuse Data protection legislation Health and safety	2 2 2 6
11.1	Data capture	7
11.2	Verification and validation	7
11.3	Organisation of data for effective retrieval	11
11.4	Software; nature, capabilities and limitations Nature and types of software Capabilities of software Upgradability Reliability	9 9 9 9 9
11.5	Manipulation and/or processing	8
11.6	Dissemination/distribution	9
11.7	Hardware; nature, capabilities and limitations	7
11.8	Security of data Back up systems	2 2, 8
11.9	Network environments	4
11.10	Human/Computer Interface	10

Topic		Chapter
13.1	Organisational structure	12
13.2	Information systems and organisations	13
	Management Information System	13
	Life cycle of an information system	14
	Success or failure of an MIS	13
13.3	Corporate information systems strategy	20
	Information flow	15
	Personnel	12
13.4	Information and data	15
13.5	The management of change	19
13.6	Legal aspects	16
	Audit requirements	16
	Disaster recovery management	17
	Legislation	16
13.7	User support	18
	Training	18
13.8	Project management and effective teams	19
13.9	Information and the professional	20
	Employee code of conduct	20
14.1	Policy and strategy issues	20
	Future proofing	20
	Backup strategies	20
14.2	Software	21
	Evaluation	21
14.3	Database management concepts	25
14.4	Communication and information systems	22
	Distributed systems	22
14.5	Networks	23
	Network security, audit and accounting	23
14.6	Human/Computer Interaction	24
14.7	Human/Computer Interface	24
14.8	Software development	21
14.9	Software reliability	21
14.10	Portability of data	23
	Protocols and standards	23

Index